Spring Flora of Wisconsin

ADDITIONAL ILLUSTRATIONS BY RODERIC L. THOMSON

FAMILY CYPERACEAE: TEXT BY JAMES H. ZIMMERMAN &
ILLUSTRATIONS BY ELIZABETH HOLLISTER ZIMMERMAN

NORMAN C. FASSETT

Spring Flora of Wisconsin

A MANUAL OF PLANTS GROWING WITHOUT
CULTIVATION AND FLOWERING BEFORE JUNE 15

FOURTH EDITION REVISED AND ENLARGED BY

OLIVE S. THOMSON

THE UNIVERSITY OF WISCONSIN PRESS

Published 1976
The University of Wisconsin Press
Box 1379
Madison, Wisconsin 53701

The University of Wisconsin Press, Ltd.
1 Gower St., London WC1E 6HA, England

Fourth edition
Copyright © 1967, 1976, 1978
The Regents of the University of Wisconsin System

The first and second editions, 1931 (reprinted 1938) and 1947,
were published by the Department of Botany, University of Wisconsin.
The third edition was published by the University of Wisconsin Press
in 1957; reprinted 1959, 1963, 1967 (with revisions), 1972.

Fourth edition printings 1976, 1977, 1978

Printed in the United States of America
ISBN 0-299-06750-5 cloth; 0-299-06754-8 paper
LC 74-27307

Contents

Preface

This Fourth Edition of *Spring Flora of Wisconsin* is ready
for printing on the twentieth anniversary of the death of its
original author, Norman C. Fassett. He probably will not be
remembered by those who will use this edition as he is by
his many friends who associate the former editions with his
personality and friendship. The widespread use enjoyed by
the book has undoubtedly helped to fulfill his hopes for an
increased awareness of and desire to protect the rich and
diverse flora of Wisconsin. Although revisions and additions
have made it considerably longer, this new edition is still in
keeping with his goals as it becomes available for present and
future wildflower enthusiasts.

From 1925 to 1954 Norman Fassett was Professor of
Botany and Curator of the Herbarium at the University of
Wisconsin. In recognition of his deep concern for wildflower
protection and his pioneer efforts in this state for habitat
preservation, Parfrey's Glen in Sauk County, one of Wisconsin's first Scientific and Natural Areas, has been dedi-
cated to his memory.

ILLUSTRATIONS

The drawings in this book have been made to illustrate
certain diagnostic characters rather than to show the appearance of the whole plant. Illustrations for previous editions

were made by the following: G. O. Cooper, W. E. Dickinson, R. I. Evans, Mrs. Jane Gilbert Peterson, S. Kliman, Florence B. Livergood, W. T. McLaughlin, Ruth P. Morgan, and Mrs. Katharine Smith Snell. New illustrations were prepared for the Poaceae, Cyperaceae, and *Juncus* by Mrs. Elizabeth Hollister Zimmerman, leaf prints of *Salix* species by F. Glenn Goff, and 121 drawings for other families by Roderic L. Thomson.

ACKNOWLEDGMENTS

My first and most important acknowledgment is to my husband, John W. Thomson, who encouraged me to undertake this revision and who, by continuous use of his expertise, made it possible. We hope that this new edition of the *Spring Flora* will help perpetuate the memory of its original author and our close friend.

I am very appreciative of the hospitality always given me by the staff at the University of Wisconsin Herbarium: Hugh H. Iltis, Mrs. Katharine S. Snell, and Theodore S. Cochrane for use of the herbarium, access to recent student reports, and unpublished distribution maps prepared by Dr. Iltis and Mr. Cochrane for an "Atlas of the Vascular Flora of Wisconsin." In addition, Mr. Cochrane critically read the manuscript and provided many helpful suggestions.

I am grateful to the following contributing authors: Albert M. Fuller, H. A. Davis, and Mrs. Tyreeca Davis for providing a key and summaries of distributions for species of the Eubati section of the genus *Rubus*; James H. Zimmerman for the entire Cyperaceae and the genus *Juncus*; Peter J. Salamun, who made improvements and additions for the family Caprifoliaceae; and Mrs. Maarit Threlfall, who compiled the list of species of the Poaceae, that for the genus *Panicum* being supplied by Robert W. Freckmann. The revisions made by Margaret S. Bergseng in the third edition of *Spring Flora* (1957) are gratefully acknowledged as an integral part of the present edition.

Material used directly from previous compilations includes the key to the genus *Salix* by George W. Argus (*Trans. Wis. Acad. Sci. Arts Letters* 53: 222-27) and the key to *Viola*, which was contributed by Norman H. Russell for use at the herbarium and by University of Wisconsin taxonomy classes.

Several recent Preliminary Reports on the Flora of Wisconsin published in the *Transactions of the Wisconsin Academy of Sciences, Arts and Letters* were a considerable asset in the new treatment of many families. They are listed in the Selected References but also merit recognition here: Caryophyllaceae by Robert A. Schlising and Hugh H. Iltis, Cruciferae by Jacqueline P. Patman and Hugh H. Iltis, Rosaceae by Harriet Gale Mason and Hugh H. Iltis, Asclepiadaceae by Gottlieb K. Noamesi and Hugh H. Iltis, Labiatae by Robert C. Koeppen, Primulaceae by Hugh H. Iltis and Winslow M. Shaughnessy, Polemoniaceae by Dale M. Smith and Donald A. Levin, Plantaginaceae by Melverne F. Tessene, Compositae I by Miles F. Johnson and Hugh H. Iltis, Compositae IV by Carol J. Mickelson and Hugh H. Iltis, and Compositae V by Edward W. Beals and Ralph F. Peters. Research on these and other aspects of Wisconsin floristics has been supported by grants to Dr. Iltis by the Research Committee of the Graduate School of the University of Wisconsin.

I consider it very fortunate and am grateful that Mrs. Margaret B. Hickey was assigned the editorial task. Her careful work has really contributed to the quality of this manual.

Olive S. Thomson

October 1974

Spring Flora of Wisconsin

Introduction

The geographic position of Wisconsin provides the state with a great diversity of plant communities and, living in them, a long list of species of native plants. The list of introduced species is also very long. At the present time, with the human population at an all-time high and the rapidly shrinking heritage of our native vegetation, interest in studying plants has been surging. It is as prevalent outside the classroom as within, and guides and manuals are needed more than ever to serve the needs of all.

Natural forces to a small degree and human activities to a greater degree prevent any possible stability for a flora. Ever-changing distributions of species already present and new species continuing to enter the state provide wildflower enthusiasts with new discoveries and surprises even after they have mastered the names of our abundant flora. Manuals continually need revisions and updating to keep abreast of the new finds.

This Fourth Edition of the *Spring Flora of Wisconsin* has been prepared to meet these needs; it has also been expanded to include some of the groups omitted in previous editions and to provide upgraded and more complete treatments of family and generic descriptions. The general style follows previous editions. Although the species are limited to those blooming before June 15, family and generic descriptions are complete enough in most cases to be helpful in naming plants at least to the generic level after that date.

The arrangement of the families follows the Englerian system and embraces only the large group of seed-bearing plants, the Angiospermae. This group, characterized by having true flowers, is further divided into two groups, the Monocotyledoneae and the Dicotyledoneae. The Monocotyledoneae are characterized by having the vascular bundles scattered in the stem, no cambium, parallel-veined leaves, parts of the flowers in 3s or 6s, and one seed leaf (cotyledon). The Dicotyledoneae have the bundles in a ring in the stem, a cambium, the veins of the leaves branching, parts of the flowers in 4s and 5s, and two cotyledons. To all these characters there are exceptions in some species. In this book the Monocotyledoneae include the families through the Orchidaceae; the Dicotyledoneae include the families from the Salicaceae through the Asteraceae. The plant families are further divided into genera and the genera into species. A species is composed of all the individuals in existence which resemble each other closely enough to be considered of one kind. The full name of each plant consists of two parts, the generic name followed by the species name, both in the accepted tradition of Latin for international usage. The common names which many people prefer are also included, although some create confusion.

Because International Rules of Nomenclature now recommend that all plant families have names ending in -aceae, the names of a few families popularly known by other terms have had to be changed, for example, the Cruciferae are now recognized as the Brassicaceae. The former well-known names have been retained in parentheses. In a few places where specific nomenclature has been recently revised, the synonymy also follows the species name in parentheses. International Rules have also determined the -i or -ii endings of many specific names.

To avoid repetition, the characters differentiating species in the keys are usually omitted in the specific descriptions. Likewise, family characters representative of each genus are not repeated at the generic level, nor are generic characters for each species repeated in the specific descriptions. When numbers are placed in parentheses following (or preceding) a range of numbers for parts of a flower, they indicate that some plants may extend the usual range to the number in parentheses. Thus, "stamens 5-10 (15)" indicates that the usual number of stamens is 5-10 but occasional flowers may have up to 15 stamens.

Concise statements of range are difficult. Most ranges have been described in terms of county boundaries (Map 1), and follow certain patterns. For example, a range stating that a certain species occurs in southern Wisconsin north to Dunn to Outagamie counties means that it probably can be found within the lines those counties delimit; on the other hand, if the statement reads that a plant is found from Grant to Racine counties and north to Dodge and La Crosse counties, its range extends northward in two places rather than having a continuous distribution between Dodge and La Crosse counties. Maps 2 and 3 should be helpful in understanding such statements as "within the Driftless Area," which is the unglaciated part of southwestern Wisconsin, or "north of the Tension Zone," a zone of transition extending from northwestern to southeastern Wisconsin and dividing the northern hardwoods from the prairie forest. Map 4 shows the original zones of vegetation of the state and should be useful in determining the range of plants which are specific to certain plant communities. Bog plants provide

Map 1. Counties of Wisconsin.

a good example. Bog areas are abundant in the northern coniferous and northern hardwood forests; they are less common in southern Wisconsin on the glaciated areas and almost lacking in the Driftless Area.

Any student of plants should attempt to gain (1) a familiarity with terms necessary for classification and (2) facility and confidence in the use of keys. The complete glossary and the figures at the end of the book are intended as aids in problems of classification terminology. All keys in this

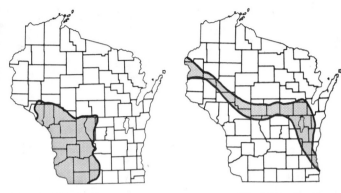

Map 2. Driftless Area.   Map 3. Tension Zone.

manual follow a uniform style, in which the reader is ex-
pected to make a selection between two choices. To reduce
the likelihood of error, the reader should always read both
choices before making a decision. After a few experiences in
following alternate choices, the procedure becomes simple.

Careful observation and understanding of the parts of a
flower and how the parts are arranged are probably the most
essential and important steps to plant identification. Each
flower is either without a stalk (sessile), or set upon a stalk,
the *pedicel*. At the tip of the stalk, the organs comprising
the flower are attached to the slightly enlarged tip, the
receptacle. The *sepals*, together making up the *calyx*, when
present are green and attached below the colored *petals*
(together called the *corolla*). When the calyx and corolla are
not differentiated, these two types of organs are called a
perianth, composed of *tepals*. If only one set is present, as in
the Marsh Marigold or the anemones, the set in most manuals is called the calyx, even if it is petallike and colored.
Above the sepals and petals are the *stamens*, composed of
filaments topped by *anthers*, within which the pollen is pro-

Map 4. Early zones of vegetation (after Curtis, 1959).

duced. Above the stamens is the *pistil* or pistils, comprised of the *stigma* to receive the pollen, the *style*, and the *ovary* containing the *ovules*, which when ripe become the seeds within the fruit developing from the ripened ovary. There may be a single or several separate pistils within a flower, as in the buttercups or the anemones, or several separate pistils may be fused to become a single compound ovary, as in the geraniums or maples. In such cases, each division of the ovary is called a *carpel* and is considered to represent a leaf which once bore ovules along its two edges, the edges infolding to meet and fuse. *Parietal placentation* is exhibited when the edges meet and fuse with each other along the outer walls of the compound ovary, providing only one cavity or *loculus* but several double rows of ovules on the inner walls of the fruit. *Axile placentation* is the term used if the edges of the carpel meet at the center of the ovary, forming partitions from the fused sides, each partition with double rows of ovules in the center, and each carpel having its own cavity or loculus. *Free central placentation* represents a further development, shown by the pinks and the primroses, in which the partitions or septa degenerate, leaving a central axis with the ovules on it freestanding within the center of the single chamber of the ovary. The number of carpels making up the pistils may often be discerned by counting the number of styles.

In flowers such as those described above, the ovary is placed above the other floral parts and is said to be a *superior* ovary. Contrariwise, the other parts are below the ovary, and the flower is termed *hypogynous*. In flowers like those of the cherries and plums where the bases of the sepals, petals, and stamens fuse to form a cup-shaped organ, the *hypanthium*, around the ovary, the flower is termed *perigynous*, but the ovary is still a superior ovary. If the hypanthium deepens and fuses with the ovary so that the sepals, petals, and stamens appear to be borne on top of the ovary, the ovary appears below and is termed an *inferior* ovary, and the flower is said to be *epigynous*. Apples, wild carrots, dogwoods, and dandelions are familiar examples of this type.

If a flower possesses both stamens and pistils, it is termed a *perfect flower*. If it lacks either, it is an *imperfect flower*. If a plant has imperfect flowers and both staminate and pistillate flowers are on the same plant, it is said to be *monoecious*. If the staminate and pistillate flowers are on separate plants, the plants are said to be *dioecious*. Occasionally plants have not developed evolutionarily to become clearly monoecious or dioecious and have partly perfect and partly imperfect flowers on the same or different plants; these are referred to as being *polygamomonoecious* or *polygamodioecious*.

It is also necessary to observe whether a flower has the petals all alike and is symmetrical along several radii, *regular*, as a rose, or whether it is bilaterally symmetrical (asymmetrical), *irregular*, as a Sweet Pea or violet. In the latter case it may be *two-lipped*, as in the Snapdragons and mints.

Some persons have difficulty distinguishing annuals and perennials. Annual plants usually have taproots which continue the main stem directly down into the soil. Perennial plants may be distinguished either by being woody or by the presence of the older dead stems of the previous year's growth around the crown of the plant at soil level. Biennials germinate their seed during the first growing season, and the plants form a basal rosette of overwintering leaves. They then produce an upright flowering and fruiting stalk the second season, dying thereafter.

Key to Families

Section A

2. Trailing or climbing vines *3*
   3. Leaves compound ....*Clematis* p. 158
   3. Leaves simple *4*
      4. Leaves palmately veined *5*
         5. Leaf blades not toothed or lobed (Fig. 74) ......
           .... *Dioscorea* p. 91
         5. Leaf blades with 3-7 coarse teeth or shallow lobes
           6
           6. Petiole attached to the blade a few mm from
              its edge .... *Menispermum* p. 165
           6. Petiole attached at the edge of the blade .....
              .... *Vitis* p. 261
      4. Leaves pinnately veined ....*Celastrus* p. 252
2. Erect plants *7*
   7. Stems woody *8*
      8. Flowers not pediceled, in heads or catkins *9*
         9. Catkins and leaves fragrant when crushed ....
           ....*Myrica* p. 115
         9. Catkins and leaves not fragrant when crushed *10*
           10. Veins of the leaves curved, not quite reaching
             the margin *10a*
             10a. Leaves usually partly lobed, sap milky;
               flowers with perianth, fruits berrylike ....
               .... MORACEAE p. 128
             10a. Leaves not lobed, sap clear; flowers lacking
               perianth, fruit capsular.................
               ....SALICACEAE p. 103
           10. Veins straight, reaching the margin ....
             .... BETULACEAE p. 117
      8. Flowers pediceled or in spherical heads, not in catkins
        11
        11. Leaves compound *12*
          12. Stems prickly .... *Xanthoxylum* p. 245
          12. Stems not prickly *13*
            13. Leaves alternate *14*
              14. Leaves glandular-dotted, stigmas 2 ....
               .... *Ptelea* p. 246
              14. Leaves not glandular-dotted, stigmas 3
               .... *Rhus* p. 250
            13. Leaves opposite *15*
              15. Petiole with a pad of hairs on the inner
               side at base ....*Acer* p. 256
              15. Petiole without a pad of hairs at base
               ....*Fraxinus* p. 302
        11. Leaves simple *16*
          16. Leaves toothed *17*
            17. Leaves palmately veined *18*
             18. Flowers pediceled or in hemispherical
              heads..................*Acer* p. 256
             18. Flowers in dense spherical heads ......
             ....PLATANACEAE p. 198
            17. Leaves pinnately veined *19*
             19. Leaf blades tapered at base evenly on
              both sides (Figs. 362-65) ....
              ....*Rhamnus* p. 258
             19. Leaf blades asymmetrical at base (Fig.
              147) ....URTICACEAE p. 129
          16. Leaves not toothed or lobed *20*

20. Leaves with a close silvery or brownish scurf .... *Shepherdia* p. 274
20. Leaves not scurfy *21*
    21. Tree .... *Nyssa* p. 288
    21. Shrub .... *Nemopanthus* p. 251
7. Stems not woody *22*
  22. Leaves compound, with 3 or more leaflets *23*
    23. Leaflets 3-11, 1 dm or more long .. *Arisaema* p. 70
    23. Leaflets very many, not over 3 cm long ....
    .... *Thalictrum* p. 161
  22. Leaves simple, sometimes deeply lobed, but without stalked leaflets *24*
    24. Stems without cobwebby hairs *25*
      25. Leaf margins not fringed with hairs *26*
        26. Leaf blades entire or with 2 lobes at base *27*
          27. Sepals 6, green, red, or yellow, the 3 outer tiny, the 3 inner much larger; leaves large (several cm long), lobed at base .... *Rumex* p. 132
          27. Sepals 6, yellow, all alike; leaves of tiny scales on green branches ....
          .... *Asparagus* p. 84
        26. Leaf blades deeply lobed (Fig. 372) ......
        .... *Napaea* p. 264
      25. Leaf margins fringed with fine hairs *28*
        28. Flowers 5 mm or less long, in a pyramidal panicle .... *Valeriana* p. 358
        28. Flowers 10 mm or more long, the inflorescence leafy-bracted ...... *Lychnis* p. 144
    24. Stems whitened with cobwebby hairs *29*
      29. Basal leaves not toothed or lobed, present at flowering time .... *Antennaria* p. 371
      29. Basal leaves toothed or lobed, usually absent at flowering time .... *Petasites* p. 367

Section B

1. Flowers in the axils of dry overlapping scales, forming spikelets without perianth *2*
  2. Stem usually cylindrical and hollow in cross section, with hard nodes; sheaths usually open; anthers attached near their middle .... **POACEAE p. 25**
  2. Stem usually triangular and solid in cross section, with soft nodes; sheaths closed; anthers attached at their base ....
  .... **CYPERACEAE p. 27**
1. Flowers not in the axils of dry overlapping scales, with sepals and petals (except in the Araceae) *3*
  3. Flowers on a fleshy common axis several or more mm thick (Figs. 36-40) .... **ARACEAE p. 70**
  3. Flowers not on a fleshy axis *4*
    4. Flowers in a head with an involucre (Figs. 540-64) ......
    .... **ASTERACEAE p. 361**
    4. Flowers not in an involucrate head *5*
      5. Petals all alike *6*

6. Filaments with long hairs . . . .
. . . . COMMELINACEAE₁ p. 73
6. Filaments without long hairs 7
   7. Perianth borne (on the receptacle) at the base
     of the ovary 8
     8. Sepals and petals papery, or stiff and scale-
       like 9
       9. Sepals and petals each 3, stiff and sharp
         . . . . JUNCACEAE p. 74
       9. Sepals and petals each 4, papery . . . . . .
         . . . . PLANTAGINACEAE p. 342
     8. At least the petals soft, white or colored
       10
       10. Flowers in separate globose heads on
         the same stalk, the staminate flower
         heads above, the pistillate below . . . . . .
         . . . . SPARGANIACEAE p. 23
       10. Flowers containing both stamens and
         pistils; inflorescences other than globose
         heads 11
         11. Perianth with 2 or 3 petals (2 or 3
          sepals) 12
          12. Ovary 1, with 3 compartments
           . . . . LILIACEAE p. 77
          12. Ovaries 3 or 6 13
           13. Inflorescence a raceme or
            spike . . . .
            . . JUNCAGINACEAE p. 23
           13. Inflorescence umbellate . . . .
            . . . . . BUTOMACEAE p. 24
         11. Petals 5 . . . . . . LINACEAE p. 244
   7. Perianth borne (on the receptacle) at the sum-
     mit of the ovary 14
     14. Stamens 6 . . . . AMARYLLIDACEAE p. 90
     14. Stamens 3 . . . . IRIDACEAE p. 91
5. One petal different from the others, and modified to
  form a lip (Figs. 80-97) . . . . . . ORCHIDACEAE p. 94

Section C

1. Calyx and stamens borne at the summit of the ovary 2
  2. Low matted herbs . . . . SAXIFRAGACEAE p. 192
  2. Stems or leaves upright 3
    3. Woody plants 4
      4. Trees; leaves lobed . . . . PLATANACEAE p. 198
      4. Shrubs; leaves simple . . . . CORNACEAE p. 286
    3. Herbs 5
      5. Leaves compound . . . . APIACEAE p. 277
      5. Leaves simple 6
        6. Leaves whorled . . . . *Galium* p. 344
        6. Leaves alternate 7
          7. Leaves narrowly oblong . . . .
          . . . . SANTALACEAE p. 130
         7. Leaves heart-shaped . . . .
          . . . . ARISTOLOCHIACEAE p. 131
1. Calyx and stamens borne at the base of the ovary 8
  8. Pistil 1 in each flower 9

9. Plants herbaceous *10*
    10. Juice milky, white or colored *11*
        11. Juice red-orange....... PAPAVERACEAE p. 165
        11. Juice white .... EUPHORBIACEAE p. 247
    10. Juice not milky *12*
        12. Styles or stigmas 2-several *13*
            13. Anthers not conspicuous *14*
                14. Styles 2-5, branched, many seeds per ovary (except in *Scleranthus*); fruit a capsule.... .... CARYOPHYLLACEAE p. 136
                14. Styles 2-3, the stigmas each with a tuft of hairs at tip; 1 seed per ovary; fruit an achene .... POLYGONACEAE p. 132
            13. Anthers conspicuous, yellow or red (Fig. 251) .... SAXIFRAGACEAE p. 192
        12. Style and stigma 1, unbranched or slightly lobed *15*
            15. Leaves divided into 2 half-leaflets (Fig. 211) .. .. *Jeffersonia* p. 163
            15. Leaves otherwise *16*
                16. Calyx about 1 cm in diameter, corollalike (Fig. 153) .... NYCTAGINACEAE p. 134
                16. Calyx much smaller *17*
                    17. Flowers in terminal racemes .... BRASSICACEAE p. 168
                    17. Flowers in axillary clusters .... URTICACEAE p. 129
9. Plants woody *18*
    18. Trees *19*
        19. Leaves compound .... OLEACEAE p. 301
        19. Leaves simple *20*
            20. Leaves opposite .... *Acer* p. 256
            20. Leaves alternate *21*
                21. Sap not milky; leaves asymmetrical at base; fruit a winged samara or drupe.......... .... ULMACEAE p. 127
                21. Sap milky; leaves symmetrical at base, lobed; fruit fleshy ....MORACEAE p. 128
    18. Shrubs *22*
        22. Leaves not toothed, coming after the flowers ....THYMELAEACEAE p. 273
        22. Leaves toothed, coming with the flowers .... RHAMNACEAE p. 258
8. Pistils several or many in each flower, or, if pistil only 1, then leaves several times compounded . RANUNCULACEAE p. 148

Section D

1. Petals not joined to each other *2*
    2. Calyx and corolla borne at the base of the ovary *3*
        3. Stamens 10 or fewer *4*
            4. Corolla symmetrical on several radii *5*
                5. Petals 3 *6*
                    6. Leaves simple........... LILIACEAE p. 77
                    6. Leaves pinnately compound *7*
                        7. Trees .... CAESALPINIACEAE p. 227
                        7. Small herbs ... LIMNANTHACEAE p. 249

5. Petals 4 or more *8*
    8. Stigmas 2-5, or each flower with several sep-
       arate pistils *9*
       9. Leaves 3-foliate *10*
          10. Herbs . . . .   OXALIDACEAE p. 242
          10. Trees or shrubs *11*
              11. Leaves opposite . . . .
                  . . . . STAPHYLEACEAE p. 253
              11. Leaves alternate . . . . . *Ptelea* p. 246
       9. Leaves simple *12*
          12. Woody plants *13*
              13. Leaves 3-5-lobed *14*
                  14. Ovary 2-lobed . . . .
                      . . . . ACERACEAE p. 254
                  14. Ovary not lobed . . . .
                      . . . . SAXIFRAGACEAE p. 192
              13. Leaves not lobed . . . .
                  . . . . RHAMNACEAE p. 258
          12. Herbs *15*
              15. Leaves lobed . . . .
                  . . . . GERANIACEAE p. 240
              15. Leaves not lobed *16*
                  16. Sepals 2; leaves on stem oppo-
                      site, unlobed, fleshy . . . .
                      . . . . PORTULACACEAE p. 134
                  16. Sepals or calyx lobes 4 or 5 *17*
                      17. Leaves opposite, or whorled
                          at several nodes on the stem
                          . . . . . CARYOPHYLLACEAE
                          . . . . p. 136
                      17. Leaves alternate, or clus-
                          tered at base of plant (in
                          which case the stem may
                          bear 1 pair of opposite
                          leaves) *18*
                          18. Leaves fleshy, much
                              thickened . . . .
                              CRASSULACEAE p. 191
                          18. Leaves flat or threadlike
                              19
                              19. Most or all of the
                                  leaves in a basal
                                  rosette . . . .
                                  . . SAXIFRAGACEAE
                                  . . . . p. 192
                              19. Leaves all on the
                                  stem . . . .
                                  . . LINACEAE p. 244
    8. Style and stigma 1, the stigma rarely 2-3-lobed
       20
       20. Herbaceous plants, or low woody peren-
           nials (not vines or tall shrubs) *21*
           21. Petals 5-8 *22*
               22. Leaves compound, with 2 or more
                   leaflets . . . . . . *Caulophyllum* p. 163
               22. Leaves simple . . . .
                   . . . . PYROLACEAE p. 289

21. Petals 4 *23*
>23. Leaves simple or pinnately compound ... BRASSICACEAE p. 168
>23. Leaves palmately compound ....CAPPARIDACEAE p. 190

20. Woody plants, climbing vines, or tall shrubs *24*
>24. Leaves compound *25*
>>25. Climbing or trailing vines *26*
>>>26. Leaflets 5 or more ....VITACEAE p. 260
>>>26. Leaflets 3 .... ANACARDIACEAE p. 250
>>25. Plants erect *27*
>>>27. Shrubs .... ANACARDIACEAE p. 250
>>>27. Trees *28*
>>>>28. Leaves opposite .... OLEACEAE p. 301
>>>>28. Leaves alternate .... CAESALPINIACEAE p. 227
>24. Leaves simple *29*
>>29. Leaves not lobed *30*
>>>30. Stems with spines .... *Berberis* p. 164
>>>30. Stems without spines *31*
>>>>31. Leaves with an abrupt short point (Fig. 351) .... AQUIFOLIACEAE p. 251
>>>>31. Leaves long-pointed (Figs. 352-53) ....CELASTRACEAE p. 252
>>29. Leaves palmately lobed and veined *32*
>>>32. Flowers in a raceme .... SAXIFRAGACEAE p. 192
>>>32. Flowers in a panicle .... VITACEAE p. 260

4. Corolla irregular *33*
>33. Leaves compound *34*
>>34. Sepals 2, small and scalelike .... FUMARIACEAE p. 166
>>34. Sepals 3-5, sometimes united, occasionally making a 2-lipped calyx *35*
>>>35. Stamens 10, leaves alternate *36*
>>>>36. Upper petal overlapped by the 2 side petals; stamens all free .... CAESALPINIACEAE p. 227
>>>>36. Upper petal overlapping the 2 side petals; stamens 10, with 9 filaments united and 1 free (except all free in *Baptisia*)......... FABACEAE p. 228
>>>35. Stamens 7; leaves opposite .... HIPPOCASTANACEAE p. 257
>33. Leaves simple, sometimes deeply cleft *37*
>>37. Flowers in heads or spikes, or solitary and erect .... POLYGALACEAE p. 246

55. Leaves 3-5-lobed ....
.... SAXIFRAGACEAE p. 192
55. Leaves not lobed *56*
    56. Leaves toothed .RHAMNACEAE p. 258
    56. Leaves not toothed ....
    .... CORNACEAE p. 286
54. Herbs *57*
  57. Leaves simple, not deeply cleft *58*
    58. Flowers in a close cluster surrounded
by 4 or more white petallike bracts
(Fig. 412) ..... CORNACEAE p. 286
    58. Flowers without petallike bracts *59*
      59. Petals narrow or fringed, dull or
whitish ....
.... SAXIFRAGACEAE p. 192
      59. Petals inverted heart-shaped, bright
yellow .... ONAGRACEAE p. 275
  57. Leaves compound, or palmately deeply
cleft *60*
    60. Petioles not winged at base (Figs.
392-94) .... ARALIACEAE p. 275
    60. Petioles winged at base, clasping the
stem (Figs. 400-401, 404) ....
.... APIACEAE p. 277
1. Petals united with each other *61*
  61. Calyx and corolla borne at the base of the ovary *62*
    62. Corolla symmetrical on several radii *63*
      63. Stamens 5-10 *64*
        64. Stigma entire; leaves simple (except *Solanum
dulcamara*) *65*
          65. Ovary not lobed *66*
            66. Corolla funnel-shaped, 12 mm or more
long; leaves with large triangular basal lobes
(Figs. 456-57) CONVOLVULACEAE p. 309
            66. Corolla shorter; leaves without triangular
basal lobes *67*
              67. Filaments united *68*
                68. Ovary 1 per flower, style separate
from stamens; juice not milky .....
.... PRIMULACEAE p. 298
                68. Ovaries 2 per flower, styles united
and fused with stamens to form a
central columnar gynostegium, juice
milky .. ASCLEPIADACEAE p. 305
              67. Filaments separate *69*
                69. Ovary 1 per flower *70*
                  70. Stems woody (or else creeping
and bearing tough evergreen
leaves) ... ERICACEAE p. 291
                  70. Stems herbaceous (except in
Lycium) and erect; leaves not
evergreen (except possibly in a
rosette at ground level) *71*
                    71. Stamens opposite the corolla
lobes...................
.... PRIMULACEAE p. 298

71. Stamens alternate with the
    corolla lobes *72*
    72. Leaf bases forming wings
        down the stem........
        SCROPHULARIACEAE
        .... p. 329
    72. Stems at most ridged,
        not winged ....
        ..SOLANACEAE p. 327
69. Ovaries 2 per flower but attached
    to a single style...............
    .... APOCYNACEAE p. 304
65. Ovary deeply 4-lobed BORAGINACEAE p. 313
64. Stigma 2-5-cleft; leaves simple or compound *73*
    73. Leaves or leaflets toothed................
        .... HYDROPHYLLACEAE p. 312
    73. Leaves or leaflets not toothed *74*
        74. Leaves simple or pinnately compound ....
            .... POLEMONIACEAE p. 310
        74. Leaves 3-foliate *75*
            75. Petals with copious wavy hairs ....
                .... GENTIANACEAE p. 303
            75. Petals without hairs...............
                .... OXALIDACEAE p. 242
63. Stamens 2 or 4 *76*
    76. Woody plants .... OLEACEAE p. 301
    76. Herbs *77*
        77. Leaves all at the base of the scapes ....
            .... PLANTAGINACEAE p. 342
        77. Leaves on the stem *78*
            78. Ovary 4-lobed ...... LAMIACEAE p. 320
            78. Ovary not lobed, or shallowly 2-lobed ....
                .... SCROPHULARIACEAE p. 329
62. Corolla irregular, without all lobes alike *79*
    79. Plants green *80*
        80. Plants terrestrial, sometimes in damp places *81*
            81. Stamens united in 1 or 2 groups ....
                .... POLYGALACEAE p. 246
            81. Stamens separate *82*
                82. Sepals usually 5, united *83*
                    83. Ovary with 4 lobes ....
                        .... LAMIACEAE p. 320
                    83. Ovary not lobed, or shallowly lobed ..
                        .... SCROPHULARIACEAE p. 329
                82. Sepals 2, separate ..FUMARIACEAE p. 166
        80. Plants submerged, with finely divided leaves......
            .... LENTIBULARIACEAE p. 341
    79. Plants not green, parasitic .OROBANCHACEAE p. 340
61. Calyx and corolla apparently borne at the summit of the
    ovary *84*
    84. Stamens free from the corolla tube *85*
        85. Corolla not blue, sometimes purple ....
            .... ERICACEAE p. 291
        85. Corolla blue .... CAMPANULACEAE p. 359
    84. Stamens borne on the corolla tube *86*
        86. Leaves without stipules (except sometimes in the
            shrub *Viburnum*) and not whorled *87*

87. Leaves simple, or, if pinnately compound, shrubs
    88
    88. Leaves alternate; corolla with a split along the
        upper side . . . . LOBELIACEAE p. 360
    88. Leaves opposite; corolla not split along the
        upper side . . . . CAPRIFOLIACEAE p. 348
87. Leaves much divided; delicate herbs . . . .
    . . . . ADOXACEAE p. 357
86. Leaves with stipules, or whorled . RUBIACEAE p. 344

Descriptive Spring Flora

SPARGANIACEAE   Bur-reed Family

Aquatic perennials; leaves long and ribbonlike, erect or limp and partly floating; inflorescence spikelike with subtended leafy bracts; flowers in spherical heads, the lower pistillate and the upper staminate; perianth of 3-6 sepals; stamens 3; pistil with 1 style and a lateral stigma or 2 stigmas; fruit an achene.

Sparganium   Bur-reed

Characters as for family.

S. eurycarpum Engelm. Plants to 12 dm high, inflorescence branched, stigmas 2.—In mud or shallow water, throughout the state.

JUNCAGINACEAE   Arrowgrass Family

Plants of wet places; leaves grasslike, basal or alternate, sheathing at the base; inflorescence a terminal spike or raceme, flowers with perianth in 1 or 2 whorls; stamens as many as the perianth parts and attached at the base; ovules 1 or 2, style lacking or very short; fruit 3 or 6 follicles.

a. Raceme subtended by bracts; follicles widely spreading ....
.... 1. *Scheuzeria*
a. Raceme bractless; follicles erect .... 2. *Triglochin*

1. Scheuchzeria

Perennial bog herbs, with alternate leaves and bracted acemes; perianth of 6 segments in 2 whorls; stamens 6; pistils 3, divergent.

S. palustris L. (Fig. 1). Plants to 4 dm tall; leaves with dilated sheaths and a terminal pore.—Cold sphagnum bogs, in scattered localities of northern and eastern Wisconsin.

Fig. 1. *Scheuchzeria palustris,*   Fig. 2. *Triglochin elatum,* × .25.
× .40.

2. Triglochin   Arrowgrass

Perennial herbs; leaves all basal, somewhat fleshy with a conspicuous sheath; inflorescence on an erect bractless scape with an elongate raceme of small greenish flowers; perianth of 2 whorls; anthers broad, sessile; ovaries 3 or 6, a solitary ovule in each; style lacking.

T. elatum Nutt. (Fig. 2). Plants to 8 dm tall; raceme 1-4 dm; ovaries 6, attached to a central axis along the inner margin. —Wet situations or bog localities, southeastern Wisconsin and scattered localities from Burnett to Door counties in northern Wisconsin. (= *T. maritima* of other manuals.)

BUTOMACEAE   Flowering Rush Family

Aquatic herbs; leaves sometimes floating, grasslike or petiolate and dilated at base; flowers perfect; sepals 3; petals 3, usually conspicuous; stamens 9-many; pistils 3-many, in a whorl; fruit a whorl of follicles.

Butomus   Flowering Rush

Sepals and petals persistent, the sepals nearly as large as

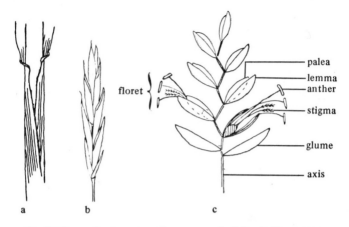

Fig. 3. Diagnostic characters of grasses: *a.* leaf sheath (*Agropyron repens*, × 2); *b.* spikelet (*Bromus inermis*, × 1.50); *c.* spikelet diagram.

the petals; stamens 9; pistils 6, united at the base.

B. umbellatus L. Perennial herbs; leaves basal, erect or floating, linear; inflorescence an umbel on a scape 1-1.5 m tall; flowers pink, 2-2.5 cm wide; pedicels 5-10 cm long; subtended by acute, lanceolate-triangular bracts. – Native of Europe; Jefferson, Waukeska, and Oconto counties.

POACEAE (GRAMINEAE)   Grass Family

Tall slender plants mostly with hollow round stems (culms); leaves consisting of 2 parts, the lower (sheath) rolled around the stem but with the edges usually not joined (Fig. 3*a*), the upper part (blade) ribbonlike and ascending or spreading; flowers minute, arranged in spikelets, with a pair of scales (glumes) at the base and 1-many florets on opposite sides of the axis (Fig. 3*b*, 3*c*); each floret subtended by 2 bracts, an outer one (lemma) and an inner one (palea); stamens 3, the anthers attached in the middle (versatile); pistil with 2 feathery stigmas, the ripened ovary becoming a grain.

A large and difficult family. See Norman C. Fassett, *Grasses of Wisconsin* (Madison, Wis.: University of Wisconsin

Press, 1951). The following list of species that flower in Wisconsin before June 15 was compiled by Mrs. Maarit Threlfall.

Agropyron dasystachyum (Hook.) Scribn.

Agropyron repens (L.) Beauv.   Quack Grass

Agropyron smithii Rydb.   Western Wheat Grass

Agrostis hyemalis (Walt.) BSP.   Tickle Grass

Alopecurus aequalis Sobol.

Alopecurus carolinianus Walt.

Alopecurus pratensis L. Meadow Foxtail

Arrhenatherum elatius (L.) Mert. & Koch   Tall Oat Grass

Bromus inermis Leyss. Smooth Brome

Bromus japonicus Thunb. Japanese Chess

Bromus kalmii Gray

Bromus purgans L.

Bromus secalinus L.   Chess

Bromus tectorum L.   Cheat

Dactylis glomerata L. Orchard Grass

Danthonia spicata (L.) Beauv.   Oat Grass

Deschampsia caespitosa (L.) Beauv.   Hair Grass

Festuca obtusa Biehler Nodding Fescue

Festuca octoflora Walt. (*Vulpia octoflora* (Walt.) Rydb.)   Six-weeks Fescue

Festuca ovina L.   Sheep Fescue

Festuca rubra L.   Red Fescue

Festuca saximontana Rydb. (*F. ovina* L., var. *saximontana* (Rydb.) Gl.)

Glyceria borealis (Nash) Batchelder   Manna Grass

Glyceria grandis S. Wats.

Glyceria striata (Lam.) Hitchc.

Hierochloë odorata (L.) Beauv.   Sweet Grass

Hordeum jubatum L. Squirrel Tail

Hystrix patula Moench Bottle-brush Grass

Koeleria cristata (L.) Pers. June Grass

Lolium perenne L. Common Rye Grass

Lolium multiflorum Lam. Italian Rye Grass

Milium effusum L.   Millet Grass

Oryzopsis asperifolia Michx. Rice Grass

Oryzopsis pungens (Torr.) Hitchc.

Panicum boreale Nash Panic Grass

Panicum columbianum Scribn.

Panicum commonsianum Ashe, var. *euchlamydeum* (Shinners) Pohl

Panicum depauperatum Muhl.

Panicum lanuginosum Ell.

Panicum leibergii (Vasey) Scribn.

Panicum linearifolium Scribn.

Panicum oligosanthes
  Schultes, var.
  scribnerianum (Nash) Fern.
Panicum perlongum Muhl.
Panicum praecocius Hitchc.
  & Chase
Panicum subvillosum Ashe
Phalaris arundinacea L.
  Reed Canary Grass
Phleum pratense L.
  Timothy
Poa alsodes Gray   Bluegrass
Poa annua L.   Annual Blue-
  grass
Poa compressa L.   Canada

Bluegrass
Poa nemoralis L.
Poa palustris L.
Poa pratensis L.   Kentucky
  Bluegrass
Poa saltuensis Fern. & Wieg.
Schizachne purpurascens
  (Torr.) Swallen
Secale cereale L.   Rye
Stipa comata Trin. & Rupr.
  Needle Grass
Stipa spartea Trin.
Stipa viridula Trin.
Triticum aestivum L.
  Wheat

CYPERACEAE   Sedge Family

Grasslike plants, mostly perennial; culms (stems) often triangular in cross section; leaves 3-ranked (Fig. 4*a*), best seen
by viewing from above; edges of the leaf sheath joined to
make a tube surrounding the culm (Fig. 9); leaf blades often
pleated (M-shaped, Fig. 8*h*), but sometimes flat or V- or
U-shaped in cross section, or reduced in size, or even absent
in some genera; flowers lacking a perianth, but often surrounded by several bristles or by a sac (perigynium), perfect
or unisexual, in spikelets with spirally arranged scales (Fig.
4*b*) hiding all but the 1-3 stamens and 2-3 threadlike stigmas
of the flowers they subtend (Fig. 4*c*); fruit an achene, bluntly to acutely triangular or lens-shaped (biconvex or plano-
convex). Spikelets arranged in simple or compound racemes,
spikes, or umbels, or combinations of these, sometimes in a
dense cluster (glomerule). In some species of *Carex* and
Scirpus, the hollow concentric sheaths of rosette leaves elongate to form "pseudoculms" that may be taller than the
flowering culms and leafy to the top. A large family with
over 200 species in Wisconsin. Identification requires
10-15 × magnification.

a. Flowers perfect; achene surrounded only by bristles; spikelets solitary, in umbels, or in crowded glomerules *b*
    b. Perianth bristles 20 or more per flower, extending conspicu-
      ously beyond the scales in whitish masses . . . . . . 1. *Eriophorum*
    b. Bristles mostly hidden, 8 or less per flower, mostly shorter
      than the scales *c*

    c. Style base enlarged and persistent at achene summit; bract-
       less spikelets solitary at culm tips; leaf blades absent ......
       .... 2. *Eleocharis*
    c. Style base not enlarged or persistent; spikelets 1-many per
       culm; leaf blades present or absent; 1-several conspicuous
       bracts subtending the inflorescence ....3. *Scirpus*
a. Flowers unisexual, borne on the same or different spikelets; pistillate flower and achene surrounded by a perigynium; spikelets in a
  spike or raceme, rarely solitary; leaf blades present ..... 4. *Carex*

1. Eriophorum   Cotton Grass

Plants of wet peaty places, with conspicuous cottony inflorescences; leaf blades not very noticeable; stamens 1-3.

a. Stems with a single spikelet .... 1. *E. spissum*
a Stems with 2 or more spikelets............. 2. *E. angustifolium*

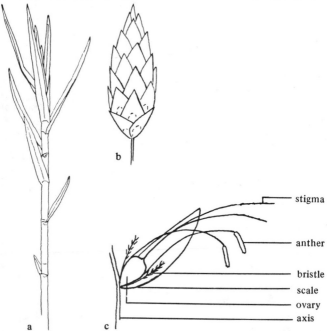

Fig. 4. Diagnostic characters of sedges: *a.* 3-ranked leaf arrangement
(*Dulichium arundinaceum,* × .60); *b.* diagram of spiral arrangement
of scales; *c.* diagram of sedge floret (except *Carex*).

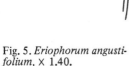

Fig. 5. *Eriophorum angusti-folium,* × 1.40.

Fig. 6. *Eleocharis palustris,* × 4.65.

1. **E. spissum** Fern. Hare's-tail. Dense tussocks of many culms like a big green pincushion; basal leaves present, about 1 mm wide. —Acid sphagnum-tamarack bogs; northern and eastern Wisconsin.

2. **E. angustifolium** Honck (Fig. 5). Culms solitary, scattered, each bearing several spikelets; basal leaves present, marked with red, up to 8 mm wide. —Open peaty meadows, fens, and bogs; throughout the state.

2. Eleocharis   Spike Rush

Our species of this genus, important components of wet meadows, shores, and shallow-water vegetation, are noted for their numerous erect, slender, green, round (rarely flat), leafless culms, some of which bear solitary bractless spikelets at the tips (Fig. 6). Some species are caespitose (forming small clumps or tufts), but many form extensive beds or sods. The leaves are reduced to a few short tightly fitting sheaths at the base of the culm. Some species are red at the plant base. Since the uniquely shaped persistent style bases are developed on only a few species by mid-June, it is best to delay identification until later in the summer.

3. Scirpus   Bulrush

A genus of mostly robust sedges occurring in shallows, shores, marshes, wet ditches, and bogs, rarely in dry places; leaves sometimes only bladeless sheaths; inflorescence of numerous spikelets in a compound umbel or dense glomerule, subtended by 1-several leafy bracts; bristles 1-6; stamens 2-3; styles 2-3-cleft; achenes ripening in summer and fall.

a. Leaves and culms numerous, submersed, long, flexible, and thread-like, abundant on a stream or lake bottom .... 1. *S. subterminalis*
a. Culms erect, emergent, or else not in water *b*
  b. Plants 1-2 dm tall, resembling *Eleocharis*, with solitary spikelets on slender leafless culms *c*
    c. Bristles protruding beyond scales, giving spikelet a whitish cast; culms in rows, in cold wet peat .... 2. *S. hudsonianus*
    c. Bristles hidden *d*
      d. Dense tufts of many slender wiry culms, in wet peat; lowest spikelet scale with prolonged tip sometimes longer than spikelet; culms round or ridged, smooth .... 3. *S. cespitosus*
      d. Loose tufts of stems, on dry sites; scale not prolonged; culms rough, triangular .... 4. *S. clintonii*
  b. Plants 2-20 dm tall; spikelets several or many at culm tip, in an umbellate cluster or crowded glomerule *e*
    e. Inflorescence borne on one side, the single erect involucral bract appearing to be a continuation of the culm; leaf blades few, mostly reduced or absent; the tall leafless green culms conspicuous *f*
      f. Culms round in section *g*
        g. Culms pale or blue green, soft when squeezed; spikelets ovoid, loosely disposed in an open inflorescence, rich orange brown .... 5. *S. validus*
        g. Culms dark olive green, firm and hard; spikelets ovoid to cylindrical, in a stiff condensed inflorescence, pale brown .... 6. *S. acutus*
      f. Culms sharply triangular .... 7. *S. americanus*
    e. Spikelets in an erect open umbellate cluster subtended by several spreading leafy bracts; leaf blades long, numerous, and conspicuous *h*
      h. Culms and tall pseudoculms sharply triangular, about 1 cm wide; spikelets 1.5-3.5 cm long, 5-10 mm thick; elongate rhizomes with tuberous swellings 2 cm in diameter .... 8. *S. fluviatilis*
      h. Culms slender, round or bluntly triangular; spikelets much smaller, often very numerous; no tubers *i*
        i. Bristles straight or with 1 or 2 kinks, minutely barbed, to one and one-half times as long as the achene *j*
          j. Leaf sheaths green; culms in rather dense clumps .... 9. *S. atrovirens*
          j. Leaf sheaths red-tinged; culms solitary or in small groups .... 10. *S. microcarpus*
        i. Bristles much kinked, smooth, longer than achene, often protruding between scales; culms clumped *k*

1. S. **subterminalis** Torr. Swaying Rush. Rarely fruiting. —In cold or soft water, in northwestern and northeastern Wisconsin, rarely to Jefferson County.

2. S. **hudsonianus** (Michx.) Fern. The whitish bristles reminiscent of *Eriophorum* but not as long or numerous. —Bogs and springs, eastern counties of the state.

3. S. **cespitosus** L. Arctic Bulrush. Wiry green pincushions atop raised tussocks. —Boggy shores, wet coniferous forests, and rock crevices near the Great Lakes, and in marly peaty fens in Dane and Jefferson counties.

4. S. **clintonii** Gray. —Sandy woods and dry slopes, rare; Adams and Brown counties.

5. S. **validus** Vahl. Soft Roundstem Bulrush (Fig. 7). —Common in lakes and marshes, forming extensive beds near shores and in shallow water to 1 m deep; throughout the state.

6. S. **acutus** Bigelow. Hard Roundstem Bulrush. More tolerant of wave action and sometimes found in deeper water than *S. validus*; hybrids between the 2 occur. —Mostly eastern and northern Wisconsin.

7. S. **americanus** Pers. Three-square Bulrush. Forming beds or colonies of stiff culms; spikelets few, sessile. —Shores,

Fig. 7. *Scirpus validus,* × 1.

especially sandy ones, and in shallow water; eastern half of the state and northwestern counties.

8. **S. fluviatilis** (Torr.) Gray. River Bulrush. Our coarsest sedge, with broad leaves, 1-3 cm wide, strongly M-shaped in section; forming extensive beds of leafy culms and pseudo-culms 1-2 m tall, with razor-sharp edges; the large tubers often floating up in numbers.—On shores and in shallow water, alone or with cattails, bur-reeds, and roundstem bul-rushes; absent from north-central Wisconsin.

9. **S. atrovirens** Willd. A common pioneer caespitose sedge of shores, wet meadows, and ditches, of husky growth; leaves M-shaped, 1-3 cm wide; inflorescences stiff, rather flat-topped, blackish green, composed of many small crowded spikelets and numerous bracts.—Throughout the state.

10. **S. microcarpus** Presl. Culms stouter than in No. 9; leaves only 4-15 mm wide.—Marshes and shores; northern and eastern Wisconsin.

11. **S. pendulus** Muhl. Caespitose; inflorescence drooping; spikelets golden brown.—Wet clearings in woods and other disturbed areas, not very common; from Dane to Brown counties and eastward, also La Crosse County.

12. **S. cyperinus** (L.) Kunth. Wool Grass. Densely caes-pitose; leaves numerous, long, slender, about 5 mm wide, C-shaped in cross section; spikelets numerous, woolly, light brown, in erect or drooping compound umbellate clus-ters.—Widespread along damp shores, in meadows, and in shallow water up to 5 cm deep; in the north dominating many acres of disturbed meadows and every roadside ditch.

4. Carex   Sedge

A large genus of perennials, with 145 species in Wiscon-sin. Leaves grasslike, 0.5-40 mm wide, often M-shaped in cross section but sometimes V- or C-shaped or flat; inflores-cence usually elevated above the leaves but sometimes hidden among the leaf bases; spikelets in racemes or spikes, or sometimes in crowded branched racemes or solitary at culm tips, each subtended by a large or small bract; flowers unisexual, each subtended by a scale; pistil surrounded by a flask-shaped sac, the perigynium (Fig. 8*c*); ovary and achene lens-shaped or triangular, the 2 or 3 stigmas protruding from the apex or the prolonged beak of the perigynium; stamens 3 per flower (Fig. 8*d*).

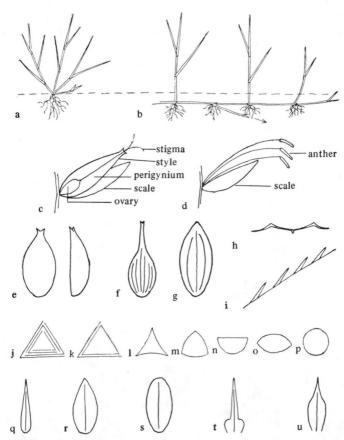

Fig. 8. Diagrams of *Carex* characters: *a.* caespitose growth; *b.* rhizomatous growth; *c.* pistillate floret; *d.* staminate floret; *e.* plano-convex perigynium; *left,* front view; *right,* side view; *f.* beaked perigynium, many nerves; *g.* beakless perigynium, few nerves; *h.* cross section of M-shaped leaf; *i.* upwardly scabrous leaf margin; *j.* cross section of pseudoculm (several leaves); *k.* cross section on culm with one leaf; *l-p.* shapes of perigynia in cross section–*l.* acutely triangular; *m.* bluntly triangular; *n.* plano-convex; *o.* biconvex; *p.* round (terete); *q-u.* scale shapes–*q.* acuminate; *r.* ovate-acute; *s.* blunt; *t.* abruptly awned; *u.* tapered into awn.

The 2 growth forms and some plant parts important in identification are shown in Figs. 8, 9. Red, pink, or purple coloring is a prominent feature of the basal sheaths of many of the species with 3 stigmas. The basal sheaths may be strongly fibrillose, either longitudinally, e.g., *C. sprengelii* and *C. pensylvanica*, or in their pinnate branching, e.g., *C. stricta* and *C. lacustris*. The perigynia and spikelets change rapidly after flowering, causing the same species at 2 stages of development to look less alike than 2 different species at the same stage. If stamens or stigmas are fresh and conspicuous, it is best to wait 1-3 weeks to obtain mature perigynia for certain identification. Most species flower in May and drop their fruits in late June. The earliest species are in the *Pensylvanica* Group, flowering in April and maturing in May, while the *Scoparia* Group is the latest, flowering through May and June and maturing its fruits in July and August. A few species in several groups retain fruits into fall. Occasionally fungi cause enlarged or distorted perigynia. Site specificity makes habitat important to *Carex* identification.

Descriptions of the species in this section are arranged in natural groupings named for the most common member species, e.g., the closely related species from the subgenera Albae, Digitatae, and Montanae of technical manuals are placed together in the *Pensylvanica* Group.

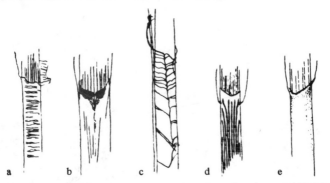

Fig. 9. *Carex* leaf sheath types as seen from inner face (opposite blade): *a.* cross-wrinkled (*C. vulpinoidea,* × 1.80); *b.* darkened at apex (*C. trichocarpa,* × 1.80); *c.* pinnate-fibrillose (*C. stricta,* × 1.80); *d.* parallel-nerved, prolonged at apex (*C. sartwellii,* × 1.80); *e.* slightly thickened at apex (*C. cephalophora,* × 1.80).

A. Culms with a single terminal spikelet (Figs. 13, 14) .... Key I
A. Culms with 2-several spikelets each *B*
    B. Stigmas 3; achene and perigynium mostly terete or triangular; spikelets mostly stalked, elongate, and of 2 sorts (Figs. 10-12) *C*
        C. Perigynia minutely pubescent even when young (Figs. 10, 17)..................................... Key II
        C. Perigynia glabrous on surface *D*
            D. Style breaking from mature achene; perigynia 2-6 mm long; teeth absent or only 0.1 mm long (Figs. 11, 12)... .... Key III
            D. Style tough and persistent; perigynia 2.5-20 mm long, often much inflated; teeth 0.2-2 mm long (Figs. 15-23) .... Key IV
    B. Stigmas 2; achene and often perigynium flattened or plano-convex *E*
        E. Upper spikelets wholly staminate, the lower ones wholly pistillate, often stalked and elongate (Fig. 25a) ..... Key V
        E. Spikelets all alike, only 1-3 times as long as wide, and sessile (Figs. 27, 29) *F*
            F. Staminate flowers at tip of spikelet (Figs. 29, 33); plants caespitose or rhizomatous (Fig. 8a, 8b)........ Key VI
            F. Staminate flowers at base of spikelet (Figs. 26, 28); plants caespitose .... Key VII

Key I

a. Stigmas 3 *b*
    b. Culms shorter than the leaves, often buried in plant base *c*
        c. Lower scales of spikelet leaflike; in deciduous forests *d*
            d. Widest leaf blades 2-3 mm .... 13. *C. jamesii*
            d. Widest leaf blades 4-5 mm ....14. *C. backii*
        c. Scales not leaflike; in dry open barrens .... 4. *C. umbellata*
    b. Culms longer than the leaves *e*
        e. Spikelet thimblelike, 30-60-flowered; in wet deciduous forests .... 73. *C. typhina*
        e. Spikelet slender, few-flowered; in bogs *f*
            f. Perigynia about 4 mm long, ascending, blunt, obovoid .. .. 50. *C. leptalea*
            f. Perigynia about 8 mm long, soon reflexed, slenderly lanceolate ....49. *C. pauciflora*
a. Stigmas 2; perigynia curved outward; tiny plants in bogs ....51. *C. gynocrates*

Key II

a. Leaves softly pubescent *b*
    b. Terminal spikelet wholly staminate and linear; in deciduous forests .... 12. *C. hirtifolia*
    b. Terminal spikelet mostly pistillate, short-cylindrical; in sand... .... 42. *C. swanii*
a. Leaves glabrous (may be rough or smooth) *c*
    c. Plants tufted, small, mostly 3-15 cm tall *d*
        d. Spikelet scales abruptly green-awned (Fig. 8u); evergreen plants of cool acid forests............. 9. *C. pedunculata*
        d. Scales blunt to acute; plants not evergreen *e*

    e. Some spikelets hidden in base of plant or on culms of
       unequal length *f*

       f. Perigynia 2-3 mm long; beaks 0.3-1 mm long *g*

          g. Pistillate spikelets 2-4, crowded at culm tip; plant
             base pink lavender; perigynia triangular, ellipsoidal;
             in dry to damp acid sand .... 6. *C. emmonsii*

          g. Pistillate spikelets 1-2 (3); perigynia globose, 2-3
             mm long, beak 0.3-1 mm long *h*

             h. Basal leaf sheaths reddish, not fibrillose; scales
                half as long as perigynia; in coniferous woods ..
                .... 8. *C. deflexa*

             h. Basal sheaths with stiff tufts of red or brown
                fibrils; scales about as long as perigynia; in dry
                prairies...................... 3. *C. abdita*

       f. Perigynia 3-4 mm long; beaks 1-2 mm long; stiff tufts
         in dry sand barrens .... 4. *C. umbellata*

    e. All spikelets elevated away from base of plant; culms
       about equal in length *i*

       i. Tiny sheathing bract hidden at base of flowering culm
         just above the leaves *j*

         j. Bract purple; in dry prairies ... 10. *C. richardsonii*

         j. Bract yellowish; in coniferous forests ....
         .... 11. *C. concinna*

       i. No bract at base of flowering culm *k*

         k. Perigynium longer than broad; spikelets crowded
           together; in cool deciduous forests .... 5. *C. peckii*

         k. Perigynium globose; spikelets spread out, the ter-
           minal (staminate) one sometimes conspicuous *l*

           l. Culms 12-54 cm tall; bases strongly red, not
             very fibrillose nor rhizomatous; widest leaves
             3-6 mm broad; in cool forests . .7. *C. communis*

           l. Culms 10-30 (35) cm tall; bases with red brown
             fibrils and elongate slender rhizomes, forming
             extensive sods or colonies of tufts; leaves 1-3
             mm broad; in all upland habitats ....
             .... 2. *C. pensylvanica*

c. Culms slender, erect, 10-50 cm tall, in small groups or scattered
   on elongate rhizomes (Fig. 8*b*) *m*

    m. Leaf blades 1-3 mm broad, often inrolled so that margins
       are hidden; in bogs .... 53. *C. lasiocarpa*

    m. Leaf blades 2.5-7 mm broad, M-shaped *n*

       n. Perigynia 3-4 mm long, finely woolly; in wet meadows,
         low prairies, fields................. 52. *C. lanuginosa*

       n. Perigynia 5-9 mm long, with sparse straight hairs, reveal-
         ing ribs *o*

         o. Purple blotch at apex of leaf sheath; in low prairies
           and wet meadows .... 59. *C. trichocarpa*

         o. No purple blotch; in dry sand... 54. *C. houghtoniana*

Key III

a. Parts of leaves obviously pilose or pubescent *b*

    b. Bract at base of inflorescence with long closed sheath *c*

       c. Terminal spikelet pistillate at tip; in wet deciduous forests
       .... 35. *C. davisii*

       c. Terminal spikelet wholly staminate; in or near bogs ....
       .... 39. *C. castanea*

 b. Bract sheathless or nearly so *d*
  d. Perigynia abrupt-beaked, ribbed; in dry prairies . . . .
  . . . . 44. *C. torreyi*
  d. Perigynia beakless, faintly nerved; in moist meadows . . . . . .
  . . . . 43. *C. pallescens*
a. Leaves glabrous (sheaths sometimes minutely hairy) *e*
 e. Leaves under 1 mm broad; plant forming tiny sods under white
  or red cedars . . . . 1. *C. eburnea*
 e. Leaves broader *f*
  f. Stems loosely scattered, often from elongate rhizomes (Fig.
   8*b*); or else bracts of the inflorescence with no sheaths *g*
   g. Wider leaf blades 3-12 mm broad, green; forest species *h*
    h. Surfaces of entire plant rough; staminate spikelet
     hidden; in cool wet woods and edges . . 38. *C. scabrata*
    h. Plant rough only on leaf margins; staminate spikelet
     long-stalked; in sugar maple forests . . . . . 28. *C. woodii*
   g. Leaf blades mostly narrower, blue green; in open places *i*
    i. Terminal spikelet pistillate at tip; in low prairies, wet
     meadows, and bogs . . . . 55. *C. buxbaumii*
    i. Terminal spikelet wholly staminate *j*
     j. Roots felted with yellow hairs; spikelets drooping;
      in bogs *k*
      k. Scales ovate, as broad as the perigynia . . . .
      . . . . 47. *C. limosa*
      k. Scales lanceolate, narrower than the perigynia
      . . . . 48. *C. paupercula*
     j. Roots not yellow; spikelets erect *l*
      l. Lowest spikelet at plant base; on limy shores
       and meadows . . . . 29. *C. crawei*
      l. Spikelets well above plant base *m*
       m. Perigynia with slight beak; in limy prairies *n*
        n. Leaf blades 3-7 mm broad; in dry sites . .
        . . . . 27. *C. meadii*
        n. Leaf blades 2-4.5 mm broad; in wet
         peaty sites . . . . 26. *C. tetanica*
       m. Perigynia with a conspicuous beak or none;
        in acid bogs *o*
        o. Perigynia far apart, beaked . . . .
        . . . . 31. *C. vaginata*
        o. Perigynia crowded, blunt; plant very
         glaucous . . . . 30. *C. livida*
  f. Plants strictly caespitose (Fig. 8*a*); bracts with short or long
   sheaths *p*
   p. At least some spikelets hidden in leaf bases; in dry sandy
    barrens . . . . 4. *C. umbellata*
   p. All spikelets elevated above the plant bases *q*
    q. Perigynia usually soon divergent to reflexed; beak
     bidentate, with erect teeth; pistillate spikelets sub-
     globose to thick-cylindrical, crowded at stem tip and
     often hiding staminate spikelets *r*
     r. Perigynia green, 1-3 mm long, 6-10-nerved; leaves
      1-2.5 mm wide; on wet sand beaches . . . .
      . . . . 46. *C. viridula*
     r. Perigynia turning yellow or orange, 3.5-6 mm long,
      8-18-nerved; leaves 1.5-5 mm wide; in wet sand,
      marl, or damp forests . . . . 45. *C. flava*

q. Perigynia ascending, orifice entire; spikelets elongate and very slender *s*
- *s.* Spikelets held well above the leaves, near tip of culm, mostly long-pedunculate and drooping; perigynia straight (Fig. 12) *t*
  - *t.* Terminal spikelet often pistillate at tip *u*
    - *u.* Perigynia 2-4 mm long *v*
      - *v.* Perigynia ellipsoidal, beakless, bluntly triangular; leaves green; in upland and wet forests . . . . 32. *C. gracillima*
      - *v.* Perigynia sharply triangular, tapering to a sharp beak; leaves blue green; in wooded springs . . . . .36. *C. prasina*
    - *u.* Perigynia 4.5-6 mm long, lanceolate, bluntly triangular or terete; achene on a tiny stalk; in wet acid woods . 37. *C. debilis,* var. *rudgei*
  - *t.* Terminal spikelet always entirely staminate *w*
    - *w.* Perigynia 10-50 per spikelet; plants 1-5 dm tall *x*
      - *x.* Perigynia globose, abruptly long-beaked; plant base very brown, very fibrillose; in deciduous woods . . . . 34. *C. sprengelii*
      - *x.* Perigynia lanceolate-ovoid, tapering to a beak; plant base red, not fibrillose; in cool acid forests. . . . . 33. *C. arctata*
    - *w.* Perigynia 1-10 per spikelet; plants 1-3 dm tall *y*
      - *y.* Plants with arching stolons; perigynia bristly, yellow; in floodplain forests . . . . 41. *C. assiniboinensis*
      - *y.* Plants with stolons absent; perigynia glabrous, green; in cool acid woods . . . . 40. *C. capillaris*
- *s.* Spikelets spaced out along the culms, which are no longer than the leaves, mostly erect, short-pedunculate; perigynia often curved (Fig. 11) *z*
  - *z.* Widest leaf blades 1.5-4 cm broad; in maple-basswood forests *a*
    - *a.* Leafless bracts, scales, and plant base red; leaves evergreen . . . . 20. *C. plantaginea*
    - *a.* Leafy bracts, scales, and bases green; leaves dying in the winter . . . . .17. *C. albursina*
  - *z.* Leaf blades narrower *b*
    - *b.* Nerves or ribs of perigynia 2-20 *c*
      - *c.* Leaves green; perigynia sharply triangular, obovoid, sharp-beaked; scales awned; in damp acid woods. . . . . 18. *C. leptonervia*
      - *c.* Leaves blue green; perigynia bluntly triangular, ellipsoidal, bluntly beaked; scales blunt; in limy wet meadows, marl, and woods . . . . 25. *C. granularis,* var. *haleana*
    - *b.* Nerves or ribs more numerous (25-50) *d*
      - *d.* Perigynia straight, terete to bluntly triangular, blunt at apex *e*
        - *e.* Staminate spikelet long-stalked, with round scales; in low prairies . . . . 24. *C. conoidea*

      e. Staminate spikelet often partly hidden; scales acute or awned; in deciduous forests and floodplains. . . . . .
. . . . 21. *C. amphibola,* var. *turgida*

   d. Perigynia curved or beaked, or both *f*
      f. Perigynia as long as wide, sharply triangular; scales acute; in maple-basswood forests . . . . 19. *C. digitalis*
      f. Perigynia longer than wide, bluntly triangular; scales with abrupt awns (Fig. 11*b*) *g*
         g. Perigynia with 30-50 impressed nerves, beaked; in maple-basswood forests *h*
            h. Bract sheaths white-hispid; perigynia 4-5 mm long, curved . . . . .
. . . . 23. *C. hitchcockiana*
            h. Bract sheaths not ciliate; perigynia 3-4 mm long, nearly straight . . . . . .22. *C. oligocarpa*
         g. Perigynia with 25-30 raised nerves *i*
            i. Perigynia crowded, curved, and mostly beakless; in southern deciduous forests . . .15. *C. blanda*
            i. Perigynia spaced out on spikelet, with short curved beak; in coniferous or mixed forests . . . . .
. . . . 16. *C. ormostachya*

Key IV

a. Culms solitary or in scattered groups from elongate rhizomes (Fig. 8*b*); in open wetlands *b*
   b. Leaves 1.5-3 mm wide, inrolled; pistillate spikelets far apart; perigynia few, 4-7 mm long; in bogs . . . . . . 61. *C. oligosperma*
   b. Leaves broader, not inrolled *c*
      c. Basal sheaths pinnate-fibrillose (Fig. 9*c*); leaves M-shaped, not conspicuously cross-veined *d*
         d. Leaves blue green, glabrous, very harsh on edges; perigynia slenderly ovoid-cylindrical, with a tiny extra tooth (under 1 mm long) next to one of the 2 apical teeth (Fig. 15); in all wetlands . . . . 56. *C. lacustris*
         d. Leaves green, parts of sheaths or blades often pubescent; ovoid perigynia tapering to a beak with 2 conspicuous teeth, 1-3 mm long (Fig. 16); in wet meadows and shores *e*
            e. Most of sheaths conspicuously pubescent . . . .
. . . . 57. *C. atherodes*
            e. Sheaths minutely pubescent only at apex, or glabrous
. . . . .58. *C. laeviconica*
      c. Basal sheaths not fibrilloses; leaves pale yellowish green, tending to be C-shaped, with numerous prominent cross-veins when dried (Fig. 18*a*) . . . . 60. *C. rostrata*
a. Plants caespitose (Fig. 8*a*); in open wetlands and wet forests *f*
   f. Spikelets very densely packed, thimble-shaped, becoming hard; staminate flowers at base of terminal spikelet; in floodplains of deciduous forests . . . . .73. *C. typhina*

   f. Perigynia loosely arranged in the spikelets; 1 or more upper
       spikelets wholly staminate *g*
      g. Spikelet scales short, abruptly narrowed to their linear awns
         (Fig. 8*t*) *h*
        h. Perigynia 1-2 mm broad, soon becoming perpendicular to
          the spikelet axis *i*
          i. Widest leaf blades 3-10 mm; perigynia 1.6-2 mm
            broad, narrowed to slender beaks; plant always red-
            based; in wet meadows and sandy shores . . . .
            . . . . 62. *C. hystericina*
          i. Widest leaf blades 7-18 mm; perigynia very slenderly
            lanceolate, 1-1.8 mm broad; rarely red at plant base;
            in floating peaty shores. . . . . 63. *C. comosa*
        h. Perigynia 2.2-3.5 mm broad, ascending-spreading; in wet
          acid woods and meadows . . . . 64. *C. lurida*
      g. Spikelet scales tapering gradually from base to apex (Fig.
         8*r*) *j*
        j. Pistallate spikelets several times as long as wide; peri-
          gynia 7-10 mm long *k*
          k. Achenes bent over on one side (Fig. 22*b*); perigynia
            becoming 5-6 mm broad; in wet woods and meadows
            . . . . .69. *C. tuckermanii*
          k. Achenes straight; perigynia only 2-3.5 mm broad at
            maturity; on wet shores and in meadows . . . .
            . . . . .70. *C. vesicaria*
        j. Pistillate spikelets at most twice as long as wide (if
          longer, perigynia 10-20 mm long) *l*
          l. Pistillate spikelets globose; each perigynium like 2
            cones fitted base to base, crowded; in floodplains of
            deciduous forests. . . . . 67. *C. grayi*
          l. Pistillate spikelets ovoid or cylindrical *m*
            m. Perigynia mostly 2-7 per spikelet, spreading, lance-
              olate-ovoid, loosely spaced; in moist to wet forests
               . . . . 66. *C. intumescens*
            m. Perigynia more numerous, somewhat crowded *n*
              n. Perigynia 5-10 mm long, curved, spreading to
                reflexed; spikelets all crowded at culm tip; in
                wet woods . . . . 68. *C. retrorsa*
              n. Perigynia 10-20 mm long, straight; spikelets not
                crowded *o*
                o. Perigynia lanceolate, divergent; in wet acid
                  soil *p*
                  p. Widest leaf blades 5-13 mm . . . .
                    . . . . 71. *C. folliculata*
                  p. Leaf blades 2-4 mm wide . . . .
                    . . . . 72. *C. michauxiana*
                o. Perigynia ovoid, appressed or ascending; in
                  wet woods . . . . 65. *C. lupulina*

Key V

a. Plants 5-50 cm tall, perigynia globose; spikelets crowded, erect;
    along shores *b*
      b. Perigynia fleshy, turning yellow . . . . 79. *C. aurea*
      b. Perigynia dry, powdery white . . . . 80. *C. garberi*
a. Plants 30-150 cm tall; perigynia flattened; spikelets arching or
    drooping *c*

    c. Spikelet scales abrupt-awned (Fig. 8*t*); root hairs yellow; spikelets pendent; in wet acid woods . . . . .78. *C. crinita*

    c. Spikelet scales blunt or acute, tapering gradually; roots not yellow; spikelets ascending or arching *d*

        d. Some of the basal leaf sheaths with pinnate fibrils (Fig. 9*c*) *e*

            e. Perigynia ovate, flattened, appressed to spikelet axis; scales ovate, purplish; in wet meadows, often forming dense raised peaty tussocks . . . . .74. *C. stricta*

            e. Perigynia ellipsoidal with tiny abrupt beak, at maturity plump, causing the acuminate scales to diverge; staminate scales brown; in wet prairies and wet sand . . . .
. . . . .76. *C. haydenii*

        d. Sheaths without pinnate fibrils *f*

            f. Perigynia ovate; in river floodplains of deciduous forests
. . . . .75. *C. emoryi*

            f. Perigynia obovate; in very wet meadows
. . . . 77. *C. aquatilis,* var. *altior*

Key VI

a. Culms solitary or in small scattered groups, arising from elongate horizontal culms *b*

    b. Culms from underground rhizomes (Fig. 8*b*) *c*

        c. Inner band of leaf sheath nerveless (Fig. 9*e*); spikelets 2-8; in dry barrens . . . . 105. *C. foenea*

        c. Inner band of leaf sheath green-nerved (Fig. 9*d*); spikelets 12-25; in wet meadows . . . . 107. *C. sartwellii*

    b. Culms arising from previous year's prostrate elongate stems, in axils of old dried leaves; in bogs . . . . .106. *C. chordorrhiza*

a. Plants caespitose (Fig. 8*a*) *d*

    d. Inflorescence compound; lower branches each bearing several tiny crowded spikelets (Fig. 33); mostly wetland species *e*

        e. Inner band of leaf sheath cross-wrinkled (Fig. 9*a*) *f*

            f. Culms soft, broadly triangular; perigynia lanceolate, plump, 3-8 mm long; in all wetlands . . . . . 97. *C. stipata*

            f. Culms wiry, only 1-2 mm thick; perigynia small, firm, and flattened *g*

                g. Culms longer than the leaves; perigynia ellipsoidal, turning yellow; scales red-tinted; in low prairies and dry sand . . . . 99. *C. annectens,* var. *xanthocarpa*

                g. Culms about as long as leaves; perigynia ovate to lanceolate, green or brown; in all wetlands . . . .
. . . . .98. *C. vulpinoidea*

        e. Inner band of leaf sheath not cross-wrinkled *h*

            h. Culms soft, broadly triangular; inner band of leaf sheath minutely orange- or dark-speckled; in floodplains . . . . . .
. . . . 100. *C. alopecoidea*

            h. Culms wiry, only about 1 mm thick; sheaths may be marked or mottled but not minutely dotted *i*

                i. Perigynia ovate, dark, shiny; spikelets all crowded together; in floating peat mats . . . . 102. *C. diandra*

                i. Perigynia lanceolate, brown; lower spikelets sometimes spread out; sheaths often yellow-tinged; in firm wet meadows . . . . .101. *C. prairea*

    d. Inflorescence rarely, if ever, compound; mostly upland species *j*

        j. Spikelets mostly spaced far apart on culm *k*

    k. Perigynia ellipsoidal, terete, minutely beaked, ascending, 1-6 per spikelet; tiny plants of wet coniferous forests...
    .... 84. *C. disperma*
    k. Perigynia plano-convex, often more numerous, divergent in starlike spikelets; in deciduous forests *l*
        l. Leaves 1-3 mm wide; in all forests ...91. *C. convoluta*
        l. Wider leaves 5-9 mm; in maple-basswood forests ....
        ....92. *C. sparganioides*
  j. Spikelets closely aggregated into an ovoid head *m*
    m. Perigynia nerved on inner face; leaves stiff, gray green; in dry barrens .... 95. *C. muhlenbergii*
    m. Perigynia nerveless on inner face *n*
        n. Leaves 2-5 mm wide; inflorescence about 1 cm long, with abundant threadlike bracts; in dry woods, openings ....94. *C. cephalophora*
        n. Leaves 3-9 mm wide; inflorescence elongate, with few or no conspicuous bracts *o*
            o. Scales acuminate or awned, reddish tinted; in fields
            .... 96. *C. gravida*
            o. Scales bluntly acute, shorter than the perigynia, green or translucent; in maple-basswood forests...
            .... 93. *C. cephaloidea*

Key VII

a. Perigynia plump to the margins (beak may be empty), without thin marginal wings (Fig. 28*b*) *b*
  b. Perigynia at maturity appressed to ascending in the spikelet (Fig. 27) *c*
    c. Perigynia 4-5.5 mm long *d*
        d. Spikelets separated on a zigzag axis; perigynia and spikelets ovate; leaves 1.5-5 mm wide; in cool forests ....
        ....103. *C. deweyana*
        d. Spikelets aggregated; perigynia and spikelets lanceolate; leaves 1-2 mm wide; in wet woods... 104. *C. bromoides*
    c. Perigynia shorter; plants of acid bogs and wet coniferous forests *e*
        e. Perigynia terete, evenly ribbed on all sides, abruptly short-beaked, 1-6 per spikelet; in cedar swamps ....
        .... 84. *C. disperma*
        e. Perigynia plano-convex or biconvex, tapered to a beak *f*
            f. Spikelets mostly spaced out on culms *g*
                g. Lower bract as long as the inflorescence; spikelets few-flowered, far apart; in bogs... 83. *C. trisperma*
                g. Bracts short; spikelets 3-20-flowered *h*
                    h. Perigynia green to brown with paler nerves and sharp margins; in bogs and coniferous forests...
                    .... 81. *C. brunnescens*
                    h. Perigynia blue green to straw-colored with darker nerves and rounded margins; in wet sand and coniferous forests .... 82. *C. canescens*
            f. Spikelets crowded into a single mass at culm tip *i*
                i. Perigynia ovoid, blunt; spikelets 2-4; in bogs .....
                ....85. *C. tenuiflora*
                i. Perigynia ovate-triangular, tapering to a short beak; spikelets 4-14; in wet openings of coniferous forests ....86. *C. arcta*

 b. Perigynia divergent in starlike spikelets (Fig. 29) *j*
  j. Perigynia slenderly lanceolate, about 1 mm wide; in bogs and wet acid sand . . . . 90. *C. angustior*
  j. Perigynia ovate, broader *k*
   k. Perigynia many-nerved on all sides; in bogs . . . . 89. *C. cephalantha*
   k. Perigynia nerveless or very weakly nerved on flat inner face *l*
    l. Spikelets 3; scales very blunt; in wet meadows and bogs . . . . 87. *C. interior*
    l. Spikelets 4; scales acute; plants sometimes dioecious; in fens . . . . 88. *C. sterilis*
a. Perigynia very thin and scalelike, with thin marginal wings that are somewhat translucent (Fig. 35*c*), closely appressed until maturity *m*
 m. Perigynia with broad wings, and beaks extending somewhat beyond the scale tips (Fig. 35*b*) *n*
  n. Plants with numerous nonflowering pseudoculms *o*
   o. Spikelets 1-3 cm long, in stiff inflorescences; perigynia 6-10 mm long, lanceolate; in floodplains of deciduous forests . . . . 112. *C. muskingumensis*
   o. Spikelets 0.5-1 cm long; plants less stiff; perigynia shorter *p*
    p. Spikelets globose, the rhombic perigynia soon spreading perpendicularly to the axis; in wet woods . . . . 116. *C. cristatella*
    p. Spikelets ovoid or broadest at middle; perigynia more or less appressed, ovate or lanceolate *q*
     q. Perigynia slenderly lanceolate, soon tan-colored *r*
      r. Inflorescence flexible; perigynium tips soon somewhat spreading; in cool forests and wooded floodplains . . . . 114. *C. projecta*
      r. Inflorescence stiff; perigynia straight or incurved, and slightly but abruptly narrowed near the middle; in floodplains . . . 113. *C. tribuloides*
     q. Perigynia ovate; green to drab; in deciduous forests . . . . 115. *C. normalis*
  n. Pseudoculms absent or inconspicuous *s*
   s. Perigynia 2.2 mm wide or wider, always broadly ovate *t*
    t. Perigynia with jagged or irregularly winged margins *u*
     u. Perigynia orange brown on margins and strongly nerved over achene on inner (concave) face; in wet to dry prairies . . . . 117. *C. bicknellii*
     u. Perigynia brown, not nerved; rocky barrens . . . . 118. *C. merritt-fernaldii*
    t. Perigynia green or tan, evenly curved at margin, not conspicuously nerved on concave face *v*
     v. Perigynia ovate; spikelets round-based, crowded at stem tip; in low meadows . . . . 120. *C. molesta*
     v. Perigynial body perfectly circular; spikelets tapering at base, separated on stem; in dry barrens . . . . 119. *C. brevior*
   s. Perigynia not over 2 mm wide, or else lanceolate *w*
    w. Spikelets well separated on a zigzag axis; perigynia ovate; in wet prairies . . . . 110. *C. tenera*
    w. Spikelets clustered near culm tip, or perigynia lanceolate *x*

Pensylvanica Group   (Albae, Montanae, Digitatae, Phyllostachyae)

Small tufted sedges 5-15 (54) cm tall; plant bases brown to red purple; flowering April-May and fruiting May-June; perigynia minutely pubescent, with only 2 ribs; stigmas 3; staminate spikelet 1.

1. **C. eburnea** Boott. Low mats; leaves 0.1-1 mm wide; staminate spikelet hidden among the 2-3 tiny pistillate spikelets; perigynia triangular, becoming black, persistent into winter, 1.5-2 mm long. —Dry to damp limy bluffs and shores, usually under red or white cedars; scattered localities in Wisconsin.

2. **C. pensylvanica** Lam. Pennsylvania Sedge (Fig. 10). Leaves yellow green, 1-3 mm wide, in tufts connected by slender rhizomes; sheaths with reddish brown longitudinal fibrils; culm erect, sometimes to 35 cm tall; staminate spikelet 5-20 mm long, tapering to each end, with purplish brown white-edged scales; pistillate spikelets 1-3, globose, sessile, close to base of the staminate one; perigynia globose, short-beaked, 2.5-4 mm long, minutely pubescent, green to gray, often attacked by a black smut fungus. —Our most abundant and widespread sedge, in all habitats except wet soil and sometimes forming the dominant ground cover in sandy oak and jack pine country; throughout the state.

3. **C. abdita** Bickn. Plant and perigynia similar to but smaller than No. 2; lacking elongate rhizomes and hiding lower spikelets in or close to the leaf bases. —Dry limestone and mesic prairies north to Chippewa and Door counties.

4. **C. umbellata** Willd. Like No. 3, but with larger sometimes glabrous perigynia (3-4 mm long, with beak 1-2 mm long) and coarser, stiffer growth; leaves 1-4 mm wide. —Common in dry open sand, differing from the robust barrens

Fig. 10. *Carex pensylvanica: a.* inflorescence, × 2.50; *b.* perigynium and scale, × 10.

form of No. 2 in not covering the ground with uniform sods; scattered localities in the state.

5. **C. peckii** Howe. Resembling No. 2, but the leaf tufts few, small; pale-scaled staminate spikelet partly hidden among the pistillate ones; perigynia longer than the scales, slenderly obovoid, slightly triangular, with coarse hairs. —Cool limy woods, more common north and east.

6. **C. emmonsii** Dew. Dense, solitary, narrow-leaved clumps; basal leaf sheaths pinkish lavender; spikelets 2-4, crowded at culm tip as in No. 5, but perigynia smaller, triangular, and very finely pubescent. —Dry to moist acid sand; central Wisconsin.

7. **C. communis** Bailey. Plants coarser than No. 2 and lacking elongate rhizomes; leaf bases not fibrillose but redder; blades soft, 2.5-6 mm wide; staminate spikelet uniformly cylindrical, pale, 4-17 mm long, sometimes hidden among the 3-5-stalked pistillate ones. —Cool coniferous or mixed forests, mostly northward.

8. **C. deflexa** Hornem. Small slender-leaved tufts; spikelets small, 1-3, crowded under a green bract, at tips of arching culms of varying length; perigynia green, globose, 2-3 mm long, longer than the scales. —In coniferous forests from Jackson to Shawano counties and northward.

9. **C. pedunculata** Willd. Solitary dense evergreen clumps with stiff flat dark green blades, 2-4.5 mm wide, dying back only near the tips in winter, with reddish band at die-back point and strongly red purple nonfibrillose basal sheaths; some spikelets on peduncles 1-12 cm long from bracts at plant base; perigynia triangular.—Cool wet to dry woods throughout the state, more common northward.

10. **C. richardsonii** R. Br. Closely resembling No. 2, but with tiny purple sheathing bract at base of flowering culm; leaves 2-4 mm wide, stiff, flat, drying to yellowish white; perigynia triangular, rarely found.—Limestone prairies and calcareous beach dunes in southern Wisconsin north to Adams and Brown counties.

11. **C. concinna** R. Br. Low mats with leaves 2-4 mm wide; a tiny yellowish green sheathing bract near the base of the flowering culm; perigynia ovoid, blunt, longer than the dark scales, and with coarse glassy hairs.—Coniferous forests; Door County.

12. **C. hirtifolia** Mackenz. Loose clumps; resembling a dark green *C. blanda*, but with softly hairy leaves and stems; perigynia triangular, pubescent.—Common in maple-basswood forests north to Barron to Shawano counties.

13. **C. jamesii** Schwein. Dense clumps of flat narrow dark green leaves, resembling No. 9 but perigynia globose, long-beaked, glabrous, with bractlike scales, and all spikelets buried in plant base.—Maple forests, southern tier of counties.

14. **C. backii** Boott. Like No. 13, but with wider leaf blades and bracts 2.5-5 mm broad.—Cool sandy woods or cliffs; sporadic.

Blanda Group   (Granulares in part, Laxiflorae, Oligocarpae)

Low, solitary, caespitose sedges, 5-20 cm tall; plant bases brown to red purple; vegetative shoots much wider-leaved than those bearing flowers; leaves short, broad, conspicuously pleated (M-shaped); sheaths loose; culms soft, triangular; spikelets slender, elongate, 5-20-flowered, borne all along leafy-bracted culms as long as the leaves; perigynia green; stigmas 3; staminate spikelet 1.

15. **C. blanda** Dew. (Fig. 11). Plant base brown; leaves 3-12 mm broad, green to blue green; perigynia 25-30-ribbed, obovoid, bluntly triangular, 3-3.5 (4.5) mm long, asymmetrical, with orifice or slight beak on one side of the blunt apex;

Fig. 11. *Carex blanda:* *a.* inflorescence, × 1; *b.* perigynium and scale, × 6.

scales with abruptly awned midribs, appressed to the perigynia; staminate spikelet often short-stalked and partly hidden among the pistillate spikelets.—Very common in dry to wet deciduous woods; southern half of the state, especially along paths and roadsides.

16. **C. ormostachya** Wieg. Very similar to No. 15, but perigynia smaller, slightly beaked, and farther apart in spikelet.—In mixed coniferous-deciduous forests; northern tier of counties. A wide-leafed form with longer spikelets and perigynia with curved beaks occurs in beech forests in counties along Lake Michigan.

17. **C. albursina** Sheldon. Like No. 15, but with wider (1-4 cm), always blue green leaf blades and floral bracts; perigynium with a distinct curved beak.—Maple-basswood forests, especially on windthrow mounds, slopes, and paths; north to Clark to Oconto counties.

18. **C. leptonervia** Fern. Like No. 15, but perigynia with only 2-15 (21) nerves and tapering gradually into a distinct curved beak; the only sedge whose leaf blade margins near the base are retrorsely (downwardly) scabrous.—Wet woods and swamps; northern half of the state and south in Wisconsin River valley.

19. **C. digitalis** Willd. Like No. 15, but the asymmetrical (slightly curved) perigynia sharply triangular and only as long as wide; pistillate scales shorter, spreading, with little or no awn, and with occasional minute dark lines or blotches. —Maple-basswood forests; sporadic.

20. **C. plantaginea** Lam. Leaves 1-4 cm wide, yellowish to dark green, evergreen; staminate spikelet long-stalked; bladeless bracts, scales, and basal sheaths strongly red purple; perigynia triangular, curved.—Climax forests, mostly northeast and west in the state.

21. **C. amphibola** Steud., var. **turgida** Fern. More erect and robust than No. 15; leaves dark green, evergreen; bases brown to reddish; perigynia 4-5.2 mm long, beakless, symmetrical, ovoid, or barrel-shaped, terete or slightly triangular, the 30-50 nerves impressed rather than raised; scales elongate, tapering, dark-speckled or streaked —Common in moist or alluvial deciduous woods in southern Wisconsin north to St. Croix to Brown counties.

22. **C. oligocarpa** Willd. Smaller than No. 20; plant bases reddish; blades only 2.5 mm wide; perigynia only 4 (2-8) per spikelet, beaked, 3-4 mm long, bluntly triangular.—Maplebasswood forests; sporadic in southwestern counties.

23. **C. hitchcockiana** Dew. Plants taller than No. 20; perigynia sharply triangular, curved, beaked, with 30-50 impressed nerves; bract sheaths minutely white-hispid.—Maplebasswood forests; sporadic in southern and eastern Wisconsin.

24. **C. conoidea** Willd. Leaves 2-5.5 mm broad; pistillate scales green-awned; perigynia ovoid, with 16-20 impressed nerves; staminate spikelet long-stalked, with rounded scales resembling those of *C. tetanica*.—Open wet meadows, fens, and low prairies; southeastern half of the state.

25. **C. granularis** Willd., var. **haleana** (Olney) Porter. Like No. 15, but leaves blue green; scales speckled; perigynia crowded, only slightly asymmetrical, broadest below middle or ellipsoidal, blue green, only 20-nerved.—Wet open calca-

reous meadows, raw marl, and wooded cliffs; southern and eastern Wisconsin.

Tetanica Group   (Granulares in part, Paniceae)

Low rhizomatous sedges with slender culms, pseudoculms, and slender V-shaped leaves; covering areas 1-20 m in diameter; bases brown except in *C. woodii* and rarely *C. tetanica*; spikelets slender, elongate, 5-20-flowered, borne all along the leafy-bracted culms as long as leaves; perigynia 14-30-nerved, green, ovoid to bluntly triangular and somewhat asymmetrical; stigmas 3; staminate spikelet long-stalked and round-scaled.

26. C. tetanica Schkuhr. Leaf blades blue green, 2-4.5 mm wide; staminate scales very round or blunt at apex.—Common in wet prairies, meadows, and fens; southeastern half of the state.

27. C. meadii Dew. Leaf blades blue green, 3-7 mm wide; perigynia 3-4.5 mm long.—Dry prairies; southwestern half of the state.

28. C. woodii Dew. Often forming nonflowering sods like *C. pensylvanica*; leaves 1-7 mm broad; basal sheaths purple; scales often pale.—Sugar maple forests; throughout the state.

29. C. crawei Dew. Plants 3-30 cm tall, resembling *C. granularis*; leaf blades 1.5-3 mm wide; lowest spikelet often close to plant base.—Limy shores and wet meadows; Waushara, Jefferson, Waukesha, and Racine counties.

30. C. livida (Wahlenb.) Willd. Plants forming beds of slender glaucous leaves; perigynia blunt, obscurely nerved. —Rare in sphagnum bogs; Rock and Bayfield counties.

31. C. vaginata Tausch. Leaves slender; perigynia with an oblique-tipped beak 1 mm long, spaced far apart on the spikelet.—Rare in sphagnum bogs; Oneida and Florence counties.

Gracillima Group   (Gracillimae, Sylvaticae, Capillares, Longirostres, Anomalae)

Culms longer than the leaves; spikelets long-pedunculate, arching or drooping, borne near the culm apex; perigynia small, mostly green, symmetrical except in No. 38, few-nerved; stigmas 3; terminal spikelet(s) wholly or partially staminate.

32. C. gracillima Schwein. Plant base red purple, older

leaf sheaths minutely hairy and finely red-dotted; leaves abundant, dark green, M-shaped, 4-11 mm broad, evergreen; terminal spikelet with some perigynia at tip; perigynia appressed, 2.5-3 mm long, 6-20-nerved, green, ellipsoidal, blunt. —Common in moist forests; throughout the state.

33. C. arctata Boott. Evergreen and similar to No. 32, perigynia lanceolate, slender-beaked, 3-4.5 mm long; terminal spikelet staminate. —Common in moist woods; northern half of the state.

34. C. sprengelii Spreng. (Fig. 12). Forming dense tussocks or fairy rings 1-5 dm across; leaves abundant, yellow green, long, 2-4.5 mm wide; old leaf sheaths forming copious tufts of gray brown fibrils at plant base; perigynia 5-7 mm long, with an abrupt linear beak as long as the globose body. —Moist woods and steep banks; throughout the state.

35. C. davisii Schwein. & Torr. Leaves 4-7 mm wide, pubescent; terminal spikelet pistillate at tip; perigynia ovoid, 4.5-6 mm long. —Wet deciduous woods along the larger rivers in southwestern Wisconsin.

36. C. prasina Wahlenb. Plants blue green; leaves 3-5 mm wide; perigynia sharply triangular, long-ovoid, beaked. —In woodland springs; Baraboo Hills, Sauk County.

37. C. debilis Michx., var. rudgei Bailey. Similar to No. 33; plant bases bright red; terminal spikelet pistillate-tipped; leaf blades 3-5 mm wide; perigynia 4.5-6 mm long; achene stalked. —Wet acid woods; scattered localities throughout the state.

38. C. scabrata Schwein. Surfaces of entire plant scabrous; leaves 5-12 mm wide; culms leafy to apex; perigynia crowded, oblique-tipped, curved, long-beaked, 6-nerved. —Wet woods; scattered localities, mostly in northern Wisconsin.

39. C. castanea Wahlenb. Leaves pubescent, 3-6 mm wide; spikelets short and pendent, resembling those of *C. limosa.* —Wet borders of coniferous forests and bogs; northern and northeastern counties in Wisconsin.

40. C. capillaris L. A diminutive of No. 33; leaves 1-3 mm wide; perigynia 2-3 mm long. —Wet coniferous woods and shores; Bayfield and Door counties.

41. C. assiniboinensis W. Boott. Plants slender, with unique arching threadlike stolons; leaves 1-3 mm wide; perigynia spaced far apart on the spikelet, becoming yellow, lanceolate, long-beaked, hispid, 4-6.5 mm long. —River floodplain forests; sporadic in northwestern Wisconsin.

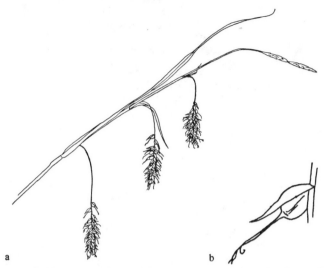

Fig. 12. *Carex sprengelii:* *a.* inflorescence, × .50; *b.* perigynium and scale, × 4.

Swanii Group (Virescentes)

Plants caespitose; leaves pubescent, 1-4 mm wide; bracts usually sheathless; spikelets crowded near culm tip, short-cylindrical, erect; perigynia small, green.

42. **C. swanii** (Fern.) Mackenz. Terminal spikelet pistillate at apex; perigynia pubescent.—Dry sand, rare; Racine and Kenosha counties.

43. **C. pallescens** L. Terminal spikelet wholly staminate; perigynia beakless, with 15-25 delicate nerves. —Meadows; Apostle Islands, Ashland County.

44. **C. torreyi** Tuckerm. Resembling No. 43; perigynia with prominent nerves, slightly beaked; scales obtuse. —Prairies; St. Croix and Trempealeau counties.

Flava Group (Extensae)

Plants slender, 2-80 cm tall, uniquely flowering all summer; leaves 1-5 mm wide; bases brown; spikelets 2-10 mm long, crowded at culm tip, only about twice as long as wide; perigynia divergent at right angles to axis or reflexed, strongly nerved, abruptly beaked; stigmas 3; staminate spikelet 1. Moist sandy or limy places.

45. **C. flava** L. Perigynia 3.5-6 mm long, becoming yellow or orange, 8-18-nerved, with conspicuous straight to decurved beaks and brown scales.—Coniferous forests, bogs, meadows, and wet sand or marl; northeastern counties of the state, rarely to Dane County.

46. **C. viridula** Michx. Leaves 1-2.5 mm wide; perigynia 1-3 mm long, green, 6-10-nerved, the beak not over half the length of the body.—Sandy lake shores.

Limosa Group   (Limosae)

Sedges with yellow-felted roots, sheathless bracts, blue green leaves; spikelets short, drooping; perigynia blunt, flattened-ovate, blue green, with contrasting purple scales (pale in shade); stigmas 3; staminate spikelet 1. In open sphagnum bogs, statewide.

47. **C. limosa** L. Leaves tending to inroll, 1-2 mm wide; scales as wide as the perigynia, persistent.

48. **C. paupercula** Michx. Leaves flattened, 1-3 mm wide; scales slenderly lanceolate, falling before the perigynia. —Widespread in northern Wisconsin, rare southward.

Pauciflora Group   (Orthocerates, Polytrichoideae, Dioicae)

Slender bog sedges with but a single apical spikelet per stem and fine leaves.

Fig. 13. *Carex pauciflora,* reflexed perigynium, × 4.

Fig. 14. *Carex leptalea,* inflorescence, × 3.

49. **C. pauciflora** Lightf. (Fig. 13). Plants caespitose; leaves 1-2 mm wide; perigynia 7-9 mm long, straight, slenderly lanceolate, soon divergent or reflexed on the spikelet.—Sphagnum bogs; Taylor to Lincoln counties and northward.

50. **C. leptalea** Wahlenb. (Fig. 14). Plants forming small mats; leaves 0.1-1.7 mm wide; perigynia oblanceolate, blunt, appressed in tiny spikelets only 5 mm long.—Common in tamarack bogs and cedar swamps, sometimes in open wet meadows; northeastern two-thirds of Wisconsin.

51. **C. gynocrates** Wormsk. Plants sometimes dioecious; leaves 0.1-1 mm wide, in small tufts; perigynia ovate, curved-reflexed.—Bogs and wet shaded sand; northeastern Wisconsin.

Lanuginosa Group   (Hirtae)

Medium-sized common rhizomatous sedges, with numerous short pseudoculms; bases reddish, with pinnate fibrillose sheaths; pistillate spikelets 1-3, ascending, remote, 1-4 cm long; perigynia small, globose, pubescent; scales often minutely fringed on margins; stigmas 3; staminate spikelets 1-3.

52. **C. lanuginosa** Michx. Plants slender; closely resembling *C. stricta* but not forming tussocks; leaf blades M-shaped, the widest 2.5-4.5 mm broad; bracts flat, somewhat divergent; perigynia 3-4 mm long and 1-2 mm thick.—Usually in calcareous mineral soil in wet meadows, low prairies, ditches, and sometimes in upland fields; throughout the state.

53. **C. lasiocarpa** Ehrh. Wiregrass. Similar to No. 52, but leaves dark green, only 0.8-2 (3) mm broad, margins inrolled; bracts ascending; inflorescence as in No. 52.—Wet peaty meadows, floating peat mats, and open sphagnum bogs, forming a tough woven mat by interconnected rhizomes, occasionally on sandy or peaty shores; throughout the state. Vegetative plants difficult to distinguish from *C. oligocarpa*.

54. **C. houghtoniana** Dew. Plants similar to No. 51, but larger; leaves 3-4.5 mm wide; perigynia 5-6 mm long, 2-3 mm thick, with coarser hairs not obscuring the 10-16 ribs, often turning reddish.—Dry open sand; northern half of the state.

Buxbaumii Group   (Atratae)

Mostly arctic sedges, distinguished by their dark purple scales and plump, pale blue green perigynia; stigmas 3.

55. **C. buxbaumii** Wahlenb. Rhizomatous; resembling *C. stricta* in having sheaths pinnate-fibrillose and bases reddish; leaves blue green, 1.8-3 mm broad; spikelets plump, the terminal one staminate only at base; scales slenderly acute, longer than the perigynia; perigynia 2.4-4.3 mm long.—Open wet meadows, floating peat mats, and wet prairies; scattered throughout the state.

Lacustris Group   (Vesicariae in part, Paludosae)

Coarse rhizomatous sedges, forming large monotypic beds to 10 m across; the numerous pseudoculms, 0.5-1.5 high, usually taller than fruiting culms; bases reddish and pinnate-fibrillose (except No. 60); spikelets ascending, cylindrical, 2-5 cm long (short-ovoid in No. 61), scattered on erect culms, the lower 1-4 pistillate, the upper 2-5 staminate; scales acute; perigynia somewhat inflated; stigmas 3. Shallow water and wet soil.

56. **C. lacustris** Willd. Sawgrass (Fig. 15). Leaves blue green, strongly M-shaped, 8-10 mm wide, margins harshly serrate; perigynia dull green, 5-7 mm long, ovoid-cylindrical, strongly many-nerved, blunt, with a tiny third tooth next to one of the 2 small apical teeth, 0.3-1 mm long.—Common in all wetlands; throughout the state.

57. **C. atherodes** Spreng. (Fig. 16). Leaves M-shaped, 4-10 mm wide; some sheaths or parts of blades finely pubescent unless inundated in deeper water; perigynia teeth 2-3 mm long.—Margins of cattail marshes; southeastern one-fourth of Wisconsin.

58. **C. laeviconica** Dew. Resembling No. 57; leaf sheaths minutely hispid at apex, or glabrous; teeth 1-2 mm long; perigynia straw-colored, dull.—Mostly near the Mississippi River.

59. **C. trichocarpa** Schkuhr (Fig. 17). Often forming dense nonflowering beds of crowded slender pseudoculms, 5-10 dm tall; leaves 4-7 mm wide, M-shaped; sheaths with a dark purple blotch at apex on side opposite blades; perigynia ovoid, pubescent, beak 6-9 mm long.—Open stream valleys and low prairies; mostly in southern half of the state.

60. **C. rostrata** Stokes (Fig. 18). Leaves pale or yellow green, when dried showing many raised crossveins more prominently than in other species, 3-8 mm wide, mostly C-shaped; new shoots green over winter; spikelets yellowish, brown, or red-tinted; perigynia 50-130 per spikelet, 16-

Fig. 15. *Carex lacustris*, × 3.
Fig. 16. *Carex atherodes*, × 3.

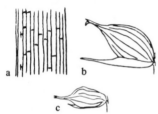

Fig. 18. *Carex rostrata:*
a. leaf section with raised crossveins, × 3;
b. large perigynium, × 3;
c. small perigynium, × 3.

Fig. 20. *Carex intumescens*, × 3.

Fig. 21. *Carex retrorsa*, × 3.
Fig. 23. *Carex vesicaria*, × 3.

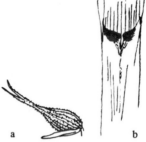

Fig. 17. *Carex trichocarpa:*
a. perigynium and scale, × 3;
b. leaf sheath, × 2.25.

Fig. 19. *Carex oligosperma:*
a. perigynium and scale, × 3;
b. end view of perigynium, × 3.

Fig. 22. *Carex tuckermanii:*
a. perigynium and scale, × 3;
b. bent achene, × 3.

Figs. 15-23. Relative shapes and positions of perigynia and scales.

nerved, variable in size and shape but tending to be broadly ovoid and more abruptly tapering into the beak than others in this group.—Open wetlands; throughout the state, but more abundant northward in cool soft water.

61. **C. oligosperma** Michx. Wiregrass (Fig. 19). With same habit as *C. lasiocarpa,* forming woven-rooted floating mats; leaves 1.5-3 (4) mm wide, strongly inrolled; pistillate spikelets 1-3, short; staminate spikelet 1; perigynia 3-18 per spikelet, persistent, leathery, broadly ovoid, dark olive green, glabrous.—Abundant in sphagnum bogs northward, less common in southeastern Wisconsin.

Hystericina Group   (Pseudo-Cypereae)

Robust mostly caespitose sedges; bases red brown, pinnate-fibrillose; staminate spikelets 1-3 above several clustered, stalked, erect or drooping pistillate spikelets which are thick-cylindrical; perigynia inflated and scales abruptly narrowed near base into a long awn; stigmas 3.

62. **C. hystericina** Willd. Sedges very variable in size, 5-30 cm tall, bases red; leaves 3-10 mm wide; perigynia numerous, divergent, lanceolate-ovoid, 13-20-nerved, 5-7 mm long, 1.6-2 mm thick, delicate, pale green to straw-colored; achene broadest above the middle.—Common in slightly disturbed wet meadows and on sandy or marly shores; throughout the state.

63. **C. comosa** Boott. Bottlebrush Sedge. Leaves 7-18 mm wide, yellowish green; perigynia divergent to reflexed, very numerous, pale green, 1.4-1.8 mm wide, the 13-20 strong nerves persisting as a "bird cage" around the achene when rolled between the fingers.—Common on wet or floating peaty shores; throughout the state.

64. **C. lurida** Wahlenb. Plant resembling a small form of *C. lupulina*; leaves 3-7 mm wide; perigynia broadly ovoid, green to brown, 7-10 mm long, 2.2-3.5 mm thick, 8-12-nerved.—Wet acid sites near streams; Lincoln to Jackson and Sauk counties.

Lupulina Group   (Vesicariae in part, Lupulinae, Folliculatae)

Robust caespitose sedges 3-13 dm tall, bases often red and pinnate-fibrillose; upper 1-4 spikelets staminate, the lower 2-5 pistillate, 1-8 cm long; perigynia large, greatly inflated, usually glabrous, persistent into autumn, well adapted for floating ashore; scales gradually tapered, acute or acuminate; stigmas 3.

65. **C. lupulina** Willd. Hops Sedge. Plant bases red; leaves 5-13 mm wide; staminate spikelet 1; pistillate spikelets 2-6, longer than thick, the larger ones with 30-60 large crowded ascending perigynia; perigynia green to brown, tough, bladderlike, broadly ovoid, 13-20 mm long, with the style looped inside.—Wet woods, throughout the state.

66. **C. intumescens** Rudge (Fig. 20). Similar habit to No. 65, but leaves evergreen; pistillate spikelets 1-3, as long as wide; perigynia 2-7 (15), dark green, lanceolate-ovoid, divergent, 10-15 mm long, not crowded; style straight. Evergreen leaves and red bases closely resemble *C. arctata* and *C. gracillima.*—Very common in cool damp woods; throughout the state.

67. **C. grayi** Carey. Plants resembling Nos. 65 and 66, but staminate spikelet sessile; perigynia shaped like 2 cones fitted base to base, numerous and crowded into 1 or 2 globose spikelets, rarely pubescent.—Deciduous floodplain forests along larger rivers; uncommon.

68. **C. retrorsa** Schwein. (Fig. 21). Variable in size of plant, spikelets, and perigynia; leaves 3-8 mm wide; pistillate spikelets crowded at culm apex, sometimes hiding the staminate ones; perigynia 5-10 mm long, green or tan, with delicate papery texture, curved back or at least divergent from the short axis.—Wet woods and adjacent lowlands; throughout the state.

69. **C. tuckermanii** Dew. (Fig. 22). Forming large tussocks of many long leaves 2-6 mm wide; pistillate spikelets short-to long-cylindrical, often arching, scattered on the culm; perigynia 7-10 mm long, 4-6 mm wide, broadly short-ovoid, shiny, becoming light brown; achene bent.—In wet hollows and meadows; common in the northern counties, southward in floodplain forests of the Wisconsin and Mississippi rivers, also along Lake Michigan.

70. **C. vesicaria** L. (Fig. 23). Plants similar to No. 69; spikelets straight and narrowly cylindrical; perigynia very shiny, ovoid, straw yellow, 2-3.5 mm wide, neatly arranged in braided pattern in side view; achene straight. The shininess of the perigynia distinguishes this species from the similar *C. rostrata* and *C. laeviconica.*—Common at pond edges and in various wetlands; throughout the state.

71. **C. folliculata** L. Leaves 6-16 mm wide; perigynia pale

green, 10-15 mm long, 2.5-3.5 mm thick.—Acid wet sandy or peaty woods; Jackson and Juneau counties.

72. **C. michauxiana** Boeckl. Resembling No. 71, but leaves only 2-4 mm wide and perigynia somewhat smaller.—Bogs; Apostle Islands, Ashland County.

Typhina Group   (Squarrosae)

Caespitose sedges with reddish bases and pinnate fibrils; spikelets 1-4, globose or cylindrical, 1-2.5 cm long, 1-2 cm thick, densely packed, persistent, staminate at base; stigmas 3.

73. **C. typhina** Michx. Plants about 5 dm tall; leaves 5-9 mm wide; perigynia 30-60 per spikelet, divergent, inflated, the body obovoid, abruptly beaked.—Deciduous floodplain forests of large rivers.

Stricta Group   (Acutae, Cryptocarpae)

Five very abundant species, each conspicuous or dominant in its special habitat; plants 30-150 cm tall; bases often reddish and pinnate-fibrillose; leaves long and slender; inflorescence arching, with 3-6 slender, many-flowered pistillate spikelets below several slender staminate spikelets; perigynia 2-3 mm long; stigmas 2; achenes flattened. Plants often forming dense tussocks, but in the first 4 species also spreading by elongate slender rhizomes.

74. **C. stricta** Lam. Tussock Sedge (Fig. 24). Plants usually forming very dense tussocks, but occasionally in uniform beds; on mounds of peat to 2 dm high; bases slender, reddish, pinnate-fibrillose; leaves slightly M-shaped, 2-4.5 mm wide, wiry, strongly glaucous toward the tips when young, later green; ligule acute, as high as wide; perigynia ascending, ovate, longer than broad when young, somewhat flattened, 2-3 mm long, 1-1.6 mm wide, pale green becoming brown when falling, faintly nerved, sometimes aborted by fungi. The purplish rounded scales are usually longer than the perigynia, but rarely much shorter.—Open acid or alkaline peaty shores and meadows, often where water level fluctuates slightly; throughout the state.

75. **C. emoryi** Dew. Similar to No. 74, but pinnate fibrils lacking; ligule less high than wide or straight across; leaves to 6.5 mm wide.—Wooded river floodplains; north to Polk and Lincoln counties.

76. **C. haydenii** Dew. Similar to No. 74; scales brownish,

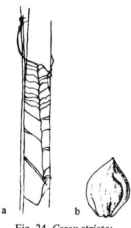
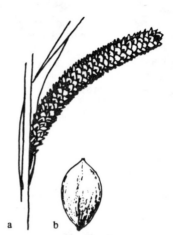

Fig. 24. *Carex stricta:*
a. leaf sheath, × 2.25;
b. perigynium, × 7.50.

Fig. 25. *Carex aquatilis:*
a. pistillate spikelet, × 1.20;
b. perigynium, × 6.

pointed, longer than perigynia; perigynia plumper and more divergent, with tiny abrupt beaks when mature, as broad as long when young.—Low prairies and wet sandy meadows; throughout the state, except eastern Wisconsin.

77. **C. aquatilis** Wahlenb., var. **altior** (Rydb.) Fern. (Fig. 25). Similar to No. 74, but coarser and lacking pinnate fibrils; leaves V-shaped, 3-8 mm wide, retaining the blue green color through autumn; perigynia 2.5-3.5 mm long, 1.5-2.6 mm wide, widest above the middle and thicker at margins, pale blue green often minutely mottled with purple brown; scales usually shorter than perigynia, sometimes longer. Occasional hybrids with No. 74 occur.—Wet shores and floating alkaline peat mats; eastern and northern Wisconsin.

78. **C. crinita** Lam. Plants with abundant yellow root hairs, rhizomes lacking; leaf sheaths smooth to rough, with minute rust-colored hairs; spikelets pendent; scales with long abrupt awns; perigynia round to ovoid; achenes bent.— Streamsides and wet woods in acid soil; northern Wisconsin and southward along the Wisconsin River.

Aurea Group   (Bicolores)

Small caespitose plants, 5-50 cm tall; pistillate spikelets

crowded around and hiding the staminate spikelet; perigynia globose, 10-20-nerved; stigmas 2.

79. **C. aurea** Nutt. Perigynia fleshy, translucent, turning bright yellow.—In damp sand; along the Great Lakes, occasionally to 50 miles inland.

80. **C. garberi** Fern. Similar to No. 79; perigynia becoming powdery white; terminal spikelet mostly pistillate.—Great Lakes shores, sporadic.

Brunnescens Group   (Heleonastes)

Plants small, slender, caespitose; leaves 0.5-4 mm wide; spikelets small, sessile, all alike, mostly as thick as long; perigynia green to pale brown, 1.6-3.8 mm long, tending to be appressed-ascending, plump, filled to the apex by the achene and sometimes splitting at apex; staminate flowers few, mostly at bases of one or more spikelets. Species of moist acid soils and with a northern distribution in Wisconsin.

81. **C. brunnescens** (Pers.) Poir. (Fig. 26). Spikelets remote to crowded; perigynia 3-10 (15) per spikelet, ovate, plano-convex, margins sharp-edged, tapered to a beak, 1.6-2.5 (3) mm long, turning light brown with a few lighter nerves of uneven length.—Common in all kinds of damp coniferous or mixed forests, wet clearings, and southward in tamarack bogs.

C. trisperma has longer bracts, *C. disperma* has terete perigynia; *C. interior* and *C. cephalantha* have perigynia

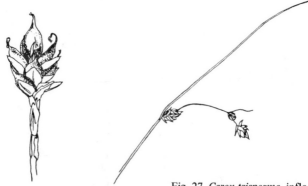

Fig. 26. *Carex brunnescens,* with staminate scales at base, × 4.50.

Fig. 27. *Carex trisperma,* inflorescence with long lower bract, × 1.

curved or bent backward, and *C. convoluta* has divergent perigynia and staminate flowers at apex of spikelets.

82. **C. canescens** L. Similar to No. 81, but leaves blue green; perigynia usually 15-20 per spikelet, round-edged, glaucous or dull, with pink to brown distinct parallel nerves of equal length on slightly convex inner face.—Bogs and wet sand.

83. **C. trisperma** Dew. (Fig. 27). Bracts threadlike, as long as the inflorescence; spikelets 2-3 (4), far apart on a zigzag culm; perigynia 2.7-3.8 mm long, strongly nerved, round-edged, slightly beaked, not quite flat on inner surface, 3 (1-5) per spikelet.—Common in sphagnum bogs, especially in the shade of spruce and tamarack.

84. **C. disperma** Dew. Very small delicate plants, mostly under 2 dm tall; spikelets mostly spaced out; perigynia terete, evenly ribbed, abrupt-beaked, ellipsoidal, often splitting at apex, 1-2 (6) per spikelet.—Damp cool woods, especially white cedar and tamarack.

85. **C. tenuiflora** Wahlenb. Spikelets 2-4, crowded within 1 cm at culm tip; perigynia few, 2.7-3.8 mm long, ovoid, beakless, biconvex.—Sporadic in sphagnum bogs.

86. **C. arcta** Boott. Spikelets (4) 6-14, crowded into an ovoid head 1-3 cm long; perigynia triangular-ovate and plano-convex. The head resembles *C. alopecoidea* and *C. cephalophora* of southern deciduous forest.—Sporadic in wet open places, south to Portage County.

Interior Group   (Stellulatae)

Plants small, slender, caespitose; leaves 1-3 mm wide; spikelets all similar, sessile, as broad as long, with staminate scales at bases; perigynia divergent or reflexed, giving spikelet a star shape when viewed from above. Mostly in wet, very acid or very alkaline soils in the sun.

87. **C. interior** Bailey. Small tufts; spikelets usually 3 (2-5) per culm; scales rounded; perigynia ovate, tapering to a short beak, almost nerveless on flat inner face, 1-2 mm wide and 2-3.5 mm long, green or yellowish.—Wet peaty places; throughout the state.

88. **C. sterilis** Willd. Similar to No. 87, forming larger, denser, more wiry tussocks; often dioecious; spikelets mostly 4 per culm; perigynia more broadly ovate and reddish, the beak half to as long as the triangular body; scales acute.—Fens and wet limy sand; throughout most of the state.

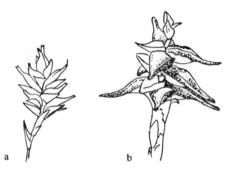

Fig. 28. *Carex cephalantha:* *a.* showing divergent immature perigynia,
× 4; *b.* mature spikelet showing staminate scales, × 6.

89. **C. cephalantha** (Bailey) Bickn. (Fig. 28). Similar to
No. 87; spikelets 4-9; perigynia more numerous, brownish,
2.7-4 mm long, with 5-20 strong nerves on flat face.—Open
sphagnum bogs; throughout the state.

90. **C. angustior** Mackenz. Perigynia only 1-1.2 (1.4) mm
wide, yellowish or pale brown, nerveless on flat face.—Bogs
and wet sandy shores; northern half of the state.

Convoluta Group   (Bracteosae)

Plants caespitose; spikelets short, sessile; staminate flowers
at apex of each spikelet, a tiny club-shaped mass of whitish
scales remaining after anthesis; perigynia plano-convex, ovate,
nerveless on flat face except No. 95, green at maturity but
becoming yellowish to red when falling, stigmas 2.

91. **C. convoluta** Mackenz. (Fig. 29). Plants solitary;
leaves 1-3 mm wide; spikelets star-shaped, far apart, with
1-10 (20) flowers; perigynia divergent, at maturity inflated
on the lower third of the flat inner face, 3-4.5 mm long,
becoming yellow. Leaves resemble those of *C. pensylvanica*
but radiate from solitary clumps.—Common in dry to wet
deciduous woods; throughout the state.

92. **C. sparganioides** Willd. Similar to No. 91, but plants
much more robust; leaves 5-9 mm wide; culms to 6 dm tall;
perigynia 12-40 per spikelet.—Maple-basswood forests; north
to Clark to Oconto counties.

93. **C. cephaloidea** (Dew.) Dew. Similar to No. 92, but
spikelets aggregated into an ovoid head, 1.5-4 cm long; pe-

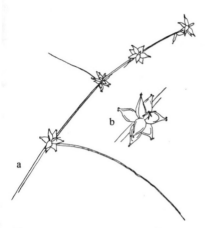

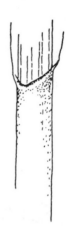

Fig. 29. *Carex convoluta:* a. inflorescence, × 1; *b.* spikelet showing staminate scales at apex, ×2.

Fig. 30. *Carex cephalophora,* leaf sheath, × 5.

rigynia 3-5 mm long, delicate glossy, becoming pale yellowish.—Maple-basswood forests; throughout the state.

94. **C. cephalophora** Willd. (Fig. 30). Plants smaller than No. 93; leaves 2-4.5 mm wide, sheaths slightly thickened at apex; head 0.7-1.8 cm long, with numerous conspicuous threadlike bracts; perigynia 2-3 mm long, becoming orange. —Common in dry deciduous forests; southern half of the state.

95. **C. muhlenbergii** Willd. Leaves sickle-shaped, wiry, gray green, 2.5-4 mm wide; spikelets aggregated into an ovoid head; perigynia gray green, becoming yellow or dark red, usually with strong nerves on slightly convex inner face.—Dry open sand barrens; common in the southern half of the state.

96. **C. gravida** Bailey. Plants to 1.5 m tall; leaves 5-8 mm wide; spikelets aggregated; perigynia 3.5-5 mm long, nerveless on flat face, red-tinged; scales red-tinged.—Old fields and roadsides; north to Chippewa and Green Lake counties.

Stipata Group   (Vulpinae, Multiflorae, Paniculatae)

Plants caespitose; inflorescence compound, of many tiny, crowded, sessile spikelets, each group of which resemble a

Fig. 31. *Carex stipata,* perigynium viewed from flat inner face, × 8.

Fig. 32. *Carex vulpinoidea,* leaf sheath, × 2.

single spikelet (Fig. 33); perigynia green to brown, plano-convex, nerveless on flat face; stigmas 2.

97. **C. stipata** Willd. (Fig. 31). Leaf blades 4-10 mm wide; sheaths cross-wrinkled and easily broken on inner side (as in No. 98); culms 2-4 mm wide, triangular, soft; inflorescence 1.5-10 cm long; scales ovate; perigynia 3-8 mm long, lanceolate, shiny green to brown, inflated, with beak as long as the body, maturing in June.—In all types of wetlands; throughout the state.

98. **C. vulpinoidea** Michx. (Fig. 32). Leaf blades 1-5 mm wide, sheaths cross-wrinkled; culms 1-2 mm thick, wiry; inflorescence 2-15 cm long, with many protruding threadlike bracts and awned scale tips; perigynia 1.7-4 mm long, ovate, dull green to brown, flattened, with 2 corky wings, beak one-half to one-third as long as body, maturing in July.—In all wetlands, rarely in old fields; throughout the state.

99. **C. annectens** Bickn., var. **xanthocarpa** (Bickn.) Wieg. Similar to No. 98; leaves shorter than the culms; perigynia with short abrupt beaks, one-third as long as the body, not winged, plump, becoming bright yellow; scales red-tinged.— Low prairies, pastures, and sandy soil; southern and western Wisconsin.

100. **C. alopecoidea** Tuckerm. Similar to No. 97; sheaths not wrinkled, but parts of inner sides minutely dotted with orange or brown; inflorescence ovoid, 0.5-4.5 cm long, soon tinted with red or yellow; perigynia 2.8-4 mm long, 1.3-2 mm wide, ovate, beak shorter than body.—Deciduous flood-plain forests and adjacent meadows; southern half of the

Fig. 33. *Carex diandra,* compound inflorescence, × 2.

state. Rare specimens of *C. sparganioides* and *C. cephaloidea* have compound inflorescences and key to this species.

101. **C. prairea** Dew. Plants in dense tussocks; leaves long and slender, resembling those of *C. stricta*, 1.5-3.5 mm wide, sheaths tinged with yellow or coppery brown; perigynia plano-convex, lanceolate, green to pale brown, soon falling. —Wet peaty meadows and streamsides; southeastern one-third of the state.

102. **C. diandra** Schrank. (Fig. 33). Similar to No. 101; sheaths not yellow, but sometimes mottled with red or brown; perigynia somewhat divergent, ovate, biconvex, shiny dark brown.—Quaky or floating peaty mats and bogs; throughout the state.

Deweyana Group   (Deweyanae)

Plants caespitose; spikelets sessile, small, all alike, longer than broad; perigynia slender, green, 4-5.5 mm long, closely appressed and hidden by the scales, fitting tightly over plump achenes, but with the beaks empty; stigmas 2.

103. **C. deweyana** Schwein. Leaves 1.5-5 mm wide; spikelets remote, on an arching zigzag inflorescence, resembling that of *C. tenera* and *C. aenea*; perigynia ovate, 1.3-1.5 (1.9) mm wide, flattened, narrowed to a beak, broad, nerveless on inner face.—Common in cool forests; throughout the state.

104. **C. bromoides** Willd. Leaves 1-2.1 (2.5) mm wide; spikelets overlapping; perigynia slenderly lanceolate, nearly terete, strongly nerved, 0.9-1.1 (1.3) mm wide.—Wet woods and river floodplains; throughout the state.

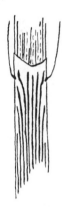

Fig. 34. *Carex sartwellii,* leaf sheath, × 3.

Foenea Group　　　(Arenariae, Chordorrhizeae)

Plants slender, rhizomatous; leaves 1.5-4.5 mm wide; spikelets all similar, short, sessile, crowded near culm apex, brown; perigynia small, brown; stigmas 2.

105. **C. foenea** Willd. Culms in rows, from slender brown fibrillose rhizomes 1-5 dm long; culms 2-6 dm tall, with 2-8 spikelets, the central ones often smaller and mostly staminate; perigynia 4.5-6 mm long, with beak as long as nearly flat ovate body, nerved on concave face.—Common in dry sand, dry prairies, and open dry woods; southern two-thirds of the state.

106. **C. chordorrhiza** L. f. Culms short, in rows, arising from the previous year's fallen-down horizontal vegetative culms; spikelets 3-5, grouped in a head, 5-15 mm long; perigynia very plump, 2-3.5 mm long, evenly nerved, abrupt-beaked.—Frequent in open sphagnum bogs and open floating peat mats; in most parts of the state.

107. **C. sartwellii** Dew. (Fig. 34). Plants 2-6 dm tall, forming dense bright green beds 1-10 m across, pseudoculms numerous, tall, slender; sheaths green-striate on all 3 faces and slightly prolonged at apex above the blade; inflorescence 1-3 cm long, of 6-30 crowded spikelets, the upper ones progressively smaller; perigynia ovate, flat, short-beaked, 2.5-4.5 mm long, 4-9-nerved on concave inner face.—Common in wet meadows; mostly in southeastern Wisconsin.

Scoparia Group   (Ovales)

Plants caespitose, slender, 1-15 dm tall; spikelets all alike, short, sessile; perigynia thin and scalelike, with translucent margins or wings, wind-dispersed, mostly appressed and all or all but the beaks hidden by the scales, becoming brown at maturity; staminate flowers confined to tapering spikelet bases; stigmas 2. The diagnostic perigynium shapes are usually not attained until July; several perigynia must be observed to determine shape and proportions.

The first 4 species are common plants 1-5 dm tall, leaf blades 1-4 mm wide; without pseudoculms; perigynia less than 2 mm wide, except rarely in No. 108.

108. **C. scoparia** Willd. Spikelets tapered to an acute point at each end, 12-14 mm long, crowded or remote on the culm; perigynia incurved, broadest at or above the upper end of achene, lanceolate to narrowly rhombic, (3 times as long as wide), 4-7 mm long, 1.2-2.5 mm wide, appearing depressed over achene on inner face.—Very common in open wetlands, sandy lake shores, and fields.

109. **C. bebbii** (Bailey) Fern. Spikelets 3-15, but usually 5 crowded at culm tip, small and rounded, warm reddish brown at maturity; perigynia ovate, 2 (rarely 3) times as long as wide, 2.5-3.3 mm long, 1-1.6 mm wide, nerveless on inner face. Occasionally seems to hybridize with *C. cristatella*.—Wet meadows and rarely in woods; throughout the state.

110. **C. tenera** Dew. Spikelets remote, on arching zigzag culm; perigynia ovate, 2-2.5 times as long as wide, 3-4 mm long, 1.5-2 mm wide, rather long-beaked, nerved on inner face. Occasionally seems to hybridize with *C. normalis*.—Wet meadows and prairies, rarely in woodlands; throughout the state.

111. **C. crawfordii** Fern. Similar to No. 109, but perigynia and scales lanceolate, 3-4.5 times as long as wide, about 1 mm wide.—Wet sandy shores; northwestern Wisconsin.

The next 5 common species are somewhat larger plants with slender pseudoculms, leaves 2-10 mm broad, and perigynia less than 2 mm wide, except for No. 112, slenderly lanceolate.

112. **C. muskingumensis** Schwein. Plants very stiff and appearing to be a giant form of *C. scoparia*, robust; pseudoculms prominent, leafy; spikelets acute at both ends, 10-30

mm long, 3-6 mm thick; perigynia incurved-appressed, lanceolate, 6-10 mm long, 1.2-2.2 mm wide.—Deciduous floodplain forests of large rivers.

113. **C. tribuloides** Wahlenb. Inflorescence stiff, rather crowded, but plants less robust than No. 112; spikelets 6-12 mm long, blunt; perigynia incurved-appressed, lanceolate, 3 times as long as wide, broadest above tip of achene, the marginal wing slightly but abruptly narrowed just below the middle.—River floodplains; north to Sawyer and Lincoln counties.

114. **C. projecta** Mackenz. Similar to No. 113, but variable in size; axis of inflorescence flexible or arching, spikelets somewhat spaced; scales short-blunt; perigynium wings not abruptly narrowed, occasionally crinkly margined; beaks tending to bend backward a little and be C-shaped in cross section, in apex of spikelet.—Very common in damp woods and clearings in northern Wisconsin and in floodplains of the Wisconsin and Mississippi rivers; also along Lake Michigan.

115. **C. normalis** Mackenz. Culms weak, up to 15 dm tall; spikelets 3-10, rounded, usually crowded near straight culm apex; perigynia ovate, beaked, 2-2.5 times as long as wide, nerved on inner face, green finally becoming pale brown, somewhat divergent. *C. tenera* has narrower leaves; *C. bebbii* has fewer spikelets; *C. projecta* has browner narrower perigynia with bent beaks; *C. cephaloidea* has delicately textured glossy perigynia.—Common in or near deciduous forests; throughout the state.

116. **C. cristatella** Britt. Spikelets 6-15 per culm, globose, hard and prickly, becoming pinkish brown, perigynia 3-4 mm long, rhombic, twice (rarely 3 times) as long as wide, broadest above achene tip, crinkly margined, divergent to slightly reflexed.—Wet places, in or near deciduous forests; mostly in southern half of the state.

The next 4 common species have perigynia ovate, beaked, and 2.2-4.2 mm wide.

117. **C. bicknellii** Britt. (Fig. 35). Spikelets ovoid, pointed at both ends when young, then resembling young *C. tenera* and *C. scoparia*, somewhat spaced on the culm; perigynia large, ovate, 4.5-7 mm long, 3-4 mm wide, with 4-7 prominent parallel nerves over achene on inner face, margins orange brown and of irregular width.—Dry to wet prairies; mostly in the southern half of the state.

118. **C. merritt-fernaldii** Mackenz. Similar to No. 117,

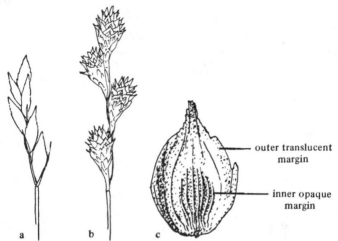

Fig. 35. *Carex bicknellii: a.* immature inflorescence, spikelets pointed, × 1.50; *b.* mature inflorescence, perigynia ripe, × 1.25; *c.* perigynium viewed from concave inner face and showing inner and outer marginal wings, × 6.

but perigynia narrower, dull brown, and nerveless.—Dry rocky barrens; northern Wisconsin.

119. **C. brevior** (Dew.) Mackenz. Spikelets 2-6, spread out or crowded, ovoid but often with a slender tapering brown base; perigynia green to pale brown, with perfectly circular nerveless body and evenly wide white wing.—Dry sand and dry prairies; mostly in the southern half of the state.

120. **C. molesta** Mackenz. Similar to No. 119; spikelets mostly 3-4, round-based and crowded at culm tip; perigynia more ovate and greener until ripe, and faintly nerved over inner face.—Fields and low meadows; southeastern Wisconsin.

The next 2 species are distinguished by longer scales which conceal even the beaks of the perigynia; perigynia plump, lanceolate-ovate, with but 1 thin marginal wing instead of 2, as shown in Fig. 35.

121. **C. adusta** Boott. Plants stiff and robust; scales pale brown or reddish tinged; spikelets ovoid, crowded within 2-3.5 cm; perigynia 4.5 mm long, 1.7-2.5 mm wide, with few or no nerves on inner face.—Dry sandy barrens; northern half of the state.

122. **C. aenea** Fern. Plants slender; spikelets in arching zigzag inflorescence, spaced far apart, attenuate-based, 3-7 cm long; perigynia pale, translucent, with 3-8 nerves near base on flat face, and the wing wider above the middle of the body; achene black; scales dull brown or yellowish.—Rocky barrens in the northern half of Wisconsin and rarely southward in cool woods.

ARACEAE   Arum Family

Perennial herbs, with acrid juice; leaves all basal or alternate; flowers minute, closely crowded on a spadix, which is subtended by a spathe; when present the sepals and petals similar; perianth parts 4-6; stamens 2-6, opposite the perianth segments; fruit a berry.

a. Leaves not grasslike, the veins netted or at least curved *b*
  b. Leaves palmately compound . . . . 1. *Arisaema*
  b. Leaves simple, the blades more or less heart-shaped *c*
    c. Spathe white, open, not rolled around the spadix . . .2. *Calla*
    c. Spathe brownish, marked with purple and green, strongly rolled and enclosing the spadix . . . . 3. *Symplocarpus*
a. Leaves grasslike, the parallel veins all essentially straight .4. *Acorus*

1. Arisaema

Monoecious or dioecious plants arising from a corm, stems somewhat fleshy, leaves compound, with long petioles; spathe rolled into a tube below, arched above; spadix long-cylindrical, covered with flowers below, the upper part sterile and smooth; perianth lacking; staminate flowers of 2-4 anthers; pistillate flowers a single-celled ovary with few ovules, only one maturing.

a. Leaves 3-foliate; spadix included in the spathe . . . 1. *A. atrorubens*
a. Leaves 7-11-foliate; spadix extending beyond spathe . . . .
. . . . 2. *A. dracontium*

1. **A. atrorubens** (Ait.) Blume. Indian Turnip; Jack-in-the-Pulpit (Fig. 36). Leaves usually 2, 3-foliate, with elliptical-ovate pointed leaflets; spadix entirely included in the spathe, the upper sterile part club-shaped; the whole plant, especially the bright red berries, which appear in late summer, pervaded by strongly acrid crystals.—Common in damp woods, throughout the state.

2. **A. dracontium** (L.) Schott. Green Dragon (Fig. 37). Usually with but 1 leaf divided into 7-11 oblong-lanceolate pointed leaflets; sterile part of the spadix prolonged into a tapering point extending far beyond the spathe.—Less common, in low rich woods; scattered localities in Wisconsin.

Fig. 36. *Arisaema atrorubens,*
× .40.

Fig. 37. *Arisaema dracontium,*
× .40.

2. Calla   Water Arum

Aquatic herbs, with heart-shaped leaves; spathe white; perianth parts lacking; stamens 6; fruit a cluster of red berries.

C. palustris L. Wild Calla (Fig. 38). Plants with long creeping stems, from which the leaves and scapes arise; leaves

Fig. 38. *Calla palustris,* × .50.

1-4 dm long, the blade usually shorter than the petiole; spadix short-cylindrical; spathe ovate, white, abruptly contracted into a point 4-10 mm long.—Quaking bogs, rather common northward, rare southward to La Crosse, Waukesha, and Milwaukee counties.

3. Symplocarpus   Skunk Cabbage

Ill-smelling herbs; flowering spadix appearing very early in the spring before the leaves, which are rolled into a cigar-shaped mass at flowering time; perianth of 4 erect segments, joined; stamens 4; style stout, elongate; seeds embedded in the enlarged spadix.

S. foetidus (L.) Nutt. (Fig. 39). Spadix globose, green or purplish, completely hidden within the hooded spathe; leaves all arising near the ground, large and flat, with thick midribs.—Wet open ground, throughout the state.

4. Acorus   Sweet Flag

Aromatic herbs, from tough rhizomes; leaves sword-shaped and stiff; inflorescence a spadix growing laterally from a triangular scape and appearing as a continuation of

Fig. 39. *Symplocarpus foetidus,* × .25.

Fig. 40. *Acorus calamus,* × .25.

the scape, cylindrical, closely covered with small yellow
flowers; perianth of 6 segments; stamens 6; ovary 2-3-celled.

A. calamus L. (Fig. 40). Leaves linear, crowded at base;
flowers yellowish brown.—Shallow muddy bogs, throughout
the state.

COMMELINACEAE   Spiderwort Family

Herbs, frequently succulent; leaves grasslike, with a short
tubular inflated sheath; inflorescence umbellike cymes or
flowers solitary; sepals 3, green, rather boat-shaped; petals 3;
stamens 6; style 1, stigma capitate; fruit a few-seeded capsule. Chiefly a tropical family, including the Inch Plant or
Wandering Jew.

Tradescantia   Spiderwort

Perennials, with alternate linear leaves; the umbellike
cymes subtended by elongate leafy bracts; flowers showy,
soon withering; sepals 3, green; petals 3, blue to pink;
stamens 6, filaments usually hairy; style slender, capitate.

a. Sepals glabrous or with nonglandular hairs . . . . 1. *T. ohiensis*
a. Sepals pubescent with glandular hairs or both glandular and nonglandular hairs *b*
   b. Sepals and pedicels densely villous, sepals 10-15 mm long . . . . .
. . . . .2. *T. bracteata*
   b. Sepals and pedicels sparsely pubescent, sepals 6-10 mm long . . .
. . . . .3. *T. occidentalis*

1. T. ohiensis Raf. (Fig. 41). Sepals without hairs except
occasionally at tips; flowers occasionally white.—Sandy soil
and along railroads, mostly southward.

Fig. 41. *Tradescantia ohiensis,* × .65.

2. **T. bracteata** Small. Sepals with glandless and gland-tipped hairs intermixed.—Rare; Dane, Iowa, St. Croix, and Oconto counties.

3. **T. occidentalis** (Britt.) Smyth. Sepals with gland-tipped hairs.—Sand terraces from Burnett to Pepin counties.

JUNCACEAE   Rush Family

Annual or perennial plants, with a grasslike aspect; both basal and alternate stem leaves with sheathing bases; sepals and petals 3, both scalelike and seldom exceeding 5 mm in length (Fig. 43*b*) stamens 3 or 6; fruit a capsule of 3 carpels.

a. Leaves without hairs ....1. *Juncus*
a. Leaves with scattered long soft hairs ....2. *Luzula*

Fig. 42. *Juncus balticus,* × 1.50.

1. Juncus   Rush

Flowers in close or loose cymes, umbels, or heads; leaves and stems slender and flexible but tough, smooth, and solid or tubular; capsules many-seeded. A large genus, mostly of wetland and shore species which flower in summer and fall.

a. Inflorescence appearing to one side of the stem, the single erect involucral bract seeming to be a continuation of the stem, 2-6 times as long as inflorescence (Fig. 42) *b*
  b. Stems numerous, in dense clumps; stamens 3 ..... 1. *J. effusus*

    b. Stems scattered, in rows from elongate horizontal rhizomes; stamens 6 ....2. *J. balticus*
a. Inflorescence appearing terminal on the stem, subtended by one or several short spreading bracts (Fig. 43*b*) *c*
    c. Leaves round in cross section, or slightly channeled; inflorescence becoming chestnut brown ....3. *J. greenei*
    c. Leaves flat, but often rolled into a quill; inflorescence straw-colored or pale brown *d*
        d. Auricles at summit of leaf sheath firm and rounded, to 0.5 mm long, yellowish or brown (Fig. 43*a*) ...... 4. *J. dudleyi*
        d. Auricles thin, papery, whitish, usually prolonged 1-3 mm above the base of blade (Fig. 44) .... 5. *J. tenuis*

1. **J. effusus** L. Soft Rush. Inflorescence appearing umbellate.—Sandy or rocky streamsides; clearings, wet pastures, and woods; throughout the state except the southern tier of counties.

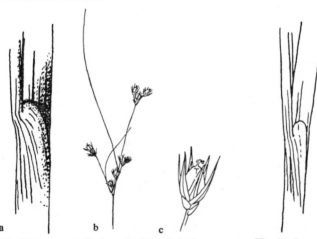

Fig. 43. *Juncus dudleyi: a.* leaf sheath, × 5; *b.* inflorescence, × .75; *c.* flower, × 3.50.

Fig. 44. *Juncus tenuis,* × 5.

2. **J. balticus** Willd. (Fig. 42). Stems dark green, often evenly spaced in rows; flowers few, ascending, becoming glossy dark purplish brown.—Wet sandy or marly shores and dunes, especially along the Great Lakes, rarely to Dane County.

3. **J. greenei** Oakes & Tuckerm. Plants very stiff; mature capsules longer than perianths.—Moist to very dry sterile

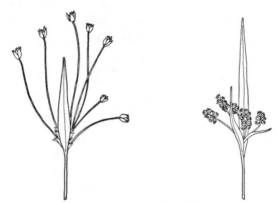

Fig. 45. *Luzula acuminata,* × 1.20.   Fig. 46. *Luzula multiflora,* × .80.

sand in open places; from Rock and Walworth counties
northwestward to Burnett County.

4. **J. dudleyi** Wieg. (Fig. 43). Appearing as a larger more
erect form of *J. tenuis*; compare several undamaged auricles;
capsule as long as the perianth.—Common in moist meadows,
low prairies, fields and shores; Dane to Sheboygan counties
north to Oconto County.

5. **J. tenuis** Willd. Yard Rush (Fig. 44). Plants usually
under 2 dm high (the species above are 3-10 dm
tall).—Probably once native to damp shores, now found on
almost every path in forest and field, along borders of roads,
and in lawns.

2. **Luzula**   Wood Rush

Leaves flat and grasslike, mostly at the base of the plant,
a few on the stem; flowers in umbellike or spikelike inflorescences; stamens 6; capsule 3-seeded.

a. Flowers solitary at the ends of the branches of the umbels . . . .
. . . . 1. *L. acuminata*
a. Flowers in glomerules or clusters at the ends of the umbel
branches . . . . 2. *L. multiflora*

1. **L. acuminata** Raf. (Fig. 45). Flowers 3-4.5 mm long, in
an umbel, with branches of the umbel about 2 cm long.—
Cool woods, southward to Iowa County and along Lake
Michigan to Racine County.

2. **L. multiflora** (Retz.) Lejeune (Fig. 46). Flowers about

2 mm long, in dense clusters at the ends of unequal branches.—In fields or open woods, more common in the southern half of the state.

LILIACEAE   Lily Family

Herbs or vines (our species); leaves various; flowers regular, ordinarily with 3 colored or white petals, 3 similar or green sepals, 6 stamens, and an ovary of 3 carpels which develops into a capsule or berry.

a. Leaves growing only from the base of the plant *b*
   b. Flowers 8 mm or less long, in an umbel . . . . 1. *Allium*
   b. Flowers 1 cm or more long, or, if shorter, in a raceme *c*
      c. Leaves not more than 10 times longer than broad *d*
         d. Flowers solitary, nodding; leaves 1-2.5 cm broad, obscurely veined . . . . 2. *Erythronium*
         d. Flowers 2 or more; leaves 3-8 cm broad, strongly veined *e*
           e. Flowers in a raceme *f*
              f. Perianth parts fused . . . . 3. *Convallaria*
              f. Perianth parts separate . . . . 4. *Scilla*
           e. Flowers in an umbel . . . . 5. *Clintonia*
      c. Leaves grasslike, many times longer than broad *g*
         g. Flowers pale blue . . . . 6. *Camassia*
         g. Flowers greenish or whitish *h*
           h. One flower in the axil of each bract; pedicels longer than the flower . . . . .7. *Zigadenus*
           h. Several flowers from each axil; pedicel at flowering time shorter than the flower . . . . 8. *Tofieldia*
a. Leaves elevated on the stem *i*
   i. Leaves scattered or in many whorls *j*
      j. Flowers solitary, or 2 in the axil of a leaf *k*
         k. Flowers 1.5 cm or more long, and at least half as wide *l*
           l. Leaves 2, opposite . . . . 2. *Erythronium*
           l. Leaves many, alternate or whorled *m*
              m. Flowers 6 cm or more long . . . . 9. *Lilium*
              m. Flowers 1.5-4.5 cm long . . . . .10. *Uvularia*
         k. Flowers 1 cm or less long, or, if longer, about one-fourth as wide *n*
           n. Leaves reduced to brownish scales . . . . .11. *Asparagus*
           n. Leaves broad and flat *o*
              o. Each peduncle appearing just under a leaf . . . . 12. *Streptopus*
              o. Each peduncle just above a leaf . . 13. *Polygonatum*
      j. Flowers in an umbel or raceme, or 2-several on a peduncle *p*
         p. Flowers 5 cm or more long . . . . 9. *Lilium*
         p. Flowers 2 cm or less long *q*
           q. Flowers in umbels, the staminate and pistillate on different plants . . . . 14. *Smilax*
           q. Flowers not in umbels, each with stamens and pistil *r*
              r. Flowers in terminal racemes *s*
                s. Stamens 4 . . . . .15. *Maianthemum*
                s. Stamens 6 . . . . 16. *Smilacina*
              r. Flowers 1-6 in each leaf axil . . . . 13. *Polygonatum*

1. **Allium** Onion

Foliage strong-scented; umbel subtended by an involucre which envelops all the buds.

1. **A. tricoccum** Ait. Wild Leek (Fig. 47). Bulb smooth; leaves flat, 1-2.3 dm long and 1-6 cm broad, appearing in the

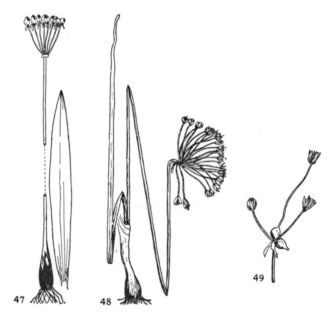

Fig. 47. *Allium tricoccum,* × .25.   Fig. 49. *Allium canadense,* × .40.
Fig. 48. *Allium cernuum,* × .40.

early spring and dying before the flowers mature; pedicels 1-1.5 cm long.—Rich woods, throughout the state.

2. **A. cernuum** Roth. Nodding Wild Onion (Fig. 48). Bulb with papery scales; leaves flattened, keeled, ribbonlike.—Prairies and along railroads, southeastern and very southern counties.

3. **A. canadense** L. Wild Garlic (Fig. 49). Bulb with fibrous coat; leaves ribbonlike, thick and fleshy; umbel erect, the pedicels reaching 5 cm in length.—Woods and meadows, north to Polk, Waupaca, and Brown counties.

4. **A. fistulosum** L. Welsh Onion. Plants stout, somewhat tufted, glaucous, up to 5 dm tall; bulbs scarcely larger in diameter than leaves or scapes.—Escape or remnant of cultivation in isolated places.

5. **A. schoenoprasum** L. Garden Chives. Plants forming tufts, up to 6 dm tall; bulbous-thickened at base; flowers many in a head, rose purple.—Occasional escape from cultivation.

2. **Erythronium**   Fawn Lily; Adder's-tongue

Plants with a deep-seated white bulb shaped like a canine tooth; leaves 2, flat and shining; flower usually solitary, nodding on a leafless scape; sepals and petals similar; stamens 6; style 1; fruit a capsule.

a. Leaves mottled; flowers yellow . . . . 1. *E. americanum*
a. Leaves less mottled; flowers pinkish white . . . . 2. *E. albidum*

1. **E. americanum** Ker. Yellow Adder's-tongue (Fig. 50). Leaves mottled with purplish and whitish; flowers light yellow; anthers yellow or reddish brown; style club-shaped, with united stigmas.—Rich woods, mostly northern, south to Sauk County and along Lake Michigan to Racine County.

2. **E. albidum** Nutt. White Adder's-tongue (Fig. 51). Leaves less mottled or not at all; flowers pinkish white; anthers yellow; style slender except at tip, with 3 spreading

Fig. 50. *Erythronium americanum,* × .25.

Fig. 51. *Erythronium albidum,* × .25.

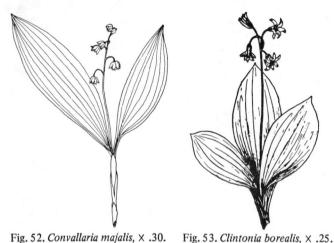

Fig. 52. *Convallaria majalis,* × .30.   Fig. 53. *Clintonia borealis,* × .25.

stigmas.—Rich woods, north to Brown to Polk counties; Ashland County.

3. Convallaria   Lily-of-the-Valley

Low plants, from horizontally branching rhizomes; leaves 2; flowers few to many on a raceme, drooping, fragrant; sepals and petals united into corollalike perianth, cup-shaped with 6 short recurved lobes; stamens 6, borne at the base of the corolla; fruit a reddish berry.

C. majalis L. (Fig. 52). Leaves 1-2 dm long, oblong-oval; flowers white.—A remnant from cultivation or escaping nearby, in scattered areas of Wisconsin.

4. Scilla   Squill

Low plants, with bulbous base; leaves narrow, basal; flowers appearing with the leaves on terminal raceme; sepals and petals similar; stamens 6; pistil 1; fruit a 3-lobed capsule.

S. sibirica Andr. Early flowering plants to 16 cm tall; flowers deep blue.—Escape from gardens; Manitowoc and Outagamie counties.

5. Clintonia   Corn Lily; Bluebead Lily

Perennials, from rhizomes, with 2-5 broad basal leaves; inflorescence an umbel on a leafless scape; sepals 3; petals 3, similar; stamens 6; style 1; fruit a berry.

C. borealis (Ait.) Raf. (Fig. 53). Leaves flat, oval, 1.5-2 dm long; flowers 3-6, straw-colored; berries blue.—In cool woods and swamps, south to the Baraboo Hills, and along Lake Michigan shore to Kenosha County.

6. Camassia   Wild Hyacinth

Perennials, from dark papery bulbs; leaves basal, linear; inflorescence a raceme, flowers subtended by filiform bracts; sepals and petals similar; stamens 6, the filaments filiform; style 1, stigma 3-lobed; fruit a capsule.

C. scilloides (Raf.) Cory. Scape 1.5-7 dm high; flowers 10-14 mm long, blue, pale violet to white, in a raceme 1-3 dm long.—Rare; Lafayette to Rock counties.

7. Zigadenus   White Camass

Perennials to 9 dm high, from papery bulbs, poisonous; basal leaves long-linear, stem leaves reduced to bractlike; inflorescence a raceme with white to greenish white flowers about 1 cm long; sepals and petals similar; stamens 6; styles 3; fruit a 3-lobed capsule subtended by the persistent perianth.

Z. elegans Pursh (Fig. 54). Leaves thin, sharp-pointed; flowers usually in a raceme with simple branches; upper bracts with papery margins and summits; capsule twice as long as the sepals.—Mesic prairies, meadows, and moist cliffs; locally northward to Door and Buffalo counties; adventive at Superior.

8. Tofieldia   False Asphodel

Perennial grasslike herbs, with 2-ranked leaves from a small rootstock; sepals and petals similar; stamens 6; styles 3; fruit a 3-lobed capsule subtended by persistent perianth and tipped by the styles.

T. glutinosa (Michx.) Pers. (Fig. 55). Scape 2-5 dm tall, covered with sticky dark glands just below the inflorescence, which is 1-5 cm long; flowers about 5 mm long.—Springy meadows, rare; Marquette and Columbia counties eastward.

9. Lilium   Lily

Erect perennials, from a scaly bulb, 4-20 dm high; leaves narrow, without petioles, whorled or alternate; flowers large, in our species reddish or orange; sepals and petals similar;

Fig. 54. *Zigadenus elegans,* × .30.   Fig. 55. *Tofieldia glutinosa,* × .25.

stamens 6, filaments elongate, anthers linear; style 1, stigma 3-lobed; fruit a 3-angled capsule.

a. Flowers erect . . . . 1. *L. philadelphicum*
a. Flowers nodding, perianth recurving . . . . 2. *L. michiganense*

1. **L. philadelphicum** L. Wood Lily (Fig. 56). Flowers 1-5, erect; perianth segments dark-spotted at the base of the broader part, much narrowed at the lower end, not recurved; leaves whorled.—Door County and perhaps elsewhere, rare.

Var. **andinum** (Nutt.) Ker, with the leaves mostly scattered, occurs throughout the state.

2. **L. michiganense** Farw. Western Turk's-cap Lily (Fig. 57). Flowers nodding; perianth segments finely dotted with dark purple, not narrowed at the lower end, strongly recurved; leaves whorled. (*L. canadense* and *L. superbum* are

Fig. 56. *Lilium philadelphicum,*
× .25.

Fig. 57. *Lilium michiganense,*
× .30.

Fig. 58. *Uvularia grandiflora,*
× .25.

Fig. 59. *Uvularia sessilifolia,*
× .25.

eastern species with which this has been confused.)—Meadowland, throughout the state.

10. Uvularia   Bellwort

Stems erect, from rhizomes, somewhat branched; flowers
yellowish, nodding; sepals and petals similar, elongate;
stamens 6; styles 3, may be partially united; fruit a 3-lobed
or winged capsule.

a. Leaves perfoliate .... 1. *U. grandiflora*
a. Leaves merely sessile .... 2. *U. sessilifolia*

1. **U. grandiflora** Sm. (Fig. 58). Stem apparently passing
through the leaves; leaves finely hairy beneath; flowers
2.5-4.5 cm long; fruit a 3-lobed capsule.—Common in woods,
throughout the state.

2. **U. sessilifolia** L. (Fig. 59). Leaves rather narrow,
whitened beneath; flowers 1.2-2 cm long; fruit 3-

Fig. 60. *Asparagus officinalis,* × .30.

winged.—Woods, south to Brown, Dane, and rarely Grant counties.

11. Asparagus   Garden Asparagus

Tall perennials, from matted rootstocks; leaves represented by small brown scales, with green branchlets from their axils; flowers in the axils of the true leaves; perianth segments 6; stamens 6; style 1; fruit a berry.

A. officinalis L. (Fig. 60). Dioecious; flowers greenish white, 3-5 mm long; berries red.—Escaping cultivation into open ground or sunny woods, throughout Wisconsin except the extreme northern counties.

12. Streptopus   Twisted Stalk

Perennial herbs, from rhizomes, 2-9 dm high; leaves alternate, 5-11 cm long, sharply pointed; flowers solitary in the leaf axils, each on a bent peduncle; sepals and petals similar; stamens 6; style 1, stigma sometimes 3-cleft; fruit a red berry.

a. Leaves clasping the stem; nodes glabrous ......1. *S. amplexifolius*
a. Leaves sessile; nodes pubescent ....2. *S. roseus*

1. S. amplexifolius   (L.) DC., var. **denticulatus** Fassett. Leaves strongly clasping the stem, whitened beneath, the margin not fringed with hairs, minutely toothed; flowers greenish white.—Cool moist woods in counties bordering on Lake Superior; Door County.

2. S. roseus Michx., var. **longipes** (Fern.) Fassett (Fig. 61). Leaves slightly clasping or merely sessile, not whitened beneath, copiously fringed with fine hairs; flowers light pink, spotted with deep pink.—Common in woods northward, and

Fig. 61. *Streptopus roseus,* var.   Fig. 62. *Polygonatum pubescens,*
longipes, × .30.   × .40, rhizome × .20.

occurring locally southward to the Baraboo Hills, to Racine
and Walworth counties, and occasionally in tamarack bogs
southwestward.

13. **Polygonatum**   Solomon's Seal

Perennials with simple erect-arching stems, from rhizomes
which bear the scars of attachment of previous stems (the
"seals"); leaves alternate, elliptic, not petioled; flowers 1-4
on axillary peduncles; sepals and petals fused for most of
their length; stamens 6; style 1; stigma capitate; fruit a black
or blue berry.

a. Leaves hairy on the veins beneath . . . . 1. *P. pubescens*
a. Leaves glabrous . . . . 2. *P. biflorum*

1. **P. pubescens** (Willd.) Pursh. Small Solomon's Seal (Fig.
62). Plants 3-9 dm high; leaves whitened and finely hairy
beneath; flowers 10-12 mm long.—Low rich woods, throughout the state.

2. **P. biflorum** (Walt.) Ell. Great Solomon's Seal. Plants
0.6-2 m high; leaves smooth, green on both sides; flowers
1.2-2 cm long.—Rich woods, throughout the state.

14. Smilax   Greenbrier

Dioecious trailing or erect perennials; stems with leaves above and scaly bracts below; leaves net-veined, the blades more or less heart-shaped; inflorescence an umbel; petals and sepals similar; fruit a berry.

a. Stem herbaceous, without prickles or bristles *b*
    b. Stem vinelike, climbing, with tendrils; umbels with 25 or more flowers *c*
        c. Leaves glabrous beneath; berries blue . . . . 1. *S. herbacea*
        c. Leaves minutely white-hairy beneath: berries black . . . .
        . . . . 2. *S. lasioneura*
    b. Stem erect, without tendrils or with tendrils from a few upper leaves; umbels with 25 or fewer flowers *d*
        d. Basal leaves broadly ovate, base cordate . . . . . 3. *S. ecirrhata*
        d. Basal leaves narrowly ovate, base mostly ovate . . . .
        . . . . 4. *S. illinoensis*
a. Stem woody, with black bristles . . . . 5. *S. hispida*

1. S. herbacea L. Carrion Flower (Fig. 63). Stem extensively climbing; flowers ill-smelling, the umbels long-peduncled from the axils of leaves; flowers many, in an umbel; berries blue.—Scattered and rare in Wisconsin.

2. S. lasioneura Hook. Similar, leaves pubescent beneath and with an abrupt transition between bracts and leaves; berries black.—In open woods and often on fences, throughout the state.

3. S. ecirrhata (Engelm.) S. Wats. Plants erect, shoots to 8 dm tall, with few or no tendrils; petioles shorter than the leaf blades; 1-3 peduncles from lower leaves and bracts; flowers few in each umbel.—Low woods, mostly in the southern half of Wisconsin but scattered localities north to Douglas County.

4. S. illinoensis Mangaly. Plants to 1 m high, with a few short tendrils; transition from bracts to leaves abrupt; petioles longer than leaf blades; 5-10 peduncles, the lower ones

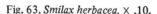

Fig. 63. *Smilax herbacea*, × .10.   Fig. 64. *Maianthemum canadense*, × .25.

long and axillary to the bracts. This recently named species is a hybrid between *S. lasioneura* and *S. ecirrhata.*—Common in southern Wisconsin.

5. **S. hispida** Torr. Greenbrier. Long thorny climbers; leaves thin, ovate to rotund; peduncles exceeding the petioles; berries black.—In woods, throughout the state.

15. Maianthemum   Canada Mayflower; Dwarf Solomon's Seal

Low perennials, from rhizomes; leaves usually 2, flat; petioles short or none; flowers 2 mm long, with 2 sepals and 2 petals; stamens 4; style 1, stigma 2-cleft; fruit a pale red berry.

M. canadense Desf. (Fig. 64). Plants 6-22 cm high; leaves heart-shaped at base.—Sandy woodlands, throughout the state, but more common northward.

Var. **interius** Fern. Stems and usually the leaves covered with short hairs.—Common southward, but also occurs in the north.

16. Smilacina   False Solomon's Seal

Plants from rhizomes; leaves alternate, many-nerved, without petioles; flowers in a terminal raceme or panicle, white; sepals 3; petals 3, similar; stamens 6; style 1; fruit a red berry.

a. Flowers 2 mm long; stamens longer than the petals .1. *S. racemosa*
a. Flowers 4-5 mm long; stamens shorter than the petals *b*
   b. Leaves 7-12, clasping the stem with their bases ... 2. *S. stellata*
   b. Leaves 2-4, sheathing at base .... 3. *S. trifolia*

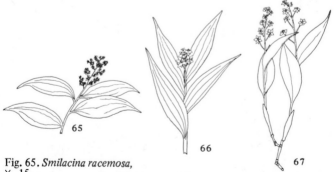

Fig. 65. *Smilacina racemosa,* × .15.
Fig. 66. *Smilacina stellata,* × .35.   Fig. 67. *Smilacina trifolia,* × .15.

1. **S. racemosa** (L.) Desf. (Fig. 65). Flowers very numerous, on pedicels about 2 mm long; leaves taper-pointed, minutely downy. Individuals with more slender panicles or racemes are treated as var. cylindrata in *Gray's Manual*, 8th ed.—Common in woods, throughout the state.

2. **S. stellata** (L.) Desf. (Fig. 66). Plants 2-5 dm high; flowers 3-14, on longer pedicels; leaves pointed but not tapered, smooth.—Open ground, throughout the state.

3. **S. trifolia** (L.) Desf. (Fig. 67). Plants 1.3-2 dm high; flowers 8-11; leaves smooth.—Peat bogs, south to Jackson and Brown counties, and along the Lake Michigan shore to Milwaukee.

17. Medeola   Indian Cucumber-root

Perennial plants, 3-9 dm high, from a white horizontal fleshy tuber; leaves in 2 whorls; inflorescence an umbel borne above the upper whorl of leaves; sepals and petals similar; stamens 6; styles 3; fruit a berry.

M. virginiana L. (Fig. 68). Plants with woolly flocculence

Fig. 68. *Medeola virginiana,* × .25.

persisting near leaf bases and lower part of stem; flowers yellowish, about 1 cm broad, with small recurved petals shorter than the styles; fruit purple.—Damp woods, eastern Wisconsin, Milwaukee to Shawano to Marinette counties.

18. **Trillium**   Trillium

Perennial plants; stem simple, from a short tuberlike rootstock; leaves 3 (rarely more), whorled, diamond-shaped, usually about as broad as long; flower solitary, terminating the stem; sepals 3; petals 3; stamens 6; style 1, stigmas 3; fruit a berry.

a. Flower sessile . . . . 1. *T. recurvatum*
a. Flower peduncled *b*
    b. Leaves without definite petioles *c*
        c. Stigmas exceeded by the stamens, usually not coiled or only
            slightly recurved at tip . . . . 2. *T. grandiflorum*
        c. Stigmas equaling or exceeding the stamens, coiled at tip *d*
            d. Peduncles 1.2-4 cm long, curved or bent downward; an-
               thers 2.5-6.5 mm long . . . . 3. *T. cernuum*
            d. Peduncles 3-12 cm long, straight; anthers 6-15 mm long
            . . . . .4. *T. flexipes*
    b. Leaves short-petioled . . . . 5. *T. nivale*

1. **T. recurvatum** Beck. Prairie Trillium (Fig. 69). Leaves often short-petioled; sepals recurved; petals 1.3-3 cm long, dark red, narrowed into a claw at base.—Moist woods and prairies, southern Wisconsin, Green to Milwaukee counties.

2. **T. grandiflorum** (Michx.) Salisb. (Fig. 70). Leaves without petioles; peduncle erect or nearly so; petals 4-6 cm long, white. turning rose color, sometimes marked with green.— Rich woods, throughout the state.

3. **T. cernuum** L. Nodding Trillium. Peduncle recurved or reflexed; filaments nearly or quite equaling anthers; petals 5-9 mm broad, white; anthers 2.5-4.5 mm long.—Rare in northern Wisconsin.

Var. **macranthum** Eames & Wieg. (Fig. 71). Petals 10-17 mm broad; anthers 4-6.5 mm long.—Rich woods, throughout the state but not common.

4. **T. flexipes** Raf. Peduncle usually horizontal; filaments less than half as long as the anthers; petals 2-3.5 cm long, white.—Rich woods, north to Monroe and Fond du Lac counties.

5. **T. nivale** Riddell. Snow Trillium (Fig. 72). Plants 5-10 cm high; leaf blades rounded at base; peduncle short, erect or recurved; petals 1.2-3 cm long, white.—Rich woods and fields, Manitowoc, Milwaukee, Ozaukee, and Pierce counties, and doubtless elsewhere in limestone areas.

Fig. 69. *Trillium recurvatum,*
× .20.

Fig. 70. *Trillium grandiflorum,*
× .20.

Fig. 71. *Trillium cernuum,* var.
macranthum, × .40.

Fig. 72. *Trillium nivale,* × .40.

AMARYLLIDACEAE　　Amaryllis Family

Perennial herbs, from bulbs, corms, or fibrous roots; flowers single or in an umbel subtended by bracts; sepals and petals each 3, similar, separate or united at base, the tube sometimes bearing a crownlike appendage (example *Narcissus*); stamens 6; style 1; ovary below the flower; fruit a capsule.

Hypoxis

Grasslike plants from bulbs or rhizomes, usually hairy; inflorescence an umbel of 2-6 flowers or flowers single, regular.

H. hirsuta (L.) Coville. Star Grass (Fig. 73). Flowers yellow, 1-4 on a scape, 1-1.5 cm broad.—Sunny woods, bluffs, and meadows, north to Pierce, Chippewa, and Oconto counties.

Fig. 73. *Hypoxis hirsuta,* × .25.   Fig. 74. *Dioscorea villosa,* × .25.

DIOSCOREACEAE   Yam Family

Herbaceous trailing vines, from thick tuberous rootstocks; leaves petioled, net-veined; flowers regular; sepals 3, petals 3, similar; stamens 6 or 3; styles 3, simple or each one branched; fruit a 3-winged capsule.—A large tropical family.

Dioscorea   Yam

Dioecious vines; flowers white to greenish yellow; pistillate flowers solitary at each node of a short spike, the staminate flowers in clusters or solitary at each node of a panicle; seeds very flat, broadly winged.

D. villosa L. Wild Yam (Fig. 74). Leaf blades heart-shaped, downy beneath; flowers very small, at intervals of about 4 mm on drooping slender branches.—Rich woods, north to Polk, Lincoln, and Oconto counties.

IRIDACEAE   Iris Family

Perennial herbs; leaves sword-shaped or grasslike; flowers from a spathe, perfect; sepals 3, petallike; petals 3; stamens 3; style 1, 3-parted, the divisions expanded, petallike; ovary inferior; fruit a capsule formed of 3 carpels.

a. Stigmas petallike; flowers 4 or more cm long . . . . 1. *Iris*
a. Stigmas threadlike; flowers about 1 cm long . . . . . 2. *Sisyrinchium*

1. Iris   Fleur-de-lis; Iris

Plants from creeping rootstocks; leaves 1-2 cm broad; flowers showy; sepals and petals united below into a tube, sepals much larger than the petals and drooping; petals erect and between the 3 branches of the style which are tipped by broad petallike crests.

Fig. 75. *Iris versicolor,*   Fig. 76. *Iris virginica,* var.
× .40.   *shrevei,* × .40.

a. Flowers blue *b*
   b. Perianth tube 5 mm or less in length *c*
      c. Petals shorter than the styles; ovary less than 2 cm in length
         (Fig. 75) . . . . 1. *I. versicolor*
      c. Petals slightly longer than the styles; ovary 2 cm or more in
         length (Fig. 76) . . . . 2. *I. virginica*
   b. Perianth tube 1 cm or more in length . . . . 3. *I. lacustris*
a. Flowers yellow . . . . 4. *I. pseudacorus*

1. **I. versicolor** L. Blue Flag (Fig. 75). Stems 1.5-5 dm
high, stout; flowers 5-8 cm long; sepals blue, sometimes with
a greenish or yellowish spot at the base of the blade; lining
of capsule shining.—Wet places, from La Crosse to Manitowoc counties northward.

2. **I. virginica** L., var. **shrevei** (Small) Anders. (Fig. 76).
Similar; sepals with a bright yellow spot at the base of the
blade; lining of capsule dull.—Wet places, north to Price and
Door counties.

3. **I. lacustris** Nutt. Dwarf Lake Iris. Stems 0.5-1.5 dm
high; free part of sepals about 2 cm long.—Rare; Lake Michigan shore, Door and Milwaukee counties.

4. **I. pseudacorus** L. Water Flag. Stems to 1 m tall; leaves
stiff and erect; flowers yellow or cream-colored; petals erect,
tongue-shaped; capsule 6-angled.—Native of Europe; shallow
water, in Douglas and Bayfield counties near Lake Superior,
in Milwaukee and Kenosha counties near Lake Michigan, and
also in Dane County.

2. Sisyrinchium   Blue-eyed Grass

Plants grasslike, tufted; stems 2-winged, longer than the
leaves; flowers blue or white, clustered, each cluster enclosed

Fig. 77. *Sisyrinchium angusti-folium*, × .15.

Fig. 78. *Sisyrinchium albidum*, × .85.

Fig. 79. *Sisyrinchium campestre*, × .90.

by 2 bracts which make up the spathe; spathe or spathes sometimes enclosed by a larger bract.—The species are separated on technical points, and are difficult of determination.

1. **S. farwellii** Bickn. Plants with tufts of hairs at the base, the hairs long but fragile and breaking to shorter tufts; flowering stems with a large bract subtending 2 long peduncles, each of which in turn bears a pair of spathes below a group of slender stalked flowers; margins of outer bract fused; capsule 3-5 mm long.—Scattered localities in Wisconsin.

2. **S. angustifolium** Mill. (Fig. 77). Stems flexuous, winged to 6 mm wide; spathes on long peduncles; inner bract 13-22 mm, outer bract longer to leaflike, margins fused.— Milwaukee County.

3. **S. albidum** Raf. (Fig. 78). Inflorescence of 2 spathes surrounded by a leaflike involucral bract, margins of bract not fused at base; capsule 2-4 mm high.—Sunny fields, from Dane and Milwaukee counties southward.

4. **S. campestre** Bickn. (Fig. 79). Plants with 1 spathe on each scape; outer bract 2.5-4.5 cm long, the inner about half as long; capsule 2-4 mm high.—Sunny fields, common from Sawyer, Waupaca, and Racine counties southward.

5. **S. montanum** Greene. Plants with inflorescence of only 1 spathe on each stem; outer bract with margin fused at base for short distance; perianth bright violet; capsule 4-6 mm high, light brown.—Scattered localities in northern Wisconsin.

ORCHIDACEAE   Orchid Family

Herbs, with thick whitish roots (or sometimes none); perianth borne at the summit of the ovary; sepals 3; petals 3, the posterior twisted so that it extends forward and downward and is modified into a lip, which sometimes has a spur extending from near its base; fertile stamen 1 (in *Cypripedium* 2) united with the style to form the column; fruit a capsule formed of 3 carpels, with innumerable minute seeds.

a. Lip a large inflated sac . . . . 1. *Cypripedium*
a. Lip hardly saclike, not inflated *b*
  b. Flowers with a spur 2 mm or more long *c*
    c. Leaves 1-2, at or near the base of the stem; each flower
      exceeded by its bract, which is about 1 cm wide . . 2. *Orchis*
    c. Leaves several, on the stem (except in 2 species which have
      narrow bracts shorter than the flowers) . . . . . . 3. *Habenaria*
  b. Flowers without conspicuous spur *d*
    d. Leaves 1 or more, green *e*
      e. Flower solitary, terminal (rarely 2) *f*
        f. Lip tonguelike, with 3 parallel lines of hairs or crests
          . . . . 4. *Arethusa*
        f. Lip scoop-shaped, with a tuft of yellow hairs . . . .
          . . . . 5. *Calypso*
      e. Flowers several or many, in a raceme *g*
        g. Leaves 2 cm or less long, a single pair halfway up the
          stem . . . . 6. *Listera*
        g. Leaves 5 cm or more long, at the base of the stem *h*
          h. Leaves 2 . . . . 7. *Liparis*
          h. Leaf solitary . . . . 8. *Aplectrum*
    d. Leaves none; plant not green . . . . 9. *Corallorhiza*

1. Cypripedium   Lady's Slipper

Leaves alternate or basal; the 2 lowest sepals, in most species, united into 1 flat structure below the lip; flowers mostly large, the lip 1.5-5 cm long, delicate and much inflated, moccasin-shaped; other petals long and spreading.

a. Leaves several, on the stem *b*
  b. Petals longer than the lip, twisted *c*
    c. Lip yellow . . . . 1. *C. calceolus*
    c. Lip white . . . . 2. *C. candidum*
  b. Petals shorter than the lip, usually not twisted *d*
    d. Lateral petals white . . . . 3. *C. reginae*
    d. Lateral petals dark purple brown, edged with green . . . .
     . . . . 4. *C. arietinum*
a. Leaves 2, or few, close to the ground . . . . 5. *C. acaule*

1. **C. calceolus** L. Leaves ovate to lanceolate-pointed; flowers 1 or 2, terminal; sepals purplish.

Var. **parviflorum** (Salisb.) Fern. Small Yellow Lady's Slipper (Fig. 80). Stems 1.9-6 dm high; lip 2-3 cm long.—Swamps; northwestern and southeastern Wisconsin, and La Crosse County.

Var. **pubescens** (Willd.) Correll. Large Yellow Lady's Slipper. Stems 2.3-7 dm high; lip 3.5-5 cm long.—Woodlands; more common, and throughout most of the state.

Fig. 80. *Cypripedium calceolus,*   Fig. 81. *Cypripedium arietinum,*
var. *parviflorum,* × .20.   × .30.
                                      Fig. 82. *Cypripedium acaule,* × .20.

C. **favillianum** Curtis and **C. andrewsii** Fuller are hybrids
of *C. candidum* with *C. calceolus* var. *pubescens* and with
var. *parviflorum*; they are intermediate between the parents.

2. **C. candidum** Muhl. Small White Lady's Slipper. Plants
1.6-3 dm high; leaves oval, folded at the base around the
stem; flower usually solitary; lip about 2 cm long, white,
purple-striped within.—Meadows and low prairies, southeastern Wisconsin.

3. **C. reginae** Walt. Showy Lady's Slipper. Plants 4-8 dm
high; stem and leaves covered with short hairs (which on the
skin of some persons cause a rash similar to that of Poison
Ivy); leaves ovate, acute, 1.5-2 dm long, folded at the base
around the stem; lip about 4 cm long, white, flushed with
wine red on face.—Woods and bogs, scattered localities in
Wisconsin.

4. **C. arietinum** R. Br. Ram's-head Lady's Slipper (Fig.
81). Stems 15-30 cm high; leaves 3-4, elliptical-lanceolate,
almost without hairs; flower 1.5-2 cm long; lip whitish with
crimson veins, prolonged at apex into a long blunt spur.—
Tamarack bogs, very rare, mostly in counties bordering Lake
Michigan.

Fig. 83. *Orchis rotundifolia,* × .30.   Fig. 84. *Orchis spectabilis,* × .30.

5. **C. acaule** Ait. Stemless Lady's Slipper; Moccasin Flower (Fig. 82). Plants 1.5-4 dm high; leaves oval, many-nerved; lip pink (rarely white), inflated, about 5 cm long and half as broad.—Common northward, south in pine woods to the Dells of the Wisconsin, and in tamarack bogs to the southeastern counties.

2. Orchis   Orchis

Leaves 1 or 2 near the base; flowers in a raceme; petals and upper sepals forming a hood over the column; lip prolonged into a conspicuous spur; disks at base of stalk of the pollen masses contained in a special pouch.

a. Leaf 1; lip 3-lobed, notched at the tip . . . . 1. *O. rotundifolia*
a. Leaves 2; lip neither lobed nor notched . . . . 2. *O. spectabilis*

1. **O. rotundifolia** Banks. Small Round-leaved Orchis (Fig. 83). Leaf solitary, 3-8 cm long; flower pinkish mauve to white.—Tamarack or white cedar bogs, rare, in counties bordering Lake Michigan.

2. **O. spectabilis** L. Showy Orchis (Fig. 84). Leaves 2, 7-17 cm long, shining; flower white and magenta.—In rich woods; many localities in southern Wisconsin, scattered localities northward to Polk and Marathon counties.

Fig. 86. *Habenaria flava,* var. *herbiola,* lip × 2.

Fig. 85. *Habenaria viridis,* var. *bracteata,* × .15, lip × 2.25.

3. Habenaria   Rein Orchis

Leaves alternate or basal; inflorescence a bracted spike or raceme, usually much longer than broad; petals (in our spring-flowering species) greenish, white, or yellowish; lip flat, entire or 3-lobed.

a. Leaves several, raised on the stem *b*
   b. Lip 2-3-toothed at apex (Fig. 85) . . . . 1. *H. viridis*
   b. Lip not toothed at apex *c*
      c. Lip cut square across at the apex, and with 2 tubercles at the base (Fig. 86) . . . . 2. *H. flava,* var. *herbiola*
      c. Lip tapered at the apex, without tubercles at the base (Fig. 87) *d*
         d. Flowers greenish *e*
            e. Lip tapered at base . 3. *H. hyperborea,* var. *huronensis*
            e. Lip abruptly wider at base . . . . .4. *H. media*
         d. Flowers white . . . . .5. *H. dilatata*
a. Leaves 1 or 2, at the base of the stem *f*
   f. Leaf 1 . . . . 6. *H. obtusata*
   f. Leaves 2 *g*
      g. Lip about 1 cm long . . . . .7. *H. hookeri*
      g. Lip 1.5-2 cm long . . . . 8. *H. orbiculata*

1. **H. viridis** (L.) R. Br. Leaves broadly lanceolate, acute or obtuse; raceme loosely flowered.

Var. **bracteata** (Muhl.) Gray. Bracted Orchis (Fig. 85). Lowest bracts divergent, 2-4 times as long as the green flowers; lip more than twice as long as the spur.—Woods, many localities in southern Wisconsin, and scattered localities in the north.

Fig. 87. *Habenaria hyperborea,* var. *huronensis,* × .10, lip × 2.

Fig. 88. *Habenaria hookeri,* × .25.

Var. interjecta Fern. Lowest bracts hardly twice as long as the flowers, the upper shorter.—Driftless Area, rare.

2. **H. flava** (L.) R. Br., var. **herbiola** (R. Br.) Ames & Correll. Pale Green Orchis (Fig. 86). Leaves lanceolate, gradually tapering to the acute apex, the upper very narrow and passing into the lower bracts; spur longer than the lip.—Wet grassy places, north to Lincoln and Brown counties.

3. **H. hyperborea** (L.) R. Br., var. **huronensis** (Nutt.) Farw. Northern Green Orchis (Fig. 87). Leaves linear-lanceolate; raceme very narrow, closely flowered; bracts about as long as the flowers.—Wet ground, northeast of a line drawn from Douglas County to Racine County, also La Crosse, Trempealeau, and Burnett counties.

4. **H. media** (Rydb.) Niles. Intermediate between *H. hyperborea* and *H. dilatata.*—Often more abundant than either; northwestern and southeastern Wisconsin, also

Shawano, Oconto, and Door counties.

5. **H. dilatata** (Pursh) Hook. Tall White Orchis. Similar to No. 3, but flowers more delicate; lip somewhat dilated at base.—Swamps in northern and eastern Wisconsin; rare.

6. **H. obtusata** (Pursh) Richards. Blunt-leaf Orchis. Flowers in a loose raceme, greenish or whitish; lip entire, 6 mm long, about equaling the slender curving spur.—Wet places, south to Taylor, Oconto, and Door counties.

7. **H. hookeri** Torr. Hooker's Orchis (Fig. 88). Leaves roundish, flat on ground; scape without bracts; much like No. 8.—Swamps and woods, very scattered localities.

8. **H. orbiculata** (Pursh) Torr. Round-leaved Orchis. Leaves 6-19 cm wide, flat on ground, shining above, silvery beneath; scape with 1-several bracts.—Swamps and rich damp woods, south to Barron, Lincoln, and Sheboygan counties.

4. Arethusa

Stems solitary, 10-25 cm high, from a bulb; a single leaf which emerges from the sheath after the flower matures; sepals and petals united at base and arching over the column; lip joined to column at base with 3 fringed-lengthwise parallel crests; column petallike.

A. bulbosa L. Swamp Pink (Fig. 89). Leaf 2-4 mm wide; flowers pink, 3-5 cm long.—Sphagnum bogs; rare, in very scattered localities in the state.

5. Calypso

Plants from a solid corm and producing a solitary basal leaf in the autumn which persists through the next season's flowering; stem with a few bladeless sheaths; sepals and lateral petals similar, spreading; lip larger than the rest of the flower, scoop-shaped, notched at the tip and with a translucent apronlike appendage; capsule erect.

C. bulbosa (L.) Oakes (Fig. 90). Plants 6-18 cm high; lip 2-2.5 cm long, whitish, becoming yellowish at the tip, irregularly streaked by red brown within, bent downward; the rest of the perianth pink, erect.—Deep mossy woods; Douglas, Price, and Burnett counties.

6. Listera   Twayblade

Plants slender and delicate, 1-2 dm high; stems with a pair of leaves near the middle; flowers small, greenish or purplish, in a loose slender raceme; sepals and petals similar,

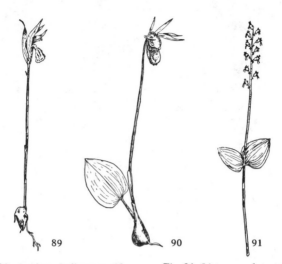

Fig. 89. *Arethusa bulbosa,* × .40. Fig. 91. *Listera cordata,* × .25.
Fig. 90. *Calypso bulbosa,* × .40.

separate; lip (in our species) much longer than the petals, shallowly to deeply notched down the center.

a. Lip cleft more than half its length . . . . 1. *L. cordata*
a. Lip cleft one-third or less of its length . . . . 2. *L. auriculata*

1. **L. cordata** (L.) R. Br. Heart-leaf Twayblade (Fig. 91). Leaves somewhat heart-shaped; lip cleft more than half of its length.—Mossy bogs, south to Burnett, Oneida, and Ozaukee counties.

2. **L. auriculata** Wieg. Eared Twayblade. Leaves rounded at base; lip cleft one-third or less of its length.— Lake Superior region.

7. Liparis Twayblade

Plants 1-2 dm high, from a solid bulb; leaves elliptical, their bases embraced by several bladeless sheaths; flowers few, in a raceme; sepals spreading; lateral petals linear, appearing threadlike.

a. Lip nearly 1 cm broad, pale purple . . . . 1. *L. lilifolia*
a. Lip about 2 mm broad, yellowish . . . . 2. *L. loeselii*

1. **L. lilifolia** (L.) Richard. Lily-leaved Twayblade (Fig.

Fig. 92. *Liparis lilifolia,* × .30.
Fig. 93. *Liparis loeselii,* × .30.

Fig. 94. *Aplectrum hyemale,*
× .25.

92). Leaves elliptical to narrowly ovate, green, lustrous; pedicels longer than the flowers, spreading; lip nearly 1 cm broad, pale purple.—Woods, southern and western Wisconsin, north to Trempealeau and Green Lake counties.

2. L. loeselii (L.) Richard. Yellow Twayblade (Fig. 93). Leaves lanceolate to lanceolate-ovate, yellow green, strongly keeled; pedicels shorter than the flowers, ascending; lip about 2 mm broad, yellowish.—Damp places in eastern half of Wisconsin, also Bayfield and Ashland counties.

8. Aplectrum   Puttyroot; Adam-and-Eve

Perennials, from a corm which bears a single leaf in late summer and a flowering scape the following year; sepals and lateral petals similar, spreading; the lip broadly obovate, 3-cleft.

A. hyemale (Muhl.) Torr. (Fig. 94). Plants about 4 dm high, from a solid corm ("Adam") near which is usually the smaller corm of last year ("Eve"); flowers 8-15, in a loose raceme, yellowish or purple.—Rich woods, north to Dunn and Brown counties.

9. **Corallorhiza** Coralroot

Brownish or yellowish saprophytes, living in leaf mold; rootless, the underground stems much branched, corallike; flowers in a raceme; sepals and petals similar, spreading; lateral sepals united with base of the column and forming a short spur on top of the ovary; lip with 1 or 2 longitudinal ridges; capsules turned sharply downward.

a. Sepals about 5 mm long; lip 3-lobed *b*
   b. Plant yellowish .... 1. *C. trifida*
   b. Plant purplish .... 2. *C. maculata*
a. Sepals about 1 cm long; lip not lobed (Fig. 97)...... 3. *C. striata*

1. **C. trifida** Chatelain, var. **verna** (Nutt.) Fern. Pale Coralroot (Fig. 95). Plants slender, 4-19 cm high, 4-12-flowered; lip white, 3-lobed.—Woods and bogs from Polk to Adams counties northward, and eastern Wisconsin from Door to Walworth counties.

 95   96   97

Fig. 95. *Corallorhiza trifida,* lip × 3.
Fig. 96. *Corallorhiza maculata,* lip × 3.

Fig. 97. *Corallorhiza striata,* lip × 2.

2. **C. maculata** Raf. Spotted Coralroot (Fig. 96). Plants stout, 2-4 dm high, 10-30 flowered; lip white, usually with crimson spots.—Dry woodlands, throughout the state.

3. **C. striata** Lindl. Striped Coralroot (Fig. 97). Plants 15-40 cm high, purplish; sepals and petals conspicuously striped with deep purple.—South to Washburn and Manitowoc counties.

SALICACEAE Willow Family

Trees or shrubs, dioecious; flowers without perianth, in catkins, often covered in bud with silky bracts (as in the Pussy Willow); fruit a capsule, the carpels curling back at maturity; seeds many, with a long tuft of silky down.

a. Buds with only 1 scale; stamens 2-8 in each flower .... 1. *Salix*
a. Buds with several overlapping scales; stamens many in each flower
  .... 2. *Populus*

1. Salix   Willow

Trees or shrubs; buds with a single outer bud scale; leaves simple, alternate, and usually stipulate; flowers in elongate catkins, insect-pollinated, each with a nectar gland and a scale subtending it; staminate flowers with 2-5 stamens; fruit a bivalved capsule.

A difficult and complex group, with 22 species in Wisconsin. Two keys are provided, one for the staminate material, the second for pistillate material. The brief descriptions which follow mainly distinguish leaf characters, but since both variation and hybridization may occur exact determinations can be difficult in this genus. The following keys, by George W. Argus, were originally published in 1964 (*Trans. Wis. Acad. Sci. Arts Letters* 53: 222-27).

Key to Staminate Specimens

a. Stamens 3 or more *b*
  b. Staminate catkins slender and loosely flowered; flowers tufted and more or less whorled along the rachis *c*
    c. Immature leaves narrowly lanceolate, green beneath; stipules prominent . . . . .1. *S. nigra*
    c. Immature leaves lanceolate, glaucous beneath; usually without stipules . . . . 2. *S. amygdaloides*
  b. Staminate catkins thickish and densely flowered; flowers spirally arranged *d*
    d. Immature leaves bearing reddish hairs which soon fall off; stipules prominently glandular . . . . 3. *S. lucida*
    d. Immature leaves glabrous; stipules minute or absent *e*
      e. Staminate catkins 3-3.5 cm long; native species . . . .
. . . . .4. *S. serissima*
      e. Staminate catkins 2-6 cm long; introduced species . . . . . .
. . . . 5. *S. pentandra*
a. Stamens 2 *f*
  f. Staminate catkins developing before the leaves *g*
    g. Staminate catkins and leaves opposite or subopposite; filaments and anthers coalescent . . . . 22. *S. purpurea*
    g. Staminate catkins and leaves alternate; filaments and anthers distinct *h*
      h. Leaves finely to densely silky beneath, margin entire or serrate; rare species in Wisconsin *i*
        i. Leaf margin serrulate; blade finely silky, at least beneath; filaments pubescent at base; native species . . . .
. . . . 18. *S. sericea*
        i. Leaf margin entire, revolute; blade densely silky beneath; filaments glabrous; introduced species . . . .
. . . . .21. *S. viminalis*
      h. Leaves pubescent when immature, but not silky; common species in Wisconsin *j*
        j. Staminate catkins 0.7-1.5 cm long . . . . . 19. *S. humilis*
        j. Staminate catkins 2-3.5 cm long . . . . . . 20. *S. discolor*
  f. Staminate catkins developing with the leaves or only slightly earlier *k*

    k. Filaments pubescent *l*
        l. Petiole glandular at distal end; introduced trees *m*
            m. Branchlets tenacious and flexible . . . . 8. *S. alba*
            m. Branchlets brittle at the base *n*
                n. Leaves silky, margin finely serrulate; branchlets pendulous; staminate catkins 3-3.5 cm long . . . . 6. *S. babylonica*
                n. Leaves glabrous or glabrate, margin coarsely serrate; branchlets not pendulous; staminate catkins 3-6 cm long . . . . 7. *S. fragilis*
        l. Petiole not glandular at distal end; native shrubs *o*
            o. Bracts black . . . . 17. *S. petiolaris*
            o. Bracts yellow or yellow green *p*
                p. Reproductive branchlets 0.8-8 cm long; staminate catkins often branched; leaves linear; margin remotely denticulate . . . . 9. *S. interior*
                p. Reproductive branchlets 0.3-0.6 cm long; staminate catkins unbranched; leaves not linear, margin entire to crenate . . . . 14. *S. bebbiana*
    k. Filaments glabrous *q*
        q. Immature leaves and branchlets dull tomentose . . . . 14. *S. candida*
        q. Immature leaves and branchlets pubescent or glabrous *r*
            r. Immature leaves thin and translucent; plants with balsamlike fragrance . . . . 13. *S. pyrifolia*
            r. Leaves or plants not as above *s*
                s. Staminate catkins few-flowered, 0.5-2 cm long; bracts yellowish; leaf margin entire, revolute . . . . 16. *S. pedicellaris*
                s. Staminate catkins many-flowered, 1.2-4 cm long; bracts dark brown to black; leaf margin serrate *t*
                    t. Inner bud scale persistent at base of catkins and vegetative shoots *u*
                        u. Immature leaves glabrous, reddish . . . . 12. *S. glaucophylloides*
                        u. Immature leaves pubescent, sometimes reddish *v*
                            v. Branchlets glabrate or velvety; immature leaves pubescent, reddish, margin serrate, not prominently glandular . . . 10. *S. rigida*
                            v. Branchlets grayish-tomentose; immature leaves densely silky, margin prominently glandular; Lake Michigan dunes, rare . . . . 11. *S. syrticola*
                    t. Inner bud scale not persistent . . 17. *S. petiolaris*

Key to Pistillate Specimens

a. Pistils and capsules pubescent *b*
    b. Pistillate catkins developing before the leaves *c*
        c. Leaves and catkins opposite or almost opposite . . . . 22. *S. purpurea*
        c. Leaves and catkins alternate *d*
            d. Capsules almost sessile, pedicels less than 1 mm long; introduced tree . . . . 21. *S. viminalis*
            d. Capsules pedicellate, pedicels 1-2.5 mm long; native species *e*

e. Pistils and capsules blunt; catkins 1-2.5 cm long; reproductive branchlets 2-10 mm long; leaves silvery silky beneath; rare . . . . 18. *S. sericea*

e. Pistils and capsules long-beaked; catkins 1.5-7 cm long; reproductive branchlets absent or very short; leaves not as above; common species *f*

    f. Pistillate catkins 1.5-4 cm long in fruit; styles 0.2-0.4 mm long; capsules 4-7 mm long . . . .
. . . . 19. *S. humilis*

    f. Pistillate catkins 4-7 cm long in fruit; styles 0.5-0.8 mm long; capsules 6-11 mm long . . .20. *S. discolor*

b. Pistillate catkins developing with the leaves or later than the leaves *g*

  g. Pistils and capsules dull white-tomentose . . . . 14. *S. candida*

  g. Pistils and capsules finely silky or glabrescent *h*

    h. Reproductive branchlets 3-12.5 cm long; bracts deciduous after flowering; capsules deciduous after dehiscence
. . . . 9. *S. interior*

    h. Reproductive branchlets 0.3-1 cm long; bracts and capsules persistent *i*

      i. Bracts brown, oblong; pistillate catkins 1.5-3.5 cm long in fruit; leaves linear-lanceolate, serrate to serrulate, sometimes with reddish pubescence . . . .
. . . . 17. *S. petiolaris*

      i. Bracts yellowish to tawny, lanceolate; pistillate catkins 3.5-6 cm long in fruit; leaves elliptical, elliptical ovate to oblanceolate, entire or crenate, lacking reddish pubescence . . . . 15. *S. bebbiana*

a. Pistils and capsules glabrous *j*

  j. Bracts deciduous after flowering, yellowish *k*

    k. Leaves green or pale beneath *l*

      l. Leaves linear to linear-lanceolate, remotely denticulate to serrulate; upper surface of blade dull *m*

        m. Leaves linear, remotely denticulate; stipules small or absent; pistillate catkins often branched; capsules slender, 4.5-7 mm long . . . . 9. *S. interior*

        m. Leaves linear-lanceolate, often falcate, serrulate; stipules large and prominent; pistillate catkins unbranched; capsules 3-4 mm long . . . . 1. *S. nigra*

      l. Leaves lanceolate or broader, serrulate; upper surface of blade glossy, often leathery or somewhat leathery *n*

        n. Immature leaves bearing early deciduous reddish trichomes; stipules prominently glandular . 3. *S. lucida*

        n. Immature leaves glabrous; stipules minute or absent *o*

          o. Pistillate catkins stout, 2-4.5 cm long; capsules 7-10 mm long; native species . . . . . 4. *S. serissima*

          o. Pistillate catkins slender, 3.5-6 cm long; capsules 1-5 mm long; introduced species . . 5. *S. pentandra*

    k. Leaves glaucous beneath *p*

      p. Pistillate catkins short and stout, 2-4.5 cm long; capsules 7-10 mm long; seeds shed late in season . . 4. *S. serissima*

      p. Pistillate catkins short or long, but slender, 2-3.5 or 4-8 cm long; capsules 1-5 mm long *q*

        q. Pistillate catkins loosely flowered; pedicels long, 1.5-2.5 mm long; leaves lanceolate to ovate-lanceolate; stipules absent or minute; native species . . . .
. . . . 2. *S. amygdaloides*

q. Pistillate catkins not as loosely flowered; pedicels short to sessile, 0-0.75 mm long; leaves linear-lanceolate to lanceolate; stipules usually small and deciduous; introduced species *r*

    r. Twigs slender and pendulous, not fragile . . . .
    . . . . 6. *S. babylonica*
    r. Twigs stout, not pendulous, fragile *s*

        s. Leaves silky, margin serrulate . . . . . . 8. *S. alba*
        s. Leaves glabrous, margin coarsely serrate . . . . . .
        . . . . .7. *S. fragilis*

j. Bracts persistent, yellow to brown *t*

  t. Leaf margin entire, revolute; bracts sparsely pubescent . . . . .
  . . . . .16. *S. pedicellaris*
  t. Leaf margin serrate to crenate; bracts pubescent to densely villous *u*

    u. Immature leaves translucent, glabrous or glabrescent; plant with balsamlike fragrance; pistillate catkins loosely flowered; pedicels 2.5-3.5 mm long . . . . . 13. *S. pyrifolia*
    u. Immature leaves opaque, glabrous to pubescent; plants lacking balsamlike fragrance; pistillate catkins densely flowered; pedicels 0.5-2(2.5) mm long *v*

      v. Immature leaves white-pubescent or densely silky, green beneath or thinly glaucous in some plants *w*

        w. Leaves oblong-lanceolate, apex gradually acuminate or attenuate, margin serrulate; immature leaves reddish purple; capsules 4-5 mm long . 10. *S. rigida*
        w. Leaves oblong-ovate, apex acute or acuminate, margin glandular serrate, teeth often prolonged; capsules 5-7 mm long . . . . 11. *S. syrticola*
      v. Immature leaves glabrous, sometimes with early deciduous reddish hairs, blade thickly glaucous beneath, often drying black . . . . .12. *S. glaucophylloides*

1. **S. nigra** Marsh. Black Willow (Fig. 98). Shrubs or trees, to 20 m tall; twigs brittle at base; leaves linear to linear-lanceolate, apex long-tapering, margins serrulate, dark green both sides; stipules prominent.—Lowland forests, throughout most of Wisconsin except the northern tier of counties.

2. **S. amygdaloides** Anderss. Peach-leaved Willow (Fig. 99). Shrubs or trees, to 20 m tall; leaves lanceolate to ovate-lanceolate, margins serrulate, glaucous beneath, petiole yellow; stipules none or minute.—Lowland forests along rivers, margins of swamps, and lakes; north to Dunn to Oconto counties (Ashland).

3. **S. lucida** Muhl. Shining Willow (Fig. 100). Shrubs or small trees, to 6 m tall; leaves lanceolate to broadly lanceolate, long-tapering, dark green above, paler below, serrulate, teeth with large glands at the tip; stipules reniform to semi-circular, 1-6 mm.—Throughout most of the state.

4. **S. serissima** Fern. Autumn Willow (Fig. 101). Shrubs, to 4 m tall; leaves lanceolate, apex acuminate, margins

Fig. 98. *Salix nigra,* × .50.   Fig. 100. *Salix lucida,* × .50.
Fig. 99. *Salix amygdaloides,* × .50.   Fig. 101. *Salix serissima,* × .50.

glandular-serrulate; stipules minute or absent.—Marshes and bogs, eastern half of Wisconsin and northwestern Wisconsin.

5. **S. pentandra** L. Bay-leaved Willow (Fig. 102). Shrubs or small trees to 7 m tall; leaves broadly lanceolate, base rounded, margins glandular-serrulate, very dark green above, paler beneath, leathery.—Introduced and cultivated.

6. **S. babylonica** L. Weeping Willow (Fig. 103). Trees, to 12 m tall; branchlets slender, pendulous, yellowish to brown; leaves linear-lanceolate, base acute, apex long-acuminate, margins serrulate.—Introduced and cultivated, also escaping along riverbanks and roadsides in scattered locations in Wisconsin.

7. **S. fragilis** L. Crack Willow (Fig. 104). Trees, to 30 m tall; branchlets yellowish to brown, very fragile at base; leaves lanceolate, margins coarsely serrate, apex long-acuminate, base acute.—Introduced and cultivated, sometimes escaping; throughout most of Wisconsin.

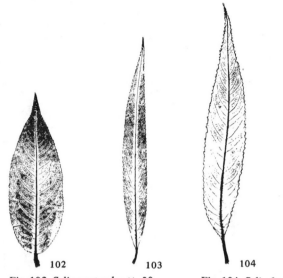

Fig. 102. *Salix pentandra,* × .50.
Fig. 103. *Salix babylonica,* × .50.
Fig. 104. *Salix fragilis,* × .50.
Fig. 105. *Salix alba,* × .50.

8. **S. alba** L. White Willow (Fig. 105). Trees, to 20 m tall; leaves lanceolate to narrowly lanceolate, margins serrulate, young leaves silky beneath, mature leaves glaucous and white-hairy on underside.—Introduced, occasionally escaping along rivers, especially in southeastern Wisconsin.

9. **S. interior** Rowlee. Sandbar Willow (Fig. 106). Shrubs, to 2 m tall, forming large patches by freely sprouting; leaf blades linear to linear-lanceolate, margins distantly denticulate.—Mostly in moist sandy situations, throughout Wisconsin.

10. **S. rigida** Muhl. Heart-leaved Willow (Fig. 107). Shrubs, to 3 m tall; leaves oblong-lanceolate, reddish purple, densely white-pubescent when immature, becoming glabrate, stipules large, lanceolate to ovate with serrulate margins. —Moist habitats, throughout most of Wisconsin.

11. **S. syrticola** Fern. Sand Dune Willow (Fig. 108). Spreading shrubs, to 3 m tall; branchlets grayish-tomentose; leaves oblong-ovate, apex acute or short-acuminate, base

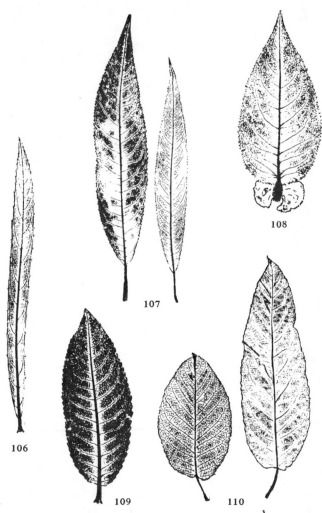

Fig. 106. *Salix interior,* × .55.
Fig. 107. *Salix rigida,* × .65.
Fig. 108. *Salix syrticola,* × .70.

Fig. 109. *Salix glaucophylloides,*
× .60.
Fig. 110. *Salix pyrifolia,* × .70.

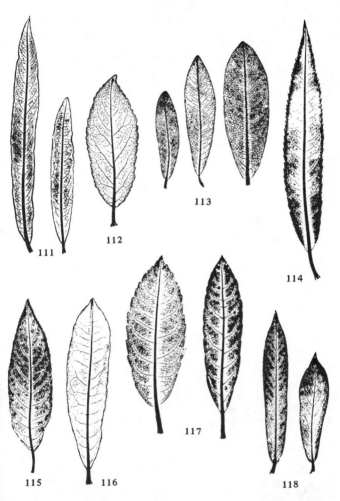

Fig. 111. *Salix candida,* × .70.
Fig. 112. *Salix bebbiana,* × .65.
Fig. 113. *Salix pedicellaris,* × .70.
Fig. 114. *Salix petiolaris,* × .70.

Fig. 115. *Salix sericea,* × .65.
Fig. 116. *Salix humilis,* × .65.
Fig. 117. *Salix discolor,* × .70.
Fig. 118. *Salix purpurea,* × .70.

cordate or broadly rounded, green on both sides; stipules prominent, semicordate to subovate.—Endemic to Great Lakes beaches; rare, Manitowoc County.

12. **S. glaucophylloides** Fern. Blue-leaved Willow (Fig. 109). Shrubs, to 2.5 m tall; leaves broadly elliptical, apex acute to short-acuminate, base obtuse, strongly glaucous beneath, often drying black; stipules prominent, ovate, to 10 mm.—Sandy areas, mostly near Lake Michigan, also wet prairies and streambanks from Dane to Portage counties.

13. **S. pyrifolia** Anderss. Balsam Willow (Fig. 110). Shrubs, to 3 m tall, with balsamlike fragrance; leaves broadly lanceolate to ovate, margins glandular-serrulate, base cordate to rounded, apex acute, mature leaves leathery, reticulate-veined and glaucous beneath; stipules small.—Most common in bogs, also in northern hardwood forests, south to Columbia County.

14. **S. candida** Flügge. Sage-leaved Willow (Fig. 111). Shrubs, to 3.5 m tall; leaves linear, margins revolute, entire, apex acute, base acuminate, dull white-tomentose on underside; stipules lanceolate, tomentose.—Alkaline shores and meadows in eastern half of Wisconsin, also Trempealeau, Washburn, and Ashland counties.

15. **S. bebbiana** Sarg. Bebb's Willow (Fig. 112). Shrubs or small trees, to 6 m tall; leaves elliptical to obovate, apex abruptly acute, base obtuse to acute, margins entire to crenate, pubescent above, underside glaucous, rough, and conspicuously veined.—Common in a wide variety of habitats.

16. **S. pedicellaris** Pursh. Bog Willow (Fig. 113). Low shrubs, to 70 cm tall; leaves oblong, apex obtuse to acute, base narrowed, margins entire, revolute, mature leaves leathery in texture, fine prominent venation above, glaucous with prominent midribs on underside.—Mostly in bogs, also in northern hardwoods, not in Driftless Area.

17. **S. petiolaris** Sm. Slender Willow (Fig. 114). Shrubs, to 3 m tall; leaves linear to lanceolate, apex acuminate, base acute, margins sharply serrate to distantly serrate to entire, silky to becoming hairless above, glaucous and variably silky below.—Common shrub of sandy to peaty low prairie habitats, throughout Wisconsin.

18. **S. sericea** Marsh. Silky Willow (Fig. 115). Shrubs, to 3 m tall; leaves narrowly lanceolate, apex acuminate, base acute, margins serrulate, immature leaves silky on both sur-

faces, mature leaves silvery silky on underside.—A rare species of boggy soils and sand terraces; Clark, Jackson, and Richland counties.

19. **S. humilis** Marsh. Prairie Willow (Fig. 116). Shrubs, to 3 m tall; leaves narrow to broadly oblanceolate, apex acute to short-acuminate, base acute, margins entire to crenate, revolute, gray green above, rough, conspicuously veined and pubescent to glabrate below.—Early flowering species of wet prairies and various upland habitats; throughout Wisconsin.

20. **S. discolor** Muhl. Pussy Willow (Fig. 117). Shrubs or small trees, to 3 m tall; leaves narrow to broadly elliptical or lanceolate, apex acute, base obtuse to acute, margins crenate to serrate, mature leaves glaucous and glabrate to pubescent beneath; stipules present.—Common species of various moist habitats, throughout Wisconsin.

21. **S. viminalis** L. Osier. Shrubs or small trees; leaves linear to linear-lanceolate, apex long-acuminate, base acute, margins entire, revolute, mature leaves densely silky on underside, midrib yellow.—Introduced cultivated species in Wisconsin.

22. **S. purpurea** L. Purple Osier (Fig. 118). Shrubs, to 2.5 m tall; leaves almost opposite, spatulate to linear, apex acute to acuminate, base obtuse, margins entire near base, irregularly serrulate above, glaucous beneath.—Introduced species; mostly along Lake Michigan as an escape in various moist habitats or waste places; also Rock, Dane, Dunn, and Barron counties.

2. **Populus** Poplar; Aspen

Trees; buds with many scales, often with a sticky, resinous coating; leaves ovate-lanceolate to almost round; staminate flowers with many stamens in a cup-shaped receptacle, pistillate flowers also with cups, wind-pollinated; fruit a 2-4-valved capsule.

a. Leaves densely white-woolly beneath . . . . 1. *P. alba*
a. Leaves not woolly beneath (except on root sprouts in No. 4) *b*
  b. Petioles not flattened *c*
    c. Leaves lance-shaped to narrowly heart-shaped, smooth or with very sparse fine hairs on midrib below 2. *P. balsamifera*
    c. Leaves broadly heart-shaped, hairy below, especially on midrib and veins . . . . 3. × *P. gileadensis*
  b. Petioles flattened, at least toward the summit *d*
    d. Buds and tips of young branches white-woolly . . . .
    . . . . 4. *P. grandidentata*
    d. Buds and twigs not woolly *e*

    e. Petiole with 2 small glands on the upper surface at its
       junction with the blade .... 5. *P. deltoides*
    e. Petiole without glands *f*
       f. Leaf blades rounded or square across the base ....
       .... 6. *P. tremuloides*
       f. Leaf blades broadly V-shaped at base ....
       .... 7. *P. nigra,* var. *italica*

1. **P. alba** L. White Poplar. Leaves somewhat 3-lobed,
coarsely and irregularly toothed, densely white-woolly beneath.—Cultivated, and frequently spreading by suckers from
the roots; Dane, Green, and Rock counties.

2. **P. balsamifera** L. Taccamahac; Balsam Poplar (Fig.
119). Trees 6-30 m high; terminal buds with fragrant gum;
leaves lighter green on lower surface with rusty blotches;
with glabrous leaves angled or rounded at base.—In low rich
soil, mostly northward and eastward, grading into var. **subcordata** Hylander, with leaves heart-shaped (Fig. 120) and
slightly pubescent on the lower surface.

3. × **P. gileadensis** Rouleau. Balm of Gilead. Leaves
heart-shaped and pubescent below.—Mostly planted trees.

4. **P. grandidentata** Michx. Large-toothed Aspen; Popple
(Fig. 121). Similar to No. 6; leaf blades not abruptly
pointed, with coarse teeth.—In somewhat moist places,
throughout the state.

5. **P. deltoides** Marsh. Cottonwood (Fig. 122). Trees
15-30 m high; leaf blades broadest near the base, which is as
if cut off almost squarely, at tip narrowed to a tapering
point about 1 cm long.—In woods and on the banks of
streams. Commonly cultivated; north to Chippewa, Lincoln,
and Brown counties.

Fig. 119. *Populus balsamifera,* × .30.
Fig. 120. *Populus balsamifera,*
var. *subcordata,* × .25.

Fig. 121. *Populus grandidentata,*
× .25.

Fig. 122. *Populus deltoides,*
× .25.

Fig. 123. *Populus tremuloides,*
× .35.

6. **P. tremuloides** Michx. Quaking Aspen; Popple (Fig. 123). Usually a small tree, with smooth greenish bark; leaf blades mostly rounded at base and abruptly short-pointed at tip, with small regular teeth.—Common, especially on cut-over and burned areas.

The common var. **tremuloides**, with slender twigs 5 mm or less thick and petioles 1 mm or less thick, grades into the rare and local var. **magnifica** Vict., with very short inter-nodes 6-12 mm thick.

7. **P. nigra** L., var. **italica** Muenchh. Lombardy Poplar. Branches all ascending; leaves much like those of the last but more finely toothed.—Sometimes persisting after cultivation.

MYRICACEAE   Sweet Gale Family

Shrubs; leaves simple, with fragrant waxy resin; flowers without perianth, the staminate and pistillate flowers in different bracted catkins; stamens 2-many; pistillate flowers subtended by 2-8 bractlets; stigmas 2; fruit a nutlet.

a. Leaves entire to serrulate; plants of shores and shallow water . . . . .
. . . . 1. *Myrica*
a. Leaves pinnatifid; plants of dry sand . . . . 2. *Comptonia*

1. **Myrica**   Sweet Gale

Shrubs; mostly dioecious; stipules lacking; fruit a resin-dotted nutlet with 2-winged scales.

M. gale L. (Fig. 124). Leaves wedge-shaped at base and toothed toward the rounded tip; flowers coming after the leaves, in short conelike catkins.—Wet places, south to Burnett, Lincoln, and Manitowoc counties.

Fig. 124. *Myrica gale*, × .30.   Fig. 125. *Comptonia peregrina*, × .30.

2. Comptonia   Sweet-fern

Shrubs; monoecious or dioecious; stipules present; staminate flowers in long cylindrical catkins, the pistillate in globular burlike heads.

C. peregrina (L.) Coult. (Fig. 125). Leaves cut almost to the midrib at intervals of 1 cm or less, fernlike, rather sticky; flowers coming before the leaves.—In dry sand, south to Columbia and Sauk counties.

JUGLANDACEAE   Walnut Family

Trees; leaves pinnately compound; staminate flowers in catkins, with an irregular calyx; pistillate flowers solitary or in clusters, with a regular 3-5-lobed calyx; fruit a large nut, enclosed in a husk developed from the calyx.

a. Staminate catkins stout, not stalked; stamens 10-40 in each flower; pith of twigs with cross partitions .... 1. *Juglans*
a. Staminate catkins slender, several at the tip of a common stalk; stamens 3-10 in each flower; pith of twigs continuous ... 2. *Carya*

1. Juglans   Walnut

Trees; pith of twigs with cross partitions; leaves pinnately compound, the median leaflets largest; surface of nuts very rough beneath the husks.

a. Pith chocolate-colored, leaf scar with a pad of down on its upper side ....1. *J. cinerea*
a. Pith tan, leaf scar without a pad of down on upper edge 2. *J. nigra*

1. J. cinerea L. Butternut (Fig. 126). Bark gray; leaf usually terminated by a single leaflet, the petiole and lower surface of leaflets covered with dark brown sticky hairs; leaf scar with a pad of down at its upper edge; nut oblong, the husk very sticky.—North to Burnett, Ashland, Lincoln, and Door counties.

Fig. 126. *Juglans cinerea,* × .25.   Fig. 127. *Carya cordiformis,* × .20.

2. **J. nigra** L. Black Walnut. Bark dark brown; leaf usually terminated by a pair of leaflets, the petiole and lower surface of leaflets downy but not sticky; leaf scar without a pad of down; nut almost spherical, the husk not sticky.— North to Pierce, Winnebago, and Door counties.

2. Carya   Hickory

Trees; pith of twigs continuous; leaves pinnately compound, the terminal leaflets larger than the others; nuts 1.5-3 cm long, the surface of the shell beneath the husk smooth.

a. Bark of trunk shaggy; buds brown; leaflets mostly 5 or 7, broadest toward tips . . . . 1. *C. ovata*
a. Bark of trunk not shaggy; buds yellow; leaflets 5-11, broadest toward centers . . . . 2. *C. cordiformis*

1. **C. ovata** (Mill.) K. Koch. Shellbark Hickory. Bark of trunk shaggy with long thick deciduous plates; bud scales becoming large and petallike as the stem starts to elongate; leaflets 5-7, with a fine tuft of hairs on each tooth; nut sweet.—Dry woods, north to Pierce and Brown counties.

2. **C. cordiformis** (Wang.) K. Koch. Bitternut Hickory (Fig. 127). Bark not shaggy; buds yellow; leaflets 5-11, the teeth without tufts of hairs; nut bitter.—Rich woods, north to Burnett, Lincoln, and Door Counties.

BETULACEAE   Birch Family

Trees or shrubs; leaves simple, alternate; monoecious; flowers in catkins, the staminate catkins many-flowered, the pistillate flowers in catkins or a few flowers in a scaly bud; fruit a nut (often small) or a samara.

a. Pistillate flowers solitary, or a few in a cluster . . . . 1. *Corylus*
a. Pistillate flowers in catkins *b*
  b. Each bract of the staminate catkin with but 1 flower, which lacks a calyx *c*
    c. Staminate catkins usually in groups of 3 (Fig. 130); bud scales with lengthwise ridges . . . . 2. *Ostrya*
    c. Staminate catkins solitary; bud scales without ridges . . . . .3. *Carpinus*
  b. Each bract of the staminate catkin with 3-6 flowers, each with a calyx *d*
    d. Stamens 2; pistillate bracts 3-lobed . . . . 4. *Betula*
    d. Stamens 3-5; pistillate bracts 5-lobed . . . . 5. *Alnus*

Since floral characters are somewhat difficult to see in this group, and last year's fruit can frequently be found, a key is also given using fruit and habit.

a. Each nut with a somewhat leafy involucre *b*
  b. Shrubs; nut 1 cm or more long . . . . .1. *Corylus*
  b. Trees; nut 6 mm or less long *c*
    c. Bark furrowed and shredding, grayish brown; involucre saclike, enclosing the nut . . . . 2. *Ostrya*
    c. Bark close, smooth, gray; involucre leafy, open, flat, coarsely toothed . . . . .3. *Carpinus*
a. Nut without involucre, in the axil of a small scaly bract *d*
  d. Bracts papery, ascending . . . . 4. *Betula*
  d. Bracts woody, at right angles to the rachis of the cone 5. *Alnus*

1. Corylus   Hazel

Shrubs or small trees with doubly serrate leaves; flowers appearing before the leaves; except for the elongated red stigmas, the tiny pistillate flowers concealed by a few scales; fruit a nut enclosed by a leafy toothed involucre.

a. Twigs and petioles glandular-bristly . . . . 1. *C. americana*
a. Twigs and petioles not bristly . . . . 2. *C. cornuta*

1. **C. americana** Walt. Hazelnut (Fig. 128). Twigs and petioles glandular-bristly; involucre of 2 broad, open, leaflike bracts, not bristly.—Thickets and pastures, throughout the state.

2. **C. cornuta** Marsh. Beaked Hazelnut (Fig. 129). Twigs and petioles not bristly; involucre enclosing the nut and pro-

Fig. 128. *Corylus americana,* × .20.

Fig. 129. *Corylus cornuta,* × .20.

Fig. 130. *Ostrya virginiana,* × .20.   Fig. 131. *Carpinus caroliniana,* × .20.

longed into a beak, covered with bristly hairs.—Common northward, less common southward to Grant and Rock counties.

2. Ostrya   Ironwood; Hop Hornbeam

Small trees; bark flaky; leaves serrate; flowers appearing with the leaves; involucre bladdery, the fruiting mass looking like that of hops; fruit a nutlet.

O. virginiana (Mill.) K. Koch (Fig. 130). Slender tree with very hard wood; young petioles and stems with a mixture of glandular and simple hairs; leaves sharply double-serrate, long-pointed, hairy on the veins.—Rich woods, throughout the state.

3. Carpinus   Ironwood; American Hornbeam; Blue Beech

Small trees or shrubs, with smooth bark; leaves double-serrate; flowers appearing with the leaves; the pistillate flowers subtended by fused bracts, which enlarge to form a 3-lobed leafy wing at maturity; fruit a small nutlet.

C. caroliniana Walt. (Fig. 131). Trunk ridged as if muscular, gray; fruiting bracts 3-toothed, 2-3 cm long.—Damp woods and along streams, throughout the state.

4. Betula   Birch

Trees or shrubs; the trees with bark which splits into fine sheets; flowers appearing before or with the leaves; staminate catkins elongate, the pistillate catkins ovoid; fruit a winged nutlet.

a. Scales of the fruiting catkins persistent; leaves with impressed veins *b*
    b. Leaf blades obliquely heart-shaped at base, more or less taper-ing at tip . . . . 1. *B. alleghaniensis*
    b. Leaf blades at base tapering to the petiole, acute at tip . . . . . .
. . . . 2. *B. nigra*

a. Scales of the fruiting catkins soon falling; veins not impressed *c*
   c. Trees; leaf blades 6-10 cm long *d*
      d. Leaf blades rounded at base . . . . 3. *B. papyrifera*
      d. Leaf blades subcordate at base . . . . 4. *B. cordifolia*
   c. Shrubs; leaf blades 2-6 cm long *e*
      e. Leaves oval to somewhat quadrangular; fruiting catkins 2-3
        cm long, with scales 4-5.5 mm long; wing as wide as or wider
        than the nutlet . . . . 5. *B. sandbergii*
      e. Leaves oval; fruiting catkins 1-2.8 cm long, with scales 3-3.5
        mm long; wing narrower than the nutlet . . 6. *B. glandulifera*

1. B. alleghaniensis Britt. Yellow Birch (Fig. 132). Bark yellowish or silvery, peeling in fine ragged fringes (rarely close and dark); twigs flavored like wintergreen; leaf blades slightly heart-shaped at base, with 5-9 pairs of veins, double-serrate.—Throughout the state.

2. B. nigra L. Red Birch; River Birch (Fig. 133). Bark reddish, peeling in coarse strips; leaf blades with 4 almost straight sides, the 2 upper the longer and doubly toothed.—Mostly following rivers; up the Mississippi to Pierce County; up the Chippewa to Chippewa County; up the Wisconsin and its tributaries to Wood and Portage counties.

3. B. papyrifera Marsh. White Birch (Fig. 134). Bark on mature trees white, peeling back in thin curled pieces; leaf blades with 6-9 pairs of veins, wedge-shaped or rounded at base.—Common northward, confined to north-facing hills southwestward, and largely to bogs southeastward.

4. B. cordifolia (Regel) Fern. Canoe Birch. Very similar to *B. papyrifera* but with the leaf bases subcordate.— A widespread species; growing with *B. papyrifera,* except in granitic areas.

5. B. sandbergii Britt. (Fig. 135). Shrubs 4-8 m high; bark dark brown, not peeling; plant much like *B. glandulifera* but larger throughout.—A hybrid between *B. glandulifera* and *B. papyrifera* and often more abundant in bogs than either parent.

6. B. glandulifera (Regel) Butler. Bog Birch (Fig. 136). Shrubs, 1-2 m high; bark dark brown, not peeling; leaf blades rounded at tip and wedge-shaped at base.—Bogs throughout Wisconsin, but rare in the Driftless Area.

5. Alnus   Alder

Shrubs; leaves alternate, serrate, about 1 dm long, oval, heart-shaped at base; staminate flowers in catkins made up of reddish bracts on which are 4-5 bractlets and 3-6 flowers

Fig. 132. *Betula alleghaniensis,* × .20.

Fig. 133. *Betula nigra,* × .40.

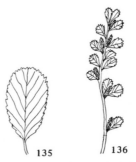

Fig. 134. *Betula papyrifera,* × .25.
Fig. 135. *Betula sandbergii,* × .40.

Fig. 136. *Betula glandulifera,* × .40.

composed of a 3-5-parted calyx and the same number of stamens; pistillate flowers in small woody conelike catkins composed of bracts which subtend 2 flowers and 4 small scales; fruits persisting through the winter.

a. Flowers developing at the same time as the leaves; leaves very finely, sharply, and regularly toothed, sticky when young . . . . 1. *A. crispa*
a. Flowers developing before the leaves; leaves coarsely double-toothed, the margins irregular, not sticky-whitened beneath . . . . 2. *A. rugosa*

1. **A** crispa (Ait.) Pursh. Green Alder (Fig. 137). Shrubs, with ascending branches; leaves round-oval, rounded or slightly cordate at base, sticky when young, 3-8 cm long, finely toothed; pistillate catkins slender, on stalks 5 mm or more long, in loosely clustered racemes, 1-2 cm long, bracts

Fig. 137. *Alnus crispa,* × .30.   Fig. 138. *Alnus rugosa,* × .30.

small and uneven; nutlets winged; winter buds not stalked, of 3-6 scales.—On sandy soils and in bogs in northern Wisconsin, south to Burnett and Lincoln counties.

Var. **mollis** Fern. Downy Alder. Young branches and leaves soft-pubescent.—Polk and Oneida counties northward.

2. **A. rugosa** (Du Roi) Spreng. Speckled Alder; Hoary Alder (Fig. 138). Shrub, with spreading branches; leaves round-oval, rounded to cordate at base, not sticky, to about 1 dm long, coarsely toothed, underside whitened; pistillate catkins sessile, drooping, bracts coarse and uneven; nutlets wingless; winter buds stalked, of 2 scales.—Low wet ground, along streams and lakes; throughout the state but rarer in southwestern Wisconsin.

FAGACEAE   Beech Family

Monoecious trees (occasionally shrubs), with alternate simple leaves; staminate flowers in catkins or heads; pistillate flowers solitary or in small clusters, each with a scaly involucre; fruit a nut surrounded by the enlarged involucre.

a. Leaves with small sharp teeth . . . . 1. *Fagus*
a. Leaves lobed or with wavy margins . . . . 2. *Quercus*

1. Fagus   Beech

Trees, with straight-veined leaves, a vein running to each tooth; flowers appearing with the leaves, the staminate flowers in small heads on drooping peduncles, the pistillate flowers paired on a short peduncle; fruit a 3-cornered nut, enclosed in a husklike involucre whose scales are somewhat fused and with their long tips free.

F. grandifolia Ehrh. Large trees, with smooth close ashy bark; leaves about 1 dm long, oblong-ovate, rounded at base.—Near Lake Michigan and the Kettle Moraine area, north to Oconto and Shawano counties.

2. **Quercus**   Oak

Large or small trees, with rough bark, or smooth when young; leaves often 2 dm long, toothed to deeply lobed; staminate flowers in catkins, appearing before the leaves; pistillate flowers solitary or in small spikes, each subtended by a bract and an involucre of many scales; fruit an acorn with the cuplike involucre enclosing its base.

a. Leaves with rounded lobes or wavy margins; stamens 6-8; fruit ripening the first year, inner surface of the nutshell without hairs *b*

  b. Leaves deeply lobed *c*

    c. Leaves with a large terminal lobe, broadest beyond the middle, finely hairy below when mature; acorn cup with fringed scales . . . . 1. *Q. macrocarpa*

    c. Leaves broadest about the middle, the tip lobes not enlarged, becoming hairless below when mature; acorn cup without fringes. . . . . 2. *Q. alba*

  b. Leaves with shallow-toothed or wavy margins *d*

    d. Leaves coarsely wavy-margined, broadened beyond the middle; acorn cups on long stalks; bark of branches peeling off in broad thin flakes . . . . 3. *Q. bicolor*

    d. Leaves with shallow-pointed (but not bristle-tipped) teeth; acorn cups on short stalks, or sessile; bark of branches not peeling *e*

      e. Leaves with 8-13 teeth per margin; tree

. . . . 4.*Q. muehlenbergii*

      e. Leaves with 3-7 teeth per margin; shrub. . .5. *Q. prinoides*

a. Leaves with pointed lobes which end in bristles; stamens 4-6; fruit ripening the second year (immature fruits may be present); inner surface of nutshell tomentose *f*

  f. Mature leaves dull above, shallowly lobed; acorn cups shallow, saucer-shaped, covering only the base of the acorn, with tightly appressed scales; winter buds 5-7 mm long, not angled, dark shining brown with very few hairs . . . . 6. *Q. borealis*

  f. Mature leaves shining above, deeply lobed; acorn cups deep, covering one-third to one-half of the acorn, the upper scales forming a fringe or else tightly appressed *g*

    g. Acorn short, oval; main trunks of the trees with few dead pendant branches *h*

      h. Scales of acorn cup forming a definite fringe on the edge of the cup, cup gray-ashy; winter buds 6-12 mm long, very gray-hairy, pointed and angular . . . . . .7. *Q. velutina*

      h. Scales of acorn cup tight along the rim, brown to chestnut brown; winter buds short, 2-4 mm, the scales

shining, only slightly hairy along the margins, short blunt-pointed, not angular . . . . 8. *Q. coccinea*
g. Acorn elongate, narrowly elliptical; cups ashy, cup scales tightly appressed; winter buds small, 3-6 mm, smooth or with the edges of the scales ciliate; small trees usually with multiple stems and many dead persistent branches extending downward . . . . 9. *Q. ellipsoidalis*

1. **Q. macrocarpa** Michx. Bur Oak (Fig. 139). Trees, with deeply ridged, dark bark and often corky ridged twigs, young twigs pubescent; leaves with rounded lobes, the tip lobe largest, tapering to smaller basal lobes, the tip half of the leaf separated from the base by a deep sinus, pubescent below, becoming glabrous above; acorn cup with thick scales, fringed to the outer tips.—Throughout the state but less common northward.

2. **Q. alba** L. White Oak (Fig. 140). Trees, with flaky light brown bark, young twigs glabrous, often red or purplish shiny; leaves with rounded deep lobes, the middle lobes longest, whitened below and pubescent, becoming green above; acorn on short stalk, the cup shallow, covering about one-third of the acorn, with rough tuberculate scales.—North to Washburn County in western Wisconsin and to Brown County in eastern Wisconsin.

3. **Q. bicolor** Willd. Swamp White Oak (Fig. 141). Trees, with flaky gray bark, that on the younger branches peeling

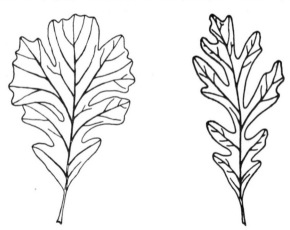

Fig. 139. *Quercus macrocarpa,* × .25.   Fig. 140. *Quercus alba,* × .30.

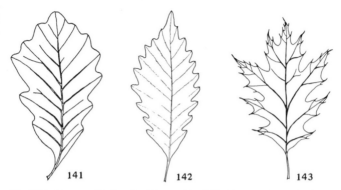

Fig. 141. *Quercus bicolor*, × .40. Fig. 143. *Quercus borealis*, × .25.
Fig. 142. *Quercus muehlenbergii*,
× .25.

off very distinctively in broad thin flakes; the twigs brown, glabrous; leaves with rounded sinuate edges, broader toward the tips, green above, whitened and downy below; acorn on an elongate stalk, the cup covering one-third to one-half of the acorn, the lower scales woody, the upper with sharp points and fringing the cup margin.—River bottoms, southern Wisconsin, north to Buffalo, Wood, and Brown counties.

4. **Q. muehlenbergii** Engelm. Yellow Oak; Chestnut Oak (Fig. 142). Small trees or large shrubs, the bark thin, flaky; twigs yellow brown to reddish glabrous; leaves broadest at

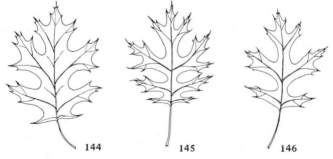

Fig. 144. *Quercus velutina*, × .25. Fig. 146. *Quercus ellipsoidalis*,
Fig. 145. *Quercus coccinea*, × .25. × .25.

middle and tapering to each end, 8-13 teeth on each margin, green above, whitened and downy below; acorn about one-half covered by the cup, its scales thin, closely appressed.— Waukesha and Crawford counties.

5. **Q. prinoides** Willd. Chinquapin Oak. Shrubs, with thin smooth bark, except at base where it is flaky; twigs yellow brown, glabrous; leaves broadest at middle, 3-7 blunted teeth on each margin, green above, whitish below; acorn as in No. 4, but smaller.—Brown County.

6. **Q. borealis** Michx. Northern Red Oak (Fig. 143). Trees, with deeply ridged bark, the younger branches smooth, inner bark reddish; twigs brown, glabrous; leaves coarsely lobed, but not as deeply as in *Q. velutina*, the lobes bristle-tipped, becoming dull green above, green and hairy in vein axils below; acorn cup shallow, covering about one-third of the acorn, with closely appressed scales.—Rich moist soils, throughout the state.

7. **Q. velutina** Lam. Black Oak (Fig. 144). Trees, with deeply ridged bark, the inner bark orange; twigs yellow brown, glabrous; leaves deeply and usually narrowly lobed, lobes bristle-tipped, upper side shiny, dark green, yellow green and hairy in vein axils below; acorn cup covering one-third to one-half of acorn (deeper than *Q. borealis*), tips of the upper scales free and spreading.—Southern and central Wisconsin, north to Eau Claire to Brown counties.

8. **Q. coccinea** Muenchh. Scarlet Oak (Fig. 145). Trees, with old bark ridged, inner bark reddish, younger branches smooth; leaves deeply and narrowly lobed, bristle-tipped, upper side shiny green, underside paler yellow green with few hairs in axils of veins below; acorn cup covering about one-third to one-half of acorn, the scales closely appressed; acorn oval, not elongate; autumn color scarlet, young leaves bright red above.—Status of this tree in Wisconsin is not clear. Its presence was denied in a paper by Wadmond (*Trans. Wis. Acad. Sci. Arts Letters* 28: 197-203, 1933). However, specimens show that it occurs in southern counties, and further field work will be necessary to determine its range.

9. **Q. ellipsoidalis** E. J. Hill. Northern Pin Oak; Jack Oak; Hill's Oak (Fig. 146). Trees, often multiple-stemmed, and often with many drooping dead branches; inner bark yellow; young branches smooth; leaves deeply and narrowly lobed, bristle-tipped, upper side shiny, bright green, lower side paler, yellow green with small tufts of hair in axils of the

veins; acorn cup narrow, covering one-third to one-half of acorn, acorn rather narrowly elongate, often with dark stripes; autumn color brown, dead leaves frequently persistent, young leaves scarcely reddened above.—Throughout the state.

ULMACEAE   Elm Family

Trees (our species) with simple alternate leaves which are asymmetrical at base; flowers perfect or unisexual; tepals 3-9, stamens the same number as the tepals and opposite them; styles 2; fruit a samara, nut, or drupe.

a. Flowers in little clusters, coming before the leaves . . . . . 1. *Ulmus*
a. Flowers solitary in each leaf axil, coming with the leaves . 2. *Celtis*

1. Ulmus   Elm

Bark in long ridges and furrows; leaves unlobed, elliptical, the bases very asymmetrical, doubly toothed, with many strong, straight, parallel veins branching from the midrib, much like the leaves of *Ostrya* and *Carpinus*; fruit a small samara with a broad circular wing.

a. Flowers on very short pedicels, 1-2.5 mm long; fruit with edges not fringed with hairs *b*
  b. Bud scales with conspicuous red brown hairs, leaves rough-hairy and folded along the midrib; new twigs 3 mm or more thick; inner bark pleasantly scented (licoricelike) . . . . 1. *U. rubra*
  b. Bud scales with gray brown hairs on their edges; leaves smooth above, flattened; new twigs less than 3 mm thick; bark not scented . . . . 2. *U. pumila*
a. Flowers on pedicels 4-10 mm long; fruit fringed with hairs on edges *c*
  c. Twigs with corky ridges; face of fruits hairy; bark with alternating brown and dull white layers . . . . 3. *U. thomasii*
  c. Twigs lacking corky ridges; face of fruits not hairy; bark with alternating brown and very white layers . . . . . . 4. *U. americana*

1. **U. rubra** Muhl. Slippery Elm. Inner bark mucilaginous when chewed; bud scales black brown with very red brown hairs; flowers in tight clusters; fruit not deeply notched, hairless except for a few hairs over the central nutlet.—Rich soils, north to Burnett, Lincoln, and Marinette counties.

2. **U. pumila** L. Siberian Elm. Branchlets very slender, drooping; buds very small, dark brown, round, the scales ciliate on the margins, glossy on surfaces; flowers in tight clusters; fruit with very shallow notch, glabrous. Often erroneously called Chinese Elm.—Escaping from cultivation.

3. **U. thomasii** Sarg. Cork Elm. Branches often with corky ridges after the previous year's growth; bud scales brown,

Fig. 146A. *Ulmus americana,*
× .25.

Fig. 147. *Celtis occidentalis,*
× .35.

ciliate on margins, surface sparingly pubescent; flowers in racemes; fruit not notched, the edges white-ciliate and the faces pubescent.—Heavy clay soil or rocky slopes, north to Forest and Marinette counties.

4. **U. americana** L. American Elm; White Elm (Fig. 146A). Inner bark not mucilaginous; bud scales brown, surface grayish-hairy, edges ciliate; flowers in umbellike clusters; fruit very deeply notched, the edges white ciliate, surfaces glabrous.— Moist woods, throughout the state.

2. **Celtis**   Hackberry

Leaves 3-5-veined from the base; flowers perfect or unisexual, in small clusters or solitary from the leaf axils of new twigs; fruit a drupe.

C. occidentalis L. (Fig. 147). Crown of tree frequently with many "witches' brooms"; bark of trunk with corky wartlike outgrowths; leaf blade tapered to the petiole, upper two-thirds of margin toothed; fruit cherrylike, reddish or yellowish when young, turning dark purple. Leaves variable in shape and texture; several varieties have been described.— Bottomlands, lakeshores, and moist woods north to Polk and Lincoln counties.

MORACEAE   Mulberry Family

Trees (our species), with milky juice; leaves simple, alternate; flowers small, in dense clusters or heads, unisexual; tepals usually 4, or united into a lobed calyx tube; stamens the same number as tepals and opposite them; styles 1-2; fruit a drupe or achene.

a. Leaves toothed or lobed; spines lacking on branches .... 1. *Morus*
a. Leaves entire; branches spiny .... 2. *Maclura*

1. Morus   Mulberry

Small trees; leaves undivided or deeply 3-lobed or pinnately several-lobed; staminate and pistillate flowers on different trees; each in catkins, the pistillate developing into a white, pink, or red aggregate fruit resembling a blackberry.

a. Leaves glabrate or pubescent in vein axils . . . . 1. *M. alba*
a. Leaves pubescent on lower surface . . . . 2. *M. rubra*

1. M. alba L. White Mulberry. Leaves smooth above, and with hairs only about the axils of the veins below; blades 6-18 cm long.—Rarely planted.

Var. tatarica (L.) Ser. with blades of leaves only 4-8 cm long. Sometimes escaping from cultivation.

2. M. rubra L. Red Mulberry. Leaves rough above, hairy below.—Bottomlands and bluffs in southwestern Wisconsin, up the Mississippi River to Pepin and up the Wisconsin River to Prairie du Sac.

2. Maclura   Osage Orange

Dioecious thorny trees; leaves alternate, entire; staminate and pistillate flowers in round heads; fruit a hard syncarp.

M. pomifera (Raf.) Schneid. Leaves lanceolate-ovate, acuminate; fruit 6-12 cm thick, yellowish green.—Persistent after cultivation; Grant, Iowa, Lafayette, and Rock counties.

URTICACEAE   Nettle Family

Herbs (our species), often with stinging hairs; flowers regular; tepals 3-5; stamens as many as the tepals and opposite them; style 1; fruit an achene or drupe.

Parietaria   Pellitory

Herbs with alternate, entire leaves; tepals 4; achene enclosed by the enlarged calyx.

P. pensylvanica Muhl. Inconspicuous herbs, with clusters of green flowers in the axils of nearly all the leaves.—Mostly about the bases of trees and on rock outcrops, north to Polk, Waupaca, and Door counties.

LORANTHACEAE   Mistletoe Family

Green half-parasitic plants attached to the branches of trees and shrubs; leaves opposite or whorled; flower parts in whorls or undifferentiated; tepals 2-6; stamens 2-6; fruit berrylike.

Arceuthobium   Dwarf Mistletoe

Dioecious woody herbs, parasitic on conifers; leaves scale-like; flowers in the leaf axils; staminate perianth 2-5-lobed, pistillate perianth 2-lobed.

A. pusillum Peck. (Fig. 148). Plants tiny, 6-20 mm tall, seldom branched, brownish; fruit a small berry which violently expels its seeds at maturity.—Parasitic on black (rarely on white) spruce in northern Wisconsin and south to Jackson and Manitowoc counties.

SANTALACEAE   Sandalwood Family

Herbs (our species), usually root parasites; leaves simple; flowers regular; sepals 3-many; stamens the same number as

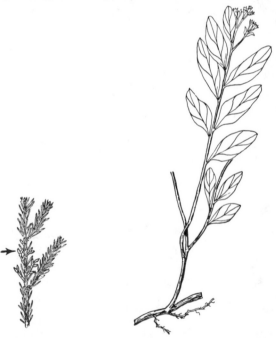

Fig. 148. *Arceuthobium pusillum* on Black Spruce, × .40.

Fig. 149. *Comandra umbellata,* × .30.

the sepals and opposite them; ovary inferior; style 1; fruit a 1-seeded nut or drupe.

a. Flowers in terminal corymbs; fruit a dry nut . . . . 1. *Comandra*
a. Flowers in axillary small cymes; fruit a fleshy drupe . 2. *Geocaulon*

1. **Comandra**   Bastard Toadflax

Stems erect, from a rhizome; leaves alternate, pale, smooth; fruit crowned by the 5 calyx lobes.

C. umbellata (L.) Nutt. (Fig. 149). Parasitic on roots of other plants, 1.5-4 dm high; leaves oblong, entire, 1.3-5 cm; flowers in dense flat-topped terminal clusters; fruits dry, greenish.—Dry sandy ground, throughout the state.

2. **Geocaulon**   Northern Comandra

Stems erect, from a dark-colored rhizome, 1-3 dm high; leaves alternate; center flower of each small cyme pistillate, others staminate; fruit a fleshy false drupe.

G. lividum (Richards.) Fern. Leaves oval; flowers few, on an axillary peduncle, only the central one developing the fleshy scarlet fruit.—Rare in mossy coniferous woods; Door County.

ARISTOLOCHIACEAE   Birthwort Family

Herbs or vines, with broad leaves; sepals fused, 3-lobed, corollalike; stamens 6-many; fruit a capsule.

Fig. 150. *Asarum canadense,* × .20.

Asarum   Wild Ginger

Stems underground, horizontal, usually with a few scalelike leaves below; foliage leaves 2, long-petioled, the blades kidney-shaped; flowers solitary, between the 2 leaves; calyx bell-shaped, purple brown inside, lobes short- to long-pointed.

A. canadense L. (Fig. 150). Plants hairy; leaves round-cordate to kidney-shaped.—In hardwoods, throughout the state.

Var. **acuminatum** Ashe. Calyx lobes with long-tapering whiplike tips.—Rich woods.

Var. **reflexum** (Bickn.) Robins. Calyx lobes short-pointed, reflexed.—Southeastern Wisconsin west to Columbia, Dane, and Green counties; Outagamie County.

POLYGONACEAE   Buckwheat Family

Herbs, usually with somewhat swollen nodes and sheathing stipules; flowers perfect or imperfect, small; sepals 3-6, or united and 3-6-lobed; corolla lacking; stamens 4-9; styles 2-3; fruit an achene.

a. Sepals 6 . . . . 1. *Rumex*
a. Sepals 4-5 . . . . 2. *Polygonum*

1. Rumex   Dock; Sorrel

Monoecious or dioecious annual or perennial herbs; leaves simple, alternate, with stipules forming tubular sheaths (ocreae) around the stem; sepals in 2 rows of 3; stamens 6, filaments very short; styles 3; fruit a 3-angled achene enclosed by the enlarged inner sepals (valves) in which (in this genus) the midrib enlarges into a protuberance known as a grain.

a. Leaves all or partly sagittate; dioecious . . . . 1. *R. acetosella*
a. Leaves not sagittate; monoecious *b*
  b. Margins of valves toothed to spiny . . . . 2. *R. obtusifolius*
  b. Margins of valves entire to wavy-edged *c*
    c. Leaves with crisped margins . . . . 3. *R. crispus*
    c. Leaves flat *d*
      d. Valves broadly round-ovate; usually only one bearing a grain . . . . 4. *R. altissimus*
      d. Valves triangular-ovate, each bearing a grain . . . . 5. *R. mexicanus*

1. **R. acetosella** L. Sheep Sorrel. Plants 1-3 dm high, from deep rootstocks; leaf blades usually arrow-shaped at base; sepals yellow on the pistillate plant, red on the staminate. —Naturalized from Europe; poor soils, throughout the state.

2. **R. obtusifolius** L. Bitter Dock. Stout perennials; lower leaves large-cordate, to 15 cm broad, upper leaves smaller; valves with 2-4 spinose teeth on each margin, one bearing a grain, the others with only a thickened midrib.—Naturalized from Europe; throughout Wisconsin.

3. **R. crispus** L. Sour Dock. Perennials, to 1 m high; leaves strongly crisped; valves thin, broadly ovate, each bearing a grain.—Naturalized from Europe; common weed, throughout the state.

4. **R. altissimus** Wood. Water Dock. Stout perennials, often branched; leaves lanceolate or lanceolate-ovate, acute; valves 4-6 mm long and wide, grains 1-3.—Wet places, north to Pierce, Brown, and Sheboygan counties.

5. **R. mexicanus** Meisn. Stout perennials, often branched; leaves pale green, narrowly lanceolate, long-tapering at both ends; valves triangular, grains 3.—Rich wet soil, north to Bayfield and Ashland counties in the west and to Juneau and Brown counties in central Wisconsin and eastward.

2. Polygonum   Knotweed; Smartweed

Erect or trailing herbs; leaves alternate, the stipules forming a tubular sheath, the ocrea, around the stem; sepals 5, equal in size; fruit a 3-angled or lenticular achene.

a. Flowers in small axillary clusters . . . . 1. *P. aviculare*
a. Flowers in terminal spikes or racemes *b*
    b. Climbing or trailing vines; sheaths small; flowers in panicled racemes *c*
        c. Nodes naked . . . . 2. *P. convolvulus*
        c. Nodes fringed . . . . 3. *P. cilinode*
    b. Erect or aquatic plants; sheaths 1 cm or more long; flowers in dense spikes . . . . 4. *P. amphibium*

1. **P. aviculare** L. Knotweed. Low, branched annual; leaves linear, 1-3 cm long; axillary flowers almost included in the ocreae; calyx lobed, green with pinkish margins; achene ovoid.—A common weed, throughout the state.

2. **P. convolvulus** L. Black Bindweed (Fig. 151). Stems roughish; sheath 1-3 mm long; leaves arrow-shaped; outer sepals narrowly winged.—Naturalized from Europe; roadsides and cultivated ground, throughout the state.

3. **P. cilinode** Michx. (Fig. 152). Stems minutely downy, usually reddish; sheath with a fringe of downward-pointing hairs at its base; leaves ovate-lanceolate with a heart-shaped base.—Fields and cutover land, south to Vernon, Dane, and Sheboygan counties.

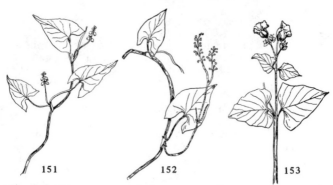

Fig. 151. *Polygonum convolvulus,*
× .30.   Fig. 153. *Mirabilis nyctaginea,*
                                      × .25.
Fig. 152. *Polygonum cilinode,* × .30.

4. **P. amphibium** L., var. **stipulaceum** (Coleman) Fern. Water Smartweed. Leaves lanceolate, short-petioled; flowers deep pink, in erect, compact, showy spikes. Plants aquatic, with floating leaves, or terrestrial with spreading green borders on the sheaths.—Throughout the state.

NYCTAGINACEAE   Four-o'clock Family

Herbs (our species) with simple, usually opposite leaves; flowers perfect, borne within a calyxlike involucre, perianth undifferentiated, calyx typically of 5 fused parts, sometimes corollalike; stamens 3-5; ovary, 1; style 1; fruit achenelike, ribbed or angled.

Mirabilis   Wild Four-o'clock

Perennial herbs; leaves opposite; flowers in clusters of 2-4, subtended by a 5-lobed involucre, perianth funnelform, 5-lobed, pink purple, opening in the morning.

M. nyctaginea (Michx.) MacM. (Fig. 153). Plants 0.3-1.5 m high; stems repeatedly forked; leaf blades heart-shaped on short petioles; bracts densely ciliate, 1 cm broad.—Common along railroad tracks; throughout the state, but less common northward.

PORTULACACEAE   Purslane Family

Small herbs, with rather fleshy entire leaves; petals 5;

sepals 2; stamens 5-many; fruit a pod, of 3 carpels, but with 1 locule and the seeds attached to a central knob.

a. Leaves flat, lanceolate; plant from a tuber . . . . 1. *Claytonia*
a. Leaves threadlike; plant from a rootstock . . . . 2. *Talinum*

1. Claytonia   Spring Beauty

Stems several, from a deep-seated globose tuber; leaves usually 2 to each stem, opposite; flowers several, rose-colored, each long-stalked from a central axis.

a. Leaf blades 5-15 times as long as wide, width 3-10 mm; petiole indistinguishable . . . . 1. *C. virginica*
a. Leaf blades 2-4.5 times as long as wide, width 6-23 mm; petiole well distinguished . . . . 2. *C. caroliniana*

1. **C. virginica** L. (Fig. 154). Leaves linear-lanceolate or ribbonlike, 3-16 cm long.—Rich woods, north to Ashland and Kewaunee counties.

2. **C. caroliniana** Michx. (Fig. 155). Leaves oval to lanceolate, 2.5-5 cm long.—Local in northern Wisconsin.

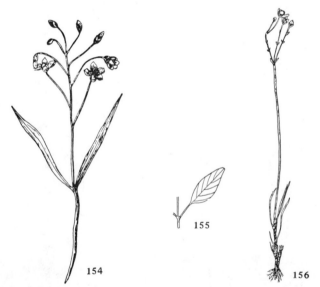

Fig. 154. *Claytonia virginica,* × .40.
Fig. 155. *Claytonia caroliniana,* × .40.

Fig. 156. *Talinum rugospermum,* × .40.

2. **Talinum**   Fameflower

Plants perennial, glabrous; leaves basal, terete; flowers in long-peduncled bracteate cymes; sepals 2, early-deciduous; petals 5; stamens 5-45; style 3-lobed.

T. rugospermum Holzinger (Fig. 156). Leaves short, much exceeded by the peduncles; peduncles branched near the summit; flowers light pink, opening but once, between 3:30 and 4 p.m., and closing at 6 p.m., when the petals shrivel; stamens 12-25, their filaments deeper pink than the petals.— Dry sand plains and sandstone ledges, southern Wisconsin, north to Burnett County along the Mississippi River, and to Waupaca and Outagamie counties in central Wisconsin.

CARYOPHYLLACEAE   Pink Family

Herbs, with opposite or whorled, entire leaves; stems usually swollen at joints; flowers in cymose inflorescences or single, mostly perfect; sepals 4-5, free, or fused and forming a tube, persistent in fruit; petals free, 4-5, or rarely lacking; stamens 1-10, usually twice the number of petals; styles 2-5; fruit a 1-seeded utricle, or many-seeded capsule with the many seeds attached to a central knob or column.

a. Sepals separate (Chickweed Tribe), or forming a small hard cup at the base and free above (Knawel Tribe) *b*
  b. Petals lacking; styles 2; fruit a 1-seeded utricle .. 1. *Scleranthus*
  b. Petals 4-5 (rarely 1-3, or lacking in *Sagina* and *Stellaria caly-cantha*); styles 3-5; fruit a few- to several-seeded capsule *c*
    c. Petals entire, not deeply notched *d*
      d. Leaves whorled *e*
        e. Petals white; styles 5 ....2. *Spergula*
        e. Petals pale rose; styles normally 3 ..... 3. *Spergularia*
      d. Leaves opposite, or some in axillary bunches *f*
        f. Sepals 4; styles 4 and alternate with sepals ..4. *Sagina*
        f. Sepals 5; styles 3, rarely 4 .... 5. *Arenaria*
    c. Petals 2-cleft, notched at tip or denticulate *g*
      g. Styles 3 *h*
        h. Petals notched; inflorescence a diffuse cyme ....
        .... 6. *Stellaria*
        h. Petals denticulate; inflorescence an umbellike cyme ..
        .... 7. *Holosteum*
      g. Styles 4 or 5 *i*
        i. Leaves pubescent; styles opposite sepals; capsule cylindrical ....8. *Cerastium*
        i. Leaves lacking hairs; styles alternate with sepals; capsule ovoid .... 9. *Myosoton*
a. Sepals united into a tube (Pink Tribe) *j*
  j. Lobes of the calyx 2-3 cm long .... 10. *Agrostemma*
  j. Lobes of the calyx shorter *k*
    k. Calyx conspicuously 10-nerved *l*
      l. Styles 5 ....11. *Lychnis*

1. Scleranthus   Knawel

Sepals 5, fused at base, becoming thick and hard in fruit; stamens 1-10, inserted on the calyx tube below the calyx teeth; styles 2, distinct; fruit a 1-seeded utricle, persistently enclosed by the hard calyx.

S. perennis L. Perennial Knawel (Fig. 157). Low herbs, with forked, pubescent, wiry stems, and numerous minute greenish flowers in tight cymes; leaves minute, stipules lacking; petals lacking.—A Eurasian weed; Juneau, Sauk, Columbia, and Marquette counties.

2. Spergula   Corn Spurrey

Annuals; leaves in whorls; stamens 5 or 10; styles 5; the 5 valves of the capsule opposite the sepals.

S. arvensis L. (Fig. 158). Stems many, radiating from the summit of a taproot, ascending or prostrate; leaves thread-like, with stipules; flowers white in much-branched cymes.—A common weed in northern Wisconsin, occasional southward; naturalized from Europe.

Fig. 157. *Scleranthus perennis,* × .50.

Fig. 158. *Spergula arvensis,* × .25.

Fig. 159. *Spergularia rubra,* × .70.

3. **Spergularia**   Sand Spurrey

Sepals 5; petals 5, entire; stamens 2-10; styles and valves of the many-seeded capsule normally 3.

S. rubra (L.) J. & C. Presl. Red Sand Spurrey (Fig. 159). Small decumbent annuals; leaves filiform, opposite, appearing whorled; stipules conspicuous, triangular-acuminate, 1.5-4 mm long; sepals 2.5-4 mm long; petals 5, entire, dull pink; stamens 6 (-10); styles and valves of capsule 3. Flowering all summer.—Naturalized from Eurasia; a rare weed of sandy areas, from Sawyer to Shawano counties in northern Wisconsin.

4. **Sagina**   Pearlwort

Sepals 4 (-5); petals 4 (-5, or lacking); styles alternate with the sepals; capsule many-seeded; valves opposite the sepals.

S. procumbens L. Matted, creeping; leaves threadlike; petals shorter than the sepals.—Damp places, sometimes found as a weed on golf greens, rare; Iron and Sheboygan counties.

5. **Arenaria**   Sandwort

Low tufted herbs; leaves without petioles; each style opposite a sepal; pod short, splitting into as many sectors as there are styles.

a. Leaves flat, 2 mm or more wide; plants pubescent and ciliate (except reduced in *A. macrophylla*); capsule with 6 teeth *b*
   b. Leaves 1-2 cm long, oblong to lanceolate-ovate; perennials with rhizomes *c*.

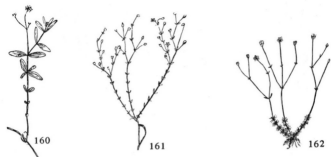

Fig. 160. *Arenaria lateriflora,* × .25.   Fig. 162. *Arenaria stricta,* × .20.
Fig. 161. *Arenaria serpyllifolia,* × .40.

   c. Leaves blunt; sepals obtuse, shorter than the petals . . . .
. . . . 1. *A. lateriflora*
   c. Leaves stiff, acute; sepals acuminate, longer than the petals
. . . . 2. *A. macrophylla*
  b. Leaves not more than 6 mm long; annuals with taproots . . . . . .
. . . . 3. *A. serpyllifolia*
a. Leaves slender, 1.5 mm or less wide, entire; plants glabrous, mat-
forming; capsule with 3 teeth . . . . 4. *A. stricta*

1. **A. lateriflora** L. Grove Sandwort (Fig. 160). Stems simple or little-branched, from a long slender rootstock; flowers 2-4 to a peduncle, with obtuse sepals.—Common in shady woods, south of the Tension Zone.

2. **A. macrophylla** Hook. Large-leaved Sandwort. Similar; sepals acuminate.—Rare, on cliffs and rocky ledges; Ashland and Iron counties.

3. **A. serpyllifolia** L. Thyme-leaved Sandwort (Fig. 161). Stems many, arising from a slender taproot; flowers many, in compound cymes.—Naturalized from Europe; wet places and gravelly hillsides; in eastern Wisconsin, especially near Lake Michigan, to Door County and Dane and Rock counties.

4. **A. stricta** Michx., ssp. **stricta.** Rock Sandwort (Fig. 162). Stems many, from a slender rootstock, with a persistent tuft of many leaves below; inflorescence much branched, often more than one-half the height of the entire plant; sepals 3.4-5.5 mm long, strongly 3-ribbed, sharply lanceolate-ovate or acuminate; petals longer than the sepals; capsules mostly shorter than the sepals.—Common and conspicuous on rocky hills and ledges in southern Wisconsin from Grant to Kenosha counties, north to Columbia to

Washington counties, and along Lake Michigan to Door
County.

Ssp. **dawsonensis** (Britt.) Maguire. Northern Rock Sandwort. Sepals blunt to acute; petals shorter than or equaling
the sepals; capsules about equaling the sepals or
longer. – Rare on rocky cliffs; La Crosse, Vernon, St. Croix,
and Waupaca counties.

6. Stellaria   Starwort; Chickweed

Mostly glabrous, delicate herbs with frequently weak and
reclining stems and with leaves lacking stipules; sepals (4-) 5;
petals (4-) 5, deeply cleft; stamens (2-) 10; styles 3; capsule
ovoid to globose, opening with as many or twice as many
valves as styles.

a. Leaves ribbonlike or narrowly lanceolate, without petioles *b*
  b. Petals shorter than the sepals or lacking . . . . . 1. *S. calycantha.*
  b. Petals mostly longer than the sepals *c*
    c. Inflorescence clearly terminal and much exceeding the
      lateral leaf shoots at its base *d*
      d. Sepals mostly over 4 mm long *e*
        e. Sepals acute, usually ciliate, strongly 3-ribbed; capsule
          pale brown to straw-colored . . . . 2. *S. graminea*
        e. Sepals blunt or subacute, without prominent ribs;
          capsule dark brown to black . . . . 3. *S. longipes*
      d. Sepals 2.3-4 (-4.4) mm long . . . . .4. *S. longifolia*
    c. Inflorescence appearing lateral, and shorter than to only
      slightly longer than the prominent leafy shoots at base;
      sepals eciliate; petals much exceeding the sepals . . . .
      . . . . .5. *S. palustris*
a. Leaves oval, middle stem leaves with distinct petioles; petals
  shorter than the sepals . . . . 6. *S. media*

1. **S. calycantha** (Ledeb.) Bong. Northern Starwort (Fig.
163). Perennials, from slender rhizomes; stems without hairs;
leaves often slightly ciliate; flowers solitary or a few together; petals minute or none. – Rather rare; northern
Wisconsin, south to Marathon County.

2. **S. graminea** L. Common Stitchwort. Leaves long-
tapered, 5-10 times as long as wide; inflorescence often
longer than leafy portion of plant; pedicels reflexed in
fruit. – Naturalized from Europe; damp open ground,
throughout the state.

3. **S. longipes** Goldie. Perennials, from slender rhizomes;
stems 1.5-2 (-3) dm tall, without hairs; leaves narrowly lanceolate, acuminate, sessile, stiff and ascending; inflorescence
terminal, with scarious bracts; sepals weakly 3-nerved; petals
longer than sepals; capsule exserted, black. – Rare; Oneida
County.

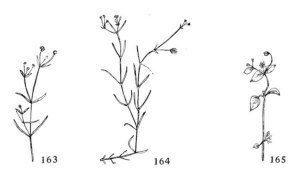

Fig. 163. *Stellaria calycantha,* × .25.

Fig. 164. *Stellaria longifolia,* × .25.

Fig. 165. *Stellaria media,* × .40.

4. **S. longifolia** Muhl. (Fig. 164). Stems without hairs, slender, branched; leaves thin, flat, acute; inflorescence few to many-flowered, with spreading pedicels; petals equaling or exceeding the sepals. Two varieties are recognized in Wisconsin, and mature specimens can usually be assigned to one of them.

Var. **longifolia**. Mature capsule straw-colored to brownish; plants dark green with leaves mostly over 1.5 mm wide; middle nodes as long or longer than the leaves; flowers on long pedicels.—In moist places; common throughout the state.

Var. **atrata** Moore. Mature capsule dark-pigmented purple brown to brownish black; leaves mostly 1.5 mm wide or less; plants usually yellowish green; middle nodes usually shorter than the leaves; flowers on short pedicels.—In moist places; mostly north of the Tension Zone, from St. Croix to Milwaukee counties.

5. **S. palustris** Retz. Marsh Stitchwort. Slender glabrous perennial, 2-4 dm tall; leaves very narrowly linear; petals conspicuously exceeding the sepals; capsule pale.—In wet places, Iowa and Dane counties, north to Wood and Portage counties.

6. **S. media** (L.) Cyrill. Common Chickweed (Fig. 165). Stems low, hairy, often reclining; leaves somewhat fleshy, the blades 5-15 mm long; flowers few, in the axils of the upper leaves.—Naturalized from Europe; a common weed about buildings, throughout Wisconsin.

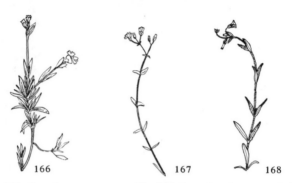

Fig. 166. *Cerastium arvense,*   Fig. 167. *Cerastium fontanum,* × .20.
× .40.   Fig. 168. *Cerastium nutans,* × .30.

7. **Holosteum**   Jagged Chickweed

Annuals or biennials, with several flowers in an umbel borne on a long terminal peduncle; sepals 5; petals 5, usually jagged or denticulate at the tip or apex; stamens 3-5 (-10); styles mostly 3; capsule ovoid.

H. umbellatum L. Glaucous or very pale herbs; leaves oblong; peduncle and upper part of stem glandular.—European; rare, Shawano County.

8. **Cerastium**   Mouse-ear Chickweed

Low, usually pubescent herbs; sepals 5; petals 5, notched or cleft; styles 5, opposite the sepals; capsule elongated, often curved, opening at the summit by twice as many teeth as there were styles.

a. Leaves with axillary bunches of smaller leaves; petals 2-3 times as long as sepals. . . . . 1. *C. arvense*
a. Leaves without axillary bunches; petals shorter or less than twice as long as sepals *b*
    b. Petals equal to or shorter than the sepals; pedicels 4-14 mm long; plants perennial . . . . 2. *C. fontanum*
    b. Petals longer than the sepals; pedicels 5-40 mm long; plants annual . . . . 3. *C. nutans*

1. **C. arvense** L. Field Chickweed (Fig. 166). Perennial, densely tufted plants 1-2 dm tall, with tufts of leaves in axils of most stem leaves; bracts and sepals with scarious margins; petals 8-12 mm long; stems sparingly to densely pubescent; leaves stiff, ribbonlike or narrowly lanceolate.—Sandy areas;

Fig. 169. *Myosoton aquaticum,* × .40.

not common, from La Crosse to Sheboygan counties, north to Dunn, Lincoln, and Douglas counties.

2. **C. fontanum** Baumg. Common Mouse-ear Chickweed (Fig. 167). Perennials, 1-5 dm tall; leaves hirsute on both surfaces; inflorescence bracts and sepals with scarious margins, sepals usually hirsute, sometimes glandular; petals slightly shorter than the sepals.—Naturalized from Europe; a common weed, throughout the state.

3. **C. nutans** Raf. Nodding Chickweed (Fig. 168). Annual, rather weak plants, 1-4 dm tall; often viscid with glandular hairs; inflorescence bracts herbaceous; petals 5-7 mm long; fruit often nodding.—Mostly south of the Tension Zone, from Barron to Kewaunee counties, south to Grant and Racine counties.

9. **Myosoton**   Giant Chickweed

Perennial, with angled stems; cordate-ovate leaves and scattered axillary flowers; sepals 5; petals 5, deeply notched; stamens 10; styles 5, alternate with the sepals; capsule broadly ovoid.

M. aquaticum (L.) Moench (Fig. 169). Similar to *Stellaria media* but much larger and coarser; leaf blades 1.5-7 cm long.—Adventive from Europe; an occasional weed in moist places; scattered throughout the state.

10. **Agrostemma**   Corn Cockle

Calyx ovoid with 10 strong ribs, teeth exceeding the

Fig. 170. *Agrostemma githago,* × .20.   Fig. 171. *Lychnis alba,* × .20.

petals; petals 5, large and unappendaged; stamens 10; capsule 1-locular.

A. githago L. (Fig. 170). Plants stout, 3-5 dm high; leaves narrowly lanceolate, 1-nerved, 5-10 cm long, covered with close hairs; calyx covered with ascending silky hairs; corolla purplish red; seeds poisonous.—Roadsides and grain fields; introduced from Europe. Rare and possibly no longer occurring in Wisconsin, although probably once common in disturbed or waste places; believed to have been associated with agricultural wheat.

11. **Lychnis**   Campion

Plants dioecious, 5-10 dm high, minutely hairy; leaves ovate, about 1 dm long, the upper without petioles, the lower sometimes petioled; sepals 5, united; petals 5; stamens 10; styles 5.

L. alba Mill. White Campion (Fig. 171). Flowers white or pink, fragrant, opening in the evening; calyx teeth about 5 mm long.—Adventive from Europe; in fields and along roadsides, throughout the state. Resembles *Silene noctiflora* from which it differs in being dioecious and in having 5 styles and shorter hairs on the stem.

12. **Silene**   Catchfly; Campion

Annual or perennial herbs; leaves opposite, entire; flowers perfect or unisexual; sepals 5, united; petals 5; stamens 10; styles 3.

Fig. 172. *Silene antirrhina,* × .20.

Fig. 173. *Silene cucubalus,* × .20; small flower, *S. cserei,* × .20.

a. Corolla pink . . . . 1. *S. antirrhina*
a. Corrolla white *b*
   b. Leaves opposite *c*
      c. Stems hairy and sticky . . . . 2. *S. noctiflora*
      c. Stems without hairs *d*
         d. Calyx subspherical to ovoid; uppermost bracts not ciliate
         . . . . 3. *S. cucubalus*
         d. Calyx narrowly ovoid-oblong to ellipsoid; uppermost
          bracts minutely ciliate . . . . 4. *S. cserei*
   b. Leaves in whorls of 4 . . . . 5. *S. stellata,* var. *scabrella*

1. **S. antirrhina** L. Sleepy Catchfly (Fig. 172). Stems usually with a sticky ring on each internode; leaves narrow; flowers many in a cyme; calyx tube 4-6 mm long.—Dry places, north to Lincoln and Douglas counties.

2. **S. noctiflora** L. Night-flowering Catchfly. Leaves broader; flowers few, fragrant, opening in the evening; calyx tube 1.5-3 cm long, cylindrical. See *Lychnis alba.*—Adventive from Europe; an occasional weed, formerly common in grain fields in eastern Wisconsin, but now very rare.

3. **S. cucubalus** Wibel. Bladder Campion (Fig. 173). Glabrous and glaucous weedy perennials with many stems; leaves ovate-lanceolate; calyx inflated and bladdery, not appressed to the much smaller capsule.—A European weed which seems to be spreading in Wisconsin and has now been found in most of the state except for the west-central part.

4. **S. cserei** Baumg. Bladder Campion (Fig. 173). Plants biennial; calyx firmer and appressed to the slightly exserted capsule.—Naturalized from Europe; mostly along railroads, and presently extending its range to most of the state.

5. **S. stellata** (L.) Ait., var. scabrella (Nieuwl.) Palmer &

Fig. 174. *Saponaria officinalis,* × .40.

Steyerm. Starry Campion. Stems swollen at the nodes; leaves long-pointed; flowers many, in a cyme; calyx tube 1-1.5 cm long, bell-shaped.—Woods, north to Dane and Milwaukee counties, and in the west to Pierce County.

13. **Saponaria**   Bouncing Bet; Soapwort

Primary leaves opposite, often with many secondary leaves in their axils; flowers pink; juice mucilaginous, making a lather with water.

S. officinalis L. (Fig. 174). Leaves oval-lanceolate, 3-nerved; flowers in 1-several close clusters. Begins flowering the last week of June.—Roadsides; introduced from Europe as a garden flower, and now found throughout the state.

14. **Vaccaria**   Cow Herb; Cow Cockle

Plants annual; inflorescence a loose open paniculate cyme; petals without auricles or appendages.

V. segetalis (Neck.) Garcke. Glabrous annuals, 2-7 dm tall; leaves lanceolate-ovate, clasping at base; calyx inflated, 1-1.6 cm long, with 5 prominent, usually green-winged ribs; petals showy, rose. 1.8-2.5 cm long, and exceeding the calyx; stamens 10; styles 2; capsule included in the inflated calyx, dehiscent by 4 teeth.—Adventive from Europe; rare, widely scattered throughout the state.

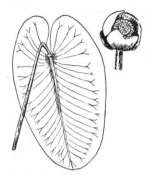

Fig. 175. *Nuphar variegatum,* × .65; flower × .40.

NYMPHAEACEAE   Water Lily Family

Aquatic perennials, with large horizontal rhizomes; leaves floating or erect; flowers solitary; sepals 4-6; petals and stamens numerous.

a. Flowers yellow . . . . 1. *Nuphar*
a. Flowers white or pinkish . . . . 2. *Nymphaea*

1. Nuphar   Yellow Pond Lily; Spatterdock

Rhizomes thick and fleshy; leaf blade rounded, the petiole attached at the base of a deep sinus; sepals 5 or more, yellow, marked with green or red, spoon-shaped; petals shorter than the stamens, thick and fleshy; style none; stigma platelike, marked with conspicuous rays.

a. Anthers at least equaling the filaments *b*
   b. Lobes of leaf spreading at an angle of 45-80 degrees . . . .
   . . . . 1. *N. advena*
   b. Lobes overlapping or nearly so . . . . 2. *N. variegatum*
a. Anthers shorter than the filaments *c*
   c. Flower about 3 cm broad . . . . 3. *N. rubrodiscum*
   c. Flower 2 cm or less broad . . . . 4. *N. microphyllum*

1. **N. advena** Ait. Leaves erect, petioles oval in cross section; sepals and fruits rarely marked with red. This species is not sharply differentiated from the next.—Rare in southeastern Wisconsin, more common northwestward, ranging from Walworth and Rock counties to Fond du Lac to Chippewa counties.

2. **N. variegatum** Engelm. (Fig. 175). Leaves mostly floating 7-28 × 11-22 cm, with petioles flattened on upper side;

sepals and fruits marked with red; flowers about 4.5 cm broad; stigma rays usually 12-14.—Widespread in quiet water, throughout the state.

3. N. rubrodiscum Morong. Leaf blades 7.5-20 × 5.5-14.5 cm, the sinus about one-half as long as the midrib, narrow or closed; stigma rays 8-13, usually 10, 11, or 12; young fruit with a ring of decayed stamens.—In the Lake Superior region south to Rusk to Shawano counties.

4. N. microphyllum (Pers.) Fern. Very slender; leaf blades 3.5-10 × 3.5-7.5 cm, the sinus two-thirds or more the length of the midrib; stigma rays 6-10; young fruit without a ring of decaying stamens.—Along our northern borders, from Douglas and Sawyer counties to Lincoln and Vilas counties.

2. Nymphaea   Water Lily

Rhizomes a few cm thick; leaf blades round, the petiole attached almost at the center at the base of a deep notch; sepals 4; petals white or pinkish, the outer longer than the sepals, the inner passing gradually into stamens.

a. Leaves purple or red beneath; petiole not striped . . . .1. *N. odorata*
a. Leaves green beneath; petiole striped . . . . 2. *N. tuberosa*

1. N. odorata Ait. Leaves 1-2 dm wide; sepals and lower surface of leaves often purple; petiole not streaked with purple; flowers fragrant, seldom more than 12 cm broad.—Throughout the state, in quiet shallow water.

2. N. tuberosa Paine. Leaves 2-3 dm wide; sepals and leaves not purple; petiole with purple streaks; flowers not fragrant, 12 cm or more wide.—Throughout the state, in quiet shallow water. Perhaps not distinct from *N. odorata*.

RANUNCULACEAE   Buttercup Family

Mostly herbs; sepals 3-20, often white or colored like petals; petals 0-15; stamens usually many; pistils 1-many; fruit an aggregate of achenes, follicles, or berries.

a. Leaves simple or once compound *b*
   b. Erect herbs *c*
      c. Petals present *d*
         d. Leaves simple, or, if compound, with leaflets longer than
            broad . . . . .1. *Ranunculus*
         d. Leaves with 3 leaflets, each as broad as long . . . 2. *Coptis*
      c. Petals absent; sepals usually petallike (in *Hepatica* 3 green
        bracts beneath the petallike sepals simulate sepals) *e*
        e. Plants leafy-stemmed *f*
          f. Flowers yellow . . . . 3. *Caltha*
          f. Flowers white or greenish *g*

```
            g. Lobes of leaves or leaflets acute h
               h. Sepals 4-20, usually silky or downy beneath ...
                  ............................4. Anemone
               h. Sepals 3, soon falling............5. Hydrastis
            g. Lobes of leaflets rounded ........ 6. Anemonella
         e. Leaves all at the base of the stem i
            i. Leaves simple, with 3 broad lobes....... 7. Hepatica
            i. Leaves compound, or finely divided into many seg-
               ments ...........................4. Anemone
      b. Climbing or trailing vines.....................8. Clematis
a. Leaves twice or more compound j
   j. Petals large and spurred k
      k. All petals spurred; flowers red and yellow..... 9. Aquilegia
      k. Two petals spurred, the petal spurs enclosed by a single
         sepal spur; flowers blue or white..........10. Delphinium
   j. Petals, when present, inconspicuous; sepals often petallike, but
      not spurred l
      l. Flowers many, in a raceme or panicle m
         m. Leaflets sharply toothed ..................11. Actaea
         m. Leaflets bluntly toothed or lobed....... 12. Thalictrum
      l. Flowers solitary or a few together n
         n. Ultimate leaf divisions ribbonlike o
            o. Plants aquatic.....................1. Ranunculus
            o. Plants terrestrial ...................  4. Anemone
         n. Ultimate leaf divisions not ribbonlike p
            p. Stem leaves compound but appearing as a whorl of
               simple leaves below the flowers ...... 6. Anemonella
            p. Stem leaves compound but not appearing whorled q
               q. Leaves with 3 blunt tips.......... 13. Isopyrum
               q. Leaves with serrate points.......... 4. Anemone
```

1. Ranunculus   Crowfoot; Buttercup

Leaves mostly palmately lobed or divided; sepals 5, green or yellowish; petals mostly 5, each with a nectariferous scale at base; stamens few to many; pistils many; fruit an achene, many making up a head.

```
a. Plants aquatic; leaves finely dissected b
   b. Petals white c
      c. Leaves without petioles, stiff, not collapsing when taken
         from the water ...................... 1. R. longirostris
      c. Leaves mostly petioled, soft, collapsing when taken from
         the water ................ 2. R. aquatilis, var. capillaceus
   b. Petals yellow d
      d. Petals 6-17 mm long...................3. R. flabellaris
      d. Petals 3.5-5 mm long.....................4. R. gmelini
a. Plants terrestrial, often in wet places e
   e. Blades of most or all of the basal leaves divided less than
      halfway to the base, or undivided; styles minute, straight or
      curved f
      f. Plants hairy .......................5. R. rhomboideus
      f. Plants not hairy; leaves somewhat shining g
         g. Plants with runners at base; flowering stems almost leaf-
            less ...............................6. R. cymbalaria
         g. Plants without runners at base; flowering stems with
            3-5-parted leaves .....................7. R. abortivus
```

 e. Blades of the basal leaves divided more than halfway to the base, or, if lowest leaves are not so deeply divided, then the styles hooked *h*
  h. Leaves compound, at least the terminal division stalked *i*
   i. Petals bright yellow, much exceeding the sepals *j*
    j. Style 1 mm or more long, slender, straight or curved *k*
     k. Lateral leaflets hardly stalked . . . .8. *R. fascicularis*
     k. Lateral leaflets on stalks about 5 mm long . . . .
     . . . . 9. *R. septentrionalis*
    j. Style short, stout, recurved *l*
     l. Stems prostrate, rooting at the nodes 10. *R. repens*
     l. Stems erect . . . . 11. *R. acris*
   i. Petals pale yellow, hardly exceeding the sepals; plant hairy . . . . 12. *R. pensylvanicus*
  h. Leaves simple, usually deeply cleft, the divisions not stalked *m*
   m. Receptacle cylindrical; head of pistils higher than broad
   . . . . 13. *R. sceleratus*
   m. Receptacle globose; head of pistils about as broad as high *n*
    n. Styles hooked . . . . 14. *R. recurvatus*
    n. Styles straight or curved, not hooked *o*
     o. Stems erect . . . . 11. *R. acris*
     o. Stems prostrate (see *d* above)

1. **R. longirostris** Godr. Stiff Water Crowfoot (Fig. 176). Leaves without petioles, with broad stipules; blades repeatedly forked, the ultimate divisions threadlike; peduncles 1-flowered, opposite the leaves, and the only part emerging from the water, raising the flower a cm or more above the surface.—In quiet or flowing shallow water, throughout the state.

2. **R. aquatilis** L., var. **capillaceus** (Thuill.) DC. White Water Crowfoot (Fig. 177). Similar to No. 1; leaves mostly petioled.—Common in quiet or slowly flowing shallow water, throughout the state.

3. **R. flabellaris** Raf. Yellow Water Crowfoot (Fig. 178). Leaves petioled, dissected, the ultimate divisions mostly flattened; flowers few, mostly terminal. Shallow water. Occasionally out of water or left on the mud of drying ponds, where it becomes forma *riparius* Fern., with firmer leaves, less cut blades, and hairy petioles.—Rock to Racine counties, north to Washburn, Bayfield, and Outagamie counties.

4. **R. gmelini** DC., var. **terrestris** (Gray) Blake. Small Yellow Water Crowfoot (Fig. 179). Similar to the last but smaller throughout.—Rare; from Lake Superior to Door County.

Forma **purshii** (Richards.) Fassett is a submersed form with the leaves cut into threadlike divisions.

5. **R. rhomboideus** Goldie. Dwarf Buttercup (Fig. 180). Plants 1-2 dm high; lowest leaves rounded; upper stem leaves scarcely petioled, 3-5-parted, the divisions ribbonlike; petals large, deep yellow.—Thinly wooded uplands, common from Pierce and Portage counties southward.

6. **R. cymbalaria** Pursh. Seaside Crowfoot (Fig. 181). Plants 4-22 cm high; leaves somewhat fleshy, at the base of the flowering stem, long-petioled, with small roundish coarsely toothed blades; petals yellow.—Rare; from Walworth to Sheboygan counties and Douglas County.

7. **R. abortivus** L. Small-flowered Crowfoot (Fig. 182). Leaf blades rounded, sometimes deeply palmately cleft, shining, slightly fleshy; petals pale yellow, shorter than the downwardly bent sepals.—Very common in clearings and woods, and from dry hillsides to low moist ground.

Var. **acrolasius** Fern. has stems with fine hairs.

8. **R. fascicularis** Muhl. Early Crowfoot (Fig. 183). Plants 1-2.5 dm high, with thickened tuberous roots; leaflets with ovate or linear divisions, which are rounded at tip; head of pistils globose; styles long and slender, straight or curved.—Common on dry hillsides, north to Pierce, Monroe, and Sheboygan counties.

9. **R. septentrionalis** Poir. Swamp Buttercup (Fig. 184). Stems 3-8 dm long, sometimes trailing; leaflets with lanceolate acute divisions; styles long, tapering.—Open ground or wet woods, throughout the state.

Var. **caricetorum** (Greene) Fern. has the petioles and lower internodes densely retrorse-hispid.—Less common.

10. **R. repens** L. Creeping Buttercup. Leaves 3-parted, the leaflets often white-spotted; flowers about 2 cm broad; sepals not reflexed; whole plant somewhat hairy.—Introduced from Europe; damp places, uncommon, from Dane and Green counties, eastward to Milwaukee and Door counties.

11. **R. acris** L. Tall Buttercup (Fig. 185). Plants 6-9 dm high; stems with fine spreading hairs; leaves deeply cut into 3 divisions, these further cut into deep lobes; inflorescence branched; flowers bright yellow, conspicuous.—Naturalized from Europe; roadsides and pastures, throughout the state.

12. **R. pensylvanicus** L. Bristly Crowfoot (Fig. 186). Plants 4-6 dm high; stems bristly with many stout spreading hairs; flowers many, axillary and terminal; pistils in a cylindrical head.—Common in wetland from Rock and Walworth counties, northwestward to Douglas and Vilas counties.

13. **R. sceleratus** L. Cursed Crowfoot (Fig. 187) Plants 1.5-4 dm high; lowest leaves 3-lobed, rounded; upper leaves 3-parted, the divisions wedge-shaped and usually 3-lobed; petals scarcely longer than the calyx.—Common in wetland, from Dane to Portage counties eastward; rare in the west, from La Crosse to Polk counties.

14. **R. recurvatus** Poir. Hooked Crowfoot (Fig. 188). Plants 3-6 dm high; stems with stiff spreading hairs; lowest leaves 3-lobed or rarely 3-parted, the divisions elliptical or wedge-shaped; petals pale, shorter than the downwardly bent sepals.—Occasional in woods; scattered localities, throughout the state except for western Wisconsin.

Fig. 176. *Ranunculus longirostris,*
× .40.

Fig. 177. *Ranunculus aquatilis,*
var. *capillaceus,* × .40.
Fig. 178. *Ranunculus flabellaris,*
× .25.

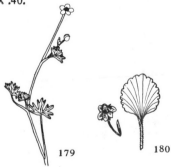

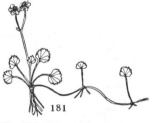

Fig. 179. *Ranunculus gmelini,*
var. *terrestris,* × .40.

Fig. 180. *Ranunculus rhomboideus,* × .50.
Fig. 181. *Ranunculus cymbalaria,*
× .55.

Fig. 182. *Ranunculus abortivus,* × .40, flower × 1.

Fig. 183. *Ranunculus fascicularis,* × .40, flower × .60.

Fig. 184. *Ranunculus septentrionalis,* × .40, flower × .30.

Fig. 185. *Ranunculus acris,* × .40.

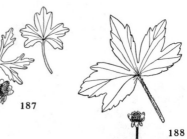

Fig. 186. *Ranunculus pensylvanicus,* flower × 1.
Fig. 187. *Ranunculus sceleratus,* × .40, flower × 1.

Fig. 188. *Ranunculus recurvatus,* × .40, flower × 1.

Fig. 189. *Coptis trifolia*, var.
groenlandica, × .50.

Fig. 190. *Caltha palustris*, × .25.

2. Coptis   Goldthread

Low evergreen perennials; leaves ternately divided; flowers
on short scapes; sepals 5-7; staminodia 5-7, nectariferous in
the hollow summit; stamens 15-25; pistils 3-7, on slender
stalks; fruit follicular.

C. trifolia (L.) Salisb., var. **groenlandica** (Oeder) Fassett
(Fig. 189). Rootstock slender, yellow; leaves all borne near
the ground; petioles long; blades 3-foliate, the leaflets
shining, wedge-shaped at base, slightly 3-lobed, toothed;
sepals white.—Frequent in woods, throughout northern
Wisconsin; southward in bogs to Racine and Green counties.

3. Caltha   Marsh Marigold; Cowslip

Glabrous perennials with alternate simple broad leaves;
sepals 5-9; stamens many, pistils 4-many; fruit, many-seeded,
somewhat flattened follicles.

C. palustris L. (Fig. 190). Plants usually growing in water
or mud; roots thickened; leaf blades kidney-shaped or round
with a deep sinus; flowers several, bright yellow.—Common
in wet places, throughout the state.

4. Anemone   Anemone

Plants with one or more palmately divided leaves at the
base of the stem, and a whorl of leaves (the involucre) at the
base of the peduncle or peduncles; sepals 4-20, petaloid;
stamens many; pistils many; fruit an achene tipped by the
persistent style, which is often woolly or feathery.

a. Leaf segments ribbonlike, not toothed on sides *b*
    b. Sepals 15-40 mm long; styles 2-4 cm in fruit, plumose . . . .
    . . . . 1. *A. patens*, var. *wolfgangiana*
    b. Sepals 5-15 mm long; styles 0.4 cm or less, glabrous . . . .
    . . . . 2. *A. multifida*, var. *hudsoniana*
a. Leaf segments toothed *c*
    c. Sepals 10-20 . . . . 3. *A. caroliniana*

    c. Sepals 4-7 *d*
        d. Sepals hairy on the back *e*
            e. Involucral leaves petioled *f*
                f. All the peduncles naked above the involucre . . . .
                . . . . 4. *A. cylindrica*
                f. All but the first peduncle with secondary involucral
                leaves halfway to the flower *g*
                    g. Anthers 0.7-1.2 mm long . . . . 5. *A. riparia*
                    g. Anthers 1.2-1.6 mm long . . . . 6. *A. virginiana*
            e. Involucral leaves not petioled . . . . .7. *A. canadensis*
        d. Sepals not hairy . . . . 8. *A. quinquefolia*

1. **A. patens** L., var. **wolfgangiana** (Bess.) Koch. Pasque Flower (Fig. 191). Plants covered with long silky hairs; stems from a woody base; sepals 15-40 mm long, purplish or blue to white; stamens accompanied by glandlike staminodia; styles plumose.—On original prairie areas only; mostly southward, but north to Douglas County in the west.

2. **A. multifida** Poir., var. **hudsoniana** DC. (Fig. 192). Plants 1-6 dm high, usually silky-villous; basal leaves with long petioles, deeply 3-parted, the segments deeply incised, entire, ribbonlike; sepals 5-9, white or yellowish, 5-15 mm long; heads of achenes subglobose; achenes densely woolly; styles short, glabrous.—Rocky places and shores, rare; Adams and Sheboygan counties.

3. **A. caroliniana** Walt. Carolina Anemone (Fig. 193). Plants 7-15 cm high, from an ovoid corm; leaves 3-parted, the leaflets often themselves cleft; involucre 3-parted; sepals about 2 cm long, ribbonlike, purple or whitish.—Rare; on prairies and sand terraces in western Wisconsin; La Crosse, Eau Claire, St. Croix, and Pierce counties.

4. **A. cylindrica** Gray. Thimbleweed (Fig. 194). Plants tall, slender, hairy; pistils densely woolly, so that only the tips of the stigmas are visible; achenes woolly, in a dense cylindrical head.—Dry hillsides and woodlands, throughout the state, north to Bayfield and Marinette counties.

5. **A. riparia** Fern. Thimbleweed (Fig. 195). Similar to No. 4; heads broader, the styles ascending.—Throughout the state.

6. **A. virginiana** L. Thimbleweed (Fig. 196). Similar to No. 4; pistils less woolly, so that the whole stigmas are plainly visible; heads ovoid, with spreading styles.—Woodlands, throughout the state.

7. **A. canadensis** L. Canada Anemone (Fig. 197). Plants 2-7 dm tall; sepals white, oval, 1.2-1.8 cm long.—Thickets, meadows, and roadside ditches, throughout the state.

Fig. 191. *Anemone patens,* var.
wolfgangiana, × .40.

Fig. 193. *Anemone caroliniana,*
× .30.

Fig. 192. *Anemone multifida,*
var. *hudsoniana,* × .25.

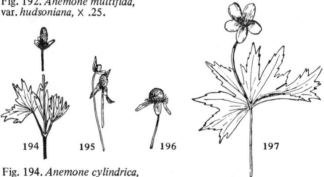

Fig. 194. *Anemone cylindrica,*
× .40.

Fig. 195. *Anemone riparia,* × .40.
Fig. 196. *Anemone virginiana,* × .40.

Fig. 197. *Anemone canadensis,*
× .40.

8. **A. quinquefolia** L., var. **interior** Fern. Wood Anemone
(Fig. 198). Plants slender and delicate, 1-1.5 dm high; leaves
3-parted, the divisions sometimes deeply cleft; flower solitary, white within, the outer surface of the sepals tinged
with red.—Common in woods and brushy pastures, throughout the state.

5. Hydrastis   Golden Seal

Perennial herbs, with thick knotty yellow rhizomes,
which send up in early spring a single leaf and a simple hairy

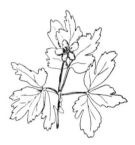

Fig. 198. *Anemone quinquefolia,* var. *interior,* × .40.

stem with 2 leaves and a solitary flower at the summit; sepals 3, petaloid; petals lacking; stamens numerous; pistils numerous with 2 ovules in each; fruit a small head of dark red 1-2-seeded berries.

H. canadensis L. (Fig. 199). Rootstocks yellow; stems 1.5-4 dm high, bearing 2 5-cleft leaves near the summit; flower solitary, greenish white.—Rich woods in areas of limestone, now rather rare; Grant to Walworth counties, north to Vernon and Sheboygan counties.

6. Anemonella   Rue Anemone

Perennial glabrous herbs; leaves ternately compound; flowers in an umbel; sepals 5-10; stamens many; pistils 4-15; fruit ovoid, strongly ribbed, sessile achenes.

A. thalictroides (L.) Spach (Fig. 200). Plants very slender and delicate, 1-3 dm high, from thickened tuberous roots;

Fig. 199. *Hydrastis canadensis,* × .30.

Fig. 200. *Anemonella thalic-troides,* × .50.

leaves 3-parted, the leaflets sometimes again 3-parted; flowers several in an umbel, a whorl of leaves sometimes arising from the pedicels; sepals 1.2 cm long, white or pinkish.—Open woods, throughout southern Wisconsin, north to Columbia and Dodge counties.

Forma **favilliana** Bergseng has many sepals.

7. **Hepatica**   Hepatica

Plants covered with spreading white hairs; blades of leaves heart-shaped at base, persisting throughout the winter, the new leaves coming after the flowers; sepals 5-12, petaloid; a calyxlike involucre of bracts beneath the flowers; flowers white, blue, or pink.

a. Leaves lobed to near the middle, lobes blunt . . . . . 1. *H. americana*
a. Leaves lobed beyond the middle, lobes acute . . . . . 2. *H. acutiloba*

1. **H. americana** (DC.) Ker (Fig. 201). Lobes of leaves, and bracts of the involucre rounded or blunt at tip, the terminal lobe often wider than long.—Woods, usually in acid soil throughout the state.

2. **H. acutiloba** DC. (Fig. 202). Very similar; lobes of leaves and bracts pointed.—Woods, preferably with calcareous soil, throughout the state to Sawyer and Oconto counties.

8. **Clematis**   Virgin's Bower; Clematis

Vines with slightly woody stems; leaves 3-foliate, the leaflets palmately veined; sepals commonly 4, petaloid; petals lacking; stamens many; pistils many; fruits achenes with feathery tails.

Fig. 201. *Hepatica americana,*
× .40.

Fig. 202. *Hepatica acutiloba,*
× .40.

a. Sepals white to yellow; staminodia lacking . . . . 1. *C. virginiana*
a. Sepals blue; staminodia present . . . . 2. *C. verticillaris*

1. **C. virginiana** L. (Fig. 203). Flowers in panicles; sepals 5-8 mm long, whitish; leaflets coarsely toothed.—Shady places, throughout the state.

2. **C. verticillaris** DC. (Fig. 204). Flowers solitary; sepals about 4 cm long, pinkish purple; leaflets slightly or not at all toothed.—Cool rocky woods, not common; scattered localities from Dane to Douglas counties.

9. **Aquilegia** Columbine

Perennial herbs, with alternate 2-3-times ternately compound leaves; sepals 5, regular, colored; petals 5, with blades prolonged backward into a hollow, straight, or curved spur, enlarged and nectar-containing at its end; fruit follicular.

A. canadensis L. (Fig. 205). Stems from a taproot; leaves chiefly at the base of the stem, 3-divided, the leaflets 3-parted, ultimate divisions 3-lobed, the lobes somewhat cut; upper leaves with branches in their axils; flowers nodding; petals scarlet, yellow within, each with a hollow, straight, or curved spur which is enlarged and nectar-containing at its end.—Rocky bluffs and in sandy soil, throughout the state.

10. **Delphinium** Larkspur

Erect annual or perennial herbs, with basal and cauline palmately divided leaves; flowers irregular; sepals 5, petallike, the upper one prolonged into a spur; petals 2 or 4, the upper pair each with long spurs which prolong backward and are enclosed in the spur of the calyx, the lower pair with short claws; stamens numerous; pistils 1-5; fruit follicular.

Fig. 203. *Clematis virginiana,*
× .20.

Fig. 204. *Clematis verticillaris,*
× .40.

Fig. 205. *Aquilegia canadensis,*
× .40.   Fig. 206. *Actaea pachypoda,* × .20.

a. Petals 4, separate; carpels 3-5; perennial . . . . .1. *D. virescens*
a. Petals 2, united to 1 body; carpel 1; annual . . . . 2. *D. ajacis*

1. **D. virescens** Nutt. Prairie Larkspur. Flowers white or bluish; spur once-and-a-half to twice the length of the upper sepal; ultimate segments of leaves narrowly linear; stems to 1.5 m high.—Native western perennial; prairies, barrens, and dry, open woods; Jackson and La Crosse counties, northwestward to St. Croix, Dunn, and Oneida counties.

2. **D. ajacis** L. Leaves with narrowly linear to filiform segments; flowers blue, violet, purple, or pink.—Escaped from cultivation; Milwaukee County.

11. Actaea   Baneberry

Stems 3-8 dm high, from stout rootstocks; leaves pinnately divided, the leaflets themselves pinnately divided, the ultimate divisions sharply and coarsely toothed or deeply cut; flowers small, in a raceme

a. Pedicels long and slender, less than 1 mm thick . . . . . . .1. *A. rubra*
a. Pedicels short, 1-2 mm thick . . . . 2. *A. pachypoda*

1. **A. rubra** (Ait.) Willd. Red Baneberry. Raceme about as broad as long; petals merely bluntly pointed; berries (which come in summer) red or rarely white.—Rich woods, throughout the state.

2. **A. pachypoda** Ell. White Baneberry (Fig. 206). Raceme much longer than broad; petals tipped by a brown antherlike body; berries (which come in summer) white or rarely pink.—Rich woods, throughout the state.

Fig. 207. *Thalictrum dioicum,* × .10.

Fig. 208. *Thalictrum dasycarpum,* × .10.

12. Thalictrum   Meadow Rue

Leaves 3-parted, the divisions again 3-parted; staminate and pistillate flowers on different plants; sepals petallike, soon dropping; petals none; stamens or pistils making up the conspicuous part of the flower.

a. Plants without rootstocks *b*
  b. Leaves with petioles ....1. *T. dioicum*
  b. Leaves without petioles, i.e., the 3 divisions arising close to the stem *c*
    c. Leaflets with fine hairs beneath ....2. *T. dasycarpum*
    c. Leaflets with waxy glands beneath ....
      ....3. *T. revolutum,* var. *glandulosior*
a. Plants with stout rootstocks (rare species) *d*
  d. Stigma 2-5 mm long; anthers 2-4 mm long with a tip 0.4-1 mm long; achene body 4-6 mm long ....4. *T. confine*
  d. Stigma 1-2.5 mm long *e*
    e. Peduncles elongate, averaging 2-3 cm, 3-5 per node; anthers with pointed tips less than 1 mm long; achene body 4-4.5 mm long ....5. *T. turneri*
    e. Peduncles averaging less than 2 cm long, 2 per node; anthers with pointed tips 0.1-0.4 mm long; achene body 3-4 mm long ....6. *T. venulosum*

1. **T. dioicum** L. Early Meadow Rue (Fig. 207). Plants 3-6 dm high, smooth and pale; leaflets drooping, the divisions rounded, 3-7 lobed.—Rich woods; throughout the state, north to Washburn, Lincoln, and Marinette counties.

2. **T. dasycarpum** Fisch. & Lall. (Fig. 208). Plants 6-12 dm high, usually purplish; divisions of leaflets oblong, mostly 3-toothed.—Wet meadows or open ground, throughout the state.

Var. **hypoglaucum** (Rydb.) Boivin. Plants glabrous.

3. **T. revolutum** DC., var. **glandulosior** Boivin. Similar to No. 2; leaflets thicker, heavy-scented, with the main branches of the compound leaves glandular.—Rare, from

Fig. 209. *Isopyrum biternatum*, × .40.

Waukesha and Walworth counties eastward.

4. **T. confine** Fern. Characters as in key above.—Brown and St. Croix counties.

5. **T. turneri** Boivin. Characters as in key above.—Woods and stream bottoms; Wisconsin Point, Douglas County.

6. **T. venulosum** Trel. Characters as in key above.—Open areas; Marinette County.

13. Isopyrum   False Rue Anemone

Perennial glabrous herbs, with basal and alternate cauline leaves, 1-3 ternately compound; flowers terminal and axillary; sepals 5; stamens numerous; pistils few.

I. biternatum (Raf.) T. & G. (Fig. 209). Plants to 4 dm high; stems several, from tufted irregularly thickened roots; flowers white.—Moist woods; Grant to Racine counties, north to Polk to Lincoln counties.

BERBERIDACEAE   Barberry Family

Sepals in 2 rows (except in *Jeffersonia*); stamens as many as the petals and opposite them (except in *Podophyllum*); anthers opening by valves or lids; style short or none; fruit a pod or berry.

a. Herbs *b*
    b. Flowers solitary, white *c*
        c. Leaves deeply palmately cleft (Fig. 210) . . 1. *Podophyllum*
        c. Leaves divided into 2 half-ovate leaflets (Fig. 211) . . . .
        . . . . 2. *Jeffersonia*
    b. Flowers in a raceme, green . . . . 3. *Caulophyllum*
a. Shrubs . . . . .4. *Berberis*

1. Podophyllum   Mayapple; Mandrake

Perennial herbs, from rhizomes, with single large peltate leaves, growing in colonies, the intermingled flowering stems with a pair of petiolate half-round or unequally peltate

Fig. 210. *Podophyllum peltatum,* × .10.

Fig. 211. *Jeffersonia diphylla,* × .30.

leaves, similarly lobed but smaller and with a terminal single flower; sepals 6, falling early; petals 6-9; stamens (in our species) twice as many as the petals; ovary thick, ovoid, with a large sessile stigma; fruit a large ovoid, fleshy, many-seeded berry.

P. peltatum L. (Fig. 210). Single white flower 3-5 cm wide on a nodding peduncle, borne between the 2 leaves; mature fruit yellow, 4-5 cm long. Herbage, seeds, and immature fruits poisonous, mature fruits edible.—Common in rich woods and pastures, north to Marathon, Oconto, and Door counties.

2. Jeffersonia   Twinleaf

Smooth perennial herbs, with long-petioled, deeply 2-parted, basal leaves; flowers solitary on leafless scape; sepals 4, early deciduous; petals 8; stamens 8; ovary with many ovules and a broad sessile stigma; fruit a capsule, splitting halfway around horizontally.

J. diphylla (L.) Pers. (Fig. 211). Leaves and peduncle all from the ground level; flowers white, 2.5 cm broad; sepals falling when buds open.—Southern Wisconsin, rare in rich woods, Kenosha to Kewaunee counties along Lake Michigan, and Grant County.

3. Caulophyllum   Blue Cohosh; Papooseroot

Smooth perennial herbs, with a large sessile triternate leaf (appearing as 3 biternate leaves) above the middle of the stem and another smaller leaf just below the panicle inflores-

Fig. 212. *Caulophyllum thalictroides,* × .20.

cence; sepals 6, petallike and subtended by 3-4 sepallike bracts; petals 6, reduced to small glandlike bodies; stamens 6; ovules 2; ovary bursting early, exposing the 2 round, stalked, blue seeds resembling berries.

C. thalictroides (L.) Michx. (Fig. 212). Plants 3-7.5 dm high, from matted rootstocks, glaucous when young; leaflets obovate-oblong, 2-5-lobed, with petioles; flowers yellowish green to greenish purple, about 1 cm wide.—Rich woods, throughout the state.

4. Berberis   Barberry

Shrubs, with yellow wood; branches armed with thorns (which are morphologically leaves); sepals 6; petals 6, yellow; stamens 6, anthers sunk in pockets in the petals, springing toward the pistil when disturbed; fruit a 1-few-seeded berry.

a. Leaf margins serrate . . . . 1. *B. vulgaris*
a. Leaf margins entire . . . . 2. *B. thunbergii*

1. **B. vulgaris** L. Common Barberry (Fig. 213). Leaves elliptical on a short jointed petiole, bristle-toothed along the margins; flowers in drooping racemes.—Fields and open woods; escaped from cultivation; a native of Europe, and the alternate host of the wheat rust. After a major eradication program, this barberry is now very rare.

2. **B. thunbergii** DC. Japanese Barberry (Fig. 214). Leaves obovate, the margins entire; flowers 1-3 in a bunch.—Widely cultivated and occasionally escaping; a native of Japan, and not a host of the wheat rust; southern Wisconsin, north to

Fig. 213. *Berberis vulgaris,* × .40.   Fig. 214. *Berberis thunbergii,* × .40.

Brown County in the east and to Pierce County in western Wisconsin.

MENISPERMACEAE   Moonseed Family

Dioecious vines, with alternate simple, mostly palmately veined leaves; sepals and petals similar; stamens 12-24 (our species); carpels separate, usually 3 with 1 ovule in each; fruit a drupe.

Menispermum   Moonseed

Leaves peltate near the margin; sepals 4-8; petals 6-8; stamens 12-24; pistils 2-4; fruit a spherical drupe, with a crescent-shaped, flattened stone.

M. canadense L. Leaf blades 0.5-1.5 dm broad, broadly heart-shaped at base, 3-7-angled; flowers in dense panicles.— Thickets and river bottoms, north to Pierce, Marathon, and Manitowoc counties.

PAPAVERACEAE   Poppy Family

Rather delicate herbs, with colored juice; sepals 2, falling as the flowers expand; petals 4-12; stamens many; fruit a capsule.

a. Plants with 1 flower appearing before the palmately lobed basal leaf . . . . 1. *Sanguinaria*
a. Flowers in an umbel, coming after the leaves; leaves pinnately divided . . . . 2. *Chelidonium*

1. Sanguinaria   Bloodroot

Low perennials, from a thick rhizome which sends up solitary leaves with stout petioles and solitary flowers on naked scapes; the entire plants with abundant red-orange juice; sepals 2, falling early; petals 8-12; stamens many; ovary 1, with a short style and capitate 2-lobed stigma; fruit a many-seeded capsule

Fig. 215. *Sanguinaria canadensis,* × .40.

S. canadensis L. (Fig. 215). Leaf blade palmately 3-9-lobed, round in outline, 2 dm wide at maturity; peduncle slender, 1-flowered, a little taller than the leaf, white, about 2 cm wide.—Rich woods, throughout the state.

2. Chelidonium   Celandine

Biennials, with yellow juice; leaves pinnately divided, the divisions with rounded teeth; flowers in an umbel; sepals 2, early deciduous; petals 4, yellow; stamens many; ovary with a 2-lobed stigma; fruit a smooth 2-valved capsule.

C. majus L. Plants 3-8 dm high; leaves glaucous; flowers about 1 cm wide; capsules 3-5 cm long.—An uncommon weed, naturalized from Europe, sometimes cultivated; southern Wisconsin from Lafayette, Green, and Dane counties, north to Vernon and Sauk counties.

FUMARIACEAE   Fumitory Family

Delicate herbs; leaves 3-divided, the leaflets pinnately divided, these secondary leaflets themselves cut into lanceolate divisions; sepals 2, small and scalelike; petals 4, in 2 pairs; stamens in 2 sets of 3 each; fruit a long pod.

a. Flowers with 2 petals prolonged into spurs (Fig. 216) . . . 1. *Dicentra*
a. Flowers with but 1 spurred petal (Fig. 218) . . . . .2. *Corydalis*

1. Dicentra

Plants about 2 dm high; flower and leaf stems arising from a cluster of small white (yellow in *D. canadensis*) bulb-

Fig. 216. *Dicentra cucullaria,*
× .40.

Fig. 217. *Dicentra canadensis,*
× .60.

like bodies that develop from thickened leaf bases; flowers 2-many, on short pedicels from a common axis; outer petals rather inflated, prolonged into spurs at base, the apex of each turned back; inner petals hidden except for their jointed tips which appear between the turned-back tips of the outer.

a. Flowers with long triangular spurs . . . . .1. *D. cucullaria*
a. Flowers with short rounded spurs (Fig. 217) . . . . . 2. *D. canadensis*

1. **D. cucullaria** (L.) Bernh. Dutchman's Breeches (Fig. 216). Plants from a cluster of white grainlike bulbs; flowers white or pink.—Woods, from Grant to Kenosha counties, north to Sawyer and Florence counties.

2. **D. canadensis** (Goldie) Walp. Squirrel Corn (Fig. 217). Plants from yellow bulblike bodies resembling grains of corn along a threadlike rhizome; flowers greenish white or rose-tinged, fragrant.—Locally abundant in rich woods, from Grant to Dane and Milwaukee counties, north to Taylor, Lincoln, and Forest counties.

2. Corydalis

Our species biennial herbs, with alternate twice pinnately dissected leaves, pale or glaucous; flowers in racemes, zygomorphic; sepals 2; petals 4, the upper spurred at base; stamens 6; fruit a slender many-seeded capsule.

a. Corolla pink or purplish, with yellow tips . . . . . 1. *C. sempervirens*
a. Corolla yellow throughout *b*
  b. Corolla 8 mm long . . . . 2. *C. micrantha*
  b. Corolla 12 mm long . . . . 3. *C. aurea*

Fig. 218. *Corydalis sempervirens,* × .40.

1. **C. sempervirens** (L.) Pers. Pale Corydalis (Fig. 218). Plants 1-6 dm high, often much branched; stems and leaves bluish green, with a whitish bloom.—Rocky places and clearings, from Iowa and Dane counties, north to Douglas to Forest counties.

2. **C. micrantha** (Engelm.) Gray. Similar to the next, but with smaller flowers; the ascending pods less than 1 cm long.—A very local adventive, in Rock and Dane counties and along the Mississippi River from Grant to Pierce counties.

3. **C. aurea** Willd. Golden Corydalis. Plants 1-3.5 dm high, much branched; pods spreading or pendulous, 1.5-2 cm long.—Dry soils, in scattered locations throughout the state, with several locations in Door County.

BRASSICACEAE (CRUCIFERAE) Mustard Family

Herbs; leaves alternate or, in *Dentaria*, apparently whorled; flowers mostly in racemes; sepals 4; petals 4 (rarely none), bent outward at the middle, the spreading parts forming a cross; stamens 6 (rarely only 2) of 2 long and 1 short pairs; fruit a pod of 2 carpels. Classification of this large family is usually based on the shape of the capsule, which occurs in 2 principal forms unique to the family: (1) a silique—an elongate capsule in which the 2 carpels are separated by outer faces split off from a thin longitudinal partition known as the replum, and (2) a silicle—a shortened silique. In certain species the fruits do not open. The following key is based on flowering or young fruiting material, and thus it is purely artificial and not always reliable; specimens with mature fruit may best be identified by the use of a more technical manual.

a. Petals white or purplish, sometimes yellow at base, or absent *b*
  b. Petals 5-15 mm long *c*
    c. Leaves or leaflets pointed *d*
      d. Leaves simple, scattered on the stem; large plants *e*
        e. Flowers purplish; leaves with regular, short, sharp-pointed teeth; no long strap-shaped basal leaves . . . . 1. *Hesperis*
        e. Flowers white; leaves with irregular, long, rounded teeth; many long strap-shaped basal leaves . . . . 2. *Armoracia*
      d. Leaves palmately compound, 2 or 3 only . . . . 3. *Dentaria*
    c. Leaves or leaflets, except the uppermost, rounded *f*
      f. Stem leaves tapered or petioled at base . . . . 4. *Cardamine*
      f. Stem leaves sessile and auricled . . . . 5. *Conringia*
  b. Petals smaller, or absent (sometimes 8 mm long in *Arabis*) *g*
    g. Leaves (at least the basal) pinnate or pinnatifid, cut at least halfway to the midrib *h*
      h. Leaves on the stem mostly simple, the pinnatifid ones confined to a basal rosette *i*
        i. Stem leaves prolonged at base into triangular auricles (Fig. 225) . . . . .6. *Capsella*
        i. Stem leaves without auricles *j*
          j. Petioles of basal leaves hairy . . . . .7. *Arabis*
          j. Petioles of basal leaves without hairs . . 8. *Lepidium*
      h. Stem leaves pinnate *k*
        k. Leaves in the inflorescence pinnate *l*
          l. Leaflets much longer than broad, often with straight sides (Fig. 224) . . . . 4. *Cardamine*
          l. Leaflets about half as broad as long, rounded (Fig. 230) . . . . 9. *Nasturtium*
        k. Leaves in the inflorescence simple, sharply toothed . . . . 10. *Rorippa*
    g. Leaves simple, sometimes toothed, but not cut halfway to the midrib *m*
      m. Plants small, with basal rosette *n*
        n. Flowering stem either leafless or with toothed leaves; fruit flattened, long-ovate . . . . 11. *Draba*
        n. Flowering stem with a few, mostly entire, linear leaves; fruit cylindrical, linear . . . . 12. *Arabidopsis*
      m. Plants larger, stems leafy *o*
        o. Leaves not toothed *p*
          p. Sepals and rachis of racemes whitened with branched or stellate hairs *q*
            q. Fruits circular, winged; styles about 1 mm long . . . . 13. *Alyssum*
            q. Fruits longer than wide, not winged; styles about 2 mm long . . . . 14. *Berteroa*
          p. Sepals and rachis not whitened, with few spreading hairs or none *r*
            r. Young fruit very narrow, the ovary hardly thicker than the stigma . . . . 7. *Arabis*

      r. Young fruit over half as broad as long, the ovary much thicker than the stigma 15. *Camelina*
   o. Leaves toothed or wavy-margined *s*
      s. Leaves roundish, wavy-margined . . . . 4. *Cardamine*
      s. Leaves with definite teeth *t*
         t. Basal rosette present, and with branched hairs *u*
            u. Anthers as broad as long . . . . 11. *Draba*
            u. Anthers much longer than broad . . 7. *Arabis*
         t. Basal rosette absent, or if present with simple hairs or none *v*
            v. Young fruit very narrow, the ovary hardly thicker than the stigma *w*
               w. Leaves much longer than wide . . 7. *Arabis*
               w. Leaves triangular, about as wide as long . . . . .16. *Alliaria*
            v. Young fruit over half as broad as long, the ovary much thicker than the stigma *x*
               x. Upper leaves prolonged at base into auricles *y*
                  y. Plants without hairs . . . . . . 17. *Thlaspi*
                  y. Plants pubescent *z*
                     z. Fruits flattened, cordate at top . . . . 8. *Lepidium*
                     z. Fruits not flattened, cordate at base . . . . 18. *Cardaria*
               x. Leaves without auricles *zz*
                 zz. Most of the leaves 1 dm or more long . . . . .10. *Rorippa*
                 zz. Leaves less than 1 dm long . . . . 8. *Lepidium*

a. Petals yellow *b*
  b. Leaves simple, not deeply cleft *c*
   c. Leaves not prolonged backward into lobes or auricles *d*
    d. Fruit elongate, elliptical to linear *e*
      e. Flowering stem arising from a small basal rosette . . . . 11. *Draba*
      e. Flowering stem with no basal rosette or very large basal rosette *f*
         f. Axis of inflorescence roughened with many short appressed hairs . . . . .19. *Erysimum*
         f. Axis of inflorescence smooth or with a few soft spreading hairs . . . . 20. *Brassica*
    d. Fruit small, orbicular, with sharp edge . . . . 13. *Alyssum*
   c. Base of leaves prolonged backward into lobes or auricles *g*
    g. Lobes rounded . . . . 5. *Conringia*
    g. Lobes pointed *h*
      h. Young fruit very narrow, the ovary hardly thicker than the stigma . . . . .7. *Arabis*
      h. Young fruit over half as broad as long, the ovary much thicker than the stigma *i*
         i. Sepals 2.5-3.5 mm long . . . . .15. *Camelina*
         i. Sepals 1-1.5 mm long . . . . .21. *Neslia*
  b. Leaves pinnately divided, or cleft at least halfway to the midrib *j*
  j. Lobes of the leaves rounded *k*
    k. Uppermost leaves clasping, stem and leaves glabrous above . . . . 22. *Barbarea*

1. Hesperis   Dame's Violet; Dame's Rocket

Erect perennials or biennials, pubescent throughout; leaves simple, lanceolate; petals up to 15 mm long; siliques up to 14 cm long.

H. matronalis L. (Fig. 219). Tall plants, with toothed leaves and many showy purple (or sometimes white) flowers. —Garden flowers, introduced from Europe, escaping and becoming locally abundant along roadsides, from Green to Racine counties, north to Pierce to Door counties.

Fig. 219. *Hesperis matronalis,* × .20.

Fig. 220. *Armoracia rusticana,* × .25.

2. Armoracia   Horseradish

Smooth perennials, with thick vertical roots; petals white; stigma large, capitate; fruit inflated, ellipsoid or obovoid; style and stigma persistent.

A. rusticana Gaertn., Mey. & Scherb. (Fig. 220). Basal leaves long-petioled, blade up to 4 dm long, wavy-margined, toothed; upper stem leaves toothed, lower stem leaves pinnately lobed; petals 5-8 mm long; fruits 3-4 mm long.—Native of Europe; often cultivated; southern and eastern Wisconsin, north to Wood and Langlade counties.

3. Dentaria   Toothwort

Perennial plants 1.5-4 dm high, from thick white rootstocks; stems with 2 or 3 leaves borne above the middle; flowers white or purplish; pods flat, lanceolate; seeds in 1 row in each carpel.

a. Leaves divided into 4-7 linear to lanceolate segments 1. *D. laciniata*
a. Leaves divided into 3 ovate segments *b*
   b. Hairs on leaflet margin 0.1 mm long, appressed; leaves usually 2 . . . . .2. *D. diphylla*
   b. Hairs on leaflet margin 0.2-0.3 mm long, spreading; leaves usually 3 . . . . .3. *D. maxima*

1. **D. laciniata** Muhl. (Fig. 221). Leaves 3, in a single whorl, each palmately 3-cleft, the 2 lateral leaflets usually

Fig. 221. *Dentaria laciniata,* × .50.   Fig. 222. *Dentaria diphylla,* × .25.

deeply 2-cleft.—Common in rich woods, north to St. Croix, Ashland, and Florence counties.

2. **D.** diphylla Michx. (Fig. 222). Leaves 2, 3-parted, the leaflets one-third to one-half as broad as long.—Rare in rich woods, south to Shawano and Sheboygan counties.

3. **D.** maxima Nutt. Very similar to *D. diphylla*; rhizome interrupted by abrupt constrictions; leaves commonly 3, alternate, the 3 leaflets more sharply, deeply, and irregularly toothed.—Rare, Ashland County.

4. Cardamine   Bitter Cress

Plants with few or no hairs; leaves simple to compound; flowers white to pink or purplish; fruit linear, flattened; seeds in 1 row in each carpel.

a. Leaves simple; plant with a bulb *b*
   b. Petals white . . . . 1. *C. bulbosa*
   b. Petals rose purple . . . . 2. *C. douglassii*
a. Leaves pinnately compound; plant without a bulb *c*
   c. Flowers 1 cm or more broad . . . . . 3. *C. pratensis*, var. *palustris*
   c. Flowers 4 mm or less broad *d*
      d. Leaflets of the stem leaves very narrow . . . .
      . . . . 4. *C. parviflora,* var. *arenicola*
      d. Leaflets ovate or roundish . . . . 5. *C. pensylvanica*

1. **C.** bulbosa (Muhl.) BSP. Spring Cress (Fig. 223). Plants about 3 dm high, unbranched; leaves 5-8, scattered, the lower sometimes petioled; petals 7-10 mm long.—Common in marshy land, north to St. Croix to Brown counties.

2. **C.** douglassii Britt. Similar to No. 1; stems somewhat hairy; petals 1-1.8 cm long.—Occasional in shady places, southeastern Wisconsin from Jefferson and Racine counties to Winnebago and Sheboygan counties; also Marinette County.

Fig. 223. *Cardamine bulbosa,*
× .25.

Fig. 224. *Cardamine parviflora,*
var. *arenicola,* × .40.

3. **C. pratensis** L., var. **palustris** Wimm. & Grab. Cuckoo Flower. Segments of the lower leaves rounded, those of the upper leaves narrow; petals white.—Rare; in swamps, or sometimes in dry places, from Dane, Waushara, and Waupaca counties, eastward to Milwaukee to Door counties; also St. Croix County.

4. **C. parviflora** L. var. **arenicola** (Britt.) Schulz (Fig. 224). Plants slender; flowers very small, white; leaflets not running together on the midrib.—Occasional in wet places, from Green north to Sauk to Clark counties and westward to Polk County.

5. **C. pensylvanica** Muhl. Similar to No. 4, but larger; leaflets tending to run together on the midrib.—Frequent in wet places, throughout the state.

5. Conringia   Hare's-ear Mustard

Annual or biennial herbs; leaves simple, entire, with auriculate basal lobes; petals yellow to white; fruit a silique.

C. orientalis (L.) Dumort. Plants reaching 7 dm in height;

Fig. 225. *Capsella bursa-pastoris,* × .20, silicle × 1.

leaves smooth and somewhat fleshy, rounded at tip; pod 4-angled, reaching 1 dm in length; seeds in 1 row in each cell. — Waste places, occasional from Dane to Waupaca counties and eastward, and Barron County north to Douglas and Ashland counties; adventive from Europe.

6. Capsella   Shepherd's Purse

Annual or biennial herbs, with basal rosettes; plants pubescent, with stellate hairs; petals white, 2-4 mm long; fruit a silicle, flattened contrary to the partition, triangular.

C. bursa-pastoris (L.) Medic. (Fig. 225). Leaves of the basal rosette variable, mostly deeply cleft; those of the stem much smaller, entire; pod 5-8 mm long, shaped like an isosceles triangle with the pedicel at the apex, and the base somewhat notched; seeds numerous in each cell. — A weed everywhere; naturalized from Europe.

7. Arabis   Rock Cress

Annuals to perennials, with basal rosettes, usually pubescent; leaves simple to pinnately cleft; flowers usually white to yellowish or purplish; fruit a linear silique, flattened (except in No. 3).

a. Stem flexuous, branching at base *b*
   b. Stem without hairs above ....1. *A. lyrata*
   b. Stem rough-hairy above ....2. *A. shortii*
a. Stem stiff, straight, unbranched below *c*
   c. Basal leaves with simple or 2-branched hairs, or none *d*

<pre>
  d. Flowers yellowish . . . . .3. A. glabra
  d. Flowers greenish, pinkish, or white e
    e. Leaves not long-pointed with large auricles at base f
      f. Stem without hairs, except sometimes at base g
        g. Stem leaves without hairs h
          h. Petals pinkish; basal leaves toothed . . . .
                . . . . 4. A. drummondii
          h. Petals creamy white; basal leaves lobed one-
              third to halfway to the midrib . . . .
                . . . . 5. A. missouriensis, var. deamii
        g. Stem leaves with short hairs . . . . . 6. A. canadensis
      f. Stem rough-hairy . . . . .7. A. hirsuta
    e. Leaves long-pointed and with large auricles at base . . . . .
          . . . . 8. A. laevigata
  c. Basal leaves with 3-parted hairs . . . . .9. A. divaricarpa
</pre>

1. **A. lyrata** L. (Fig. 226). Basal rosette of many small, sometimes hairy, pinnatifid leaves, with the terminal lobe much the largest; stem much branched, smooth; stem leaves simple, lanceolate, entire or pinnately lobed; flowers small, white; seeds very small, in 1 row in each cell.—On sandstone and sand, throughout the state.

2. **A. shortii** (Fern.) Gl. Stem leaves rather broad, shallowly toothed; pods widely spreading.—Rich woodlands, not common; Grant to Dane and Jefferson counties, also in St. Croix County.

3. **A. glabra** (L.) Bernh. Tower Mustard. Plants somewhat powdery-whitened; leaves smooth, the midrib conspicuous, but not the branch veins; pods 8 cm long; seeds in 1 row in each cell.—Dry places, throughout the state.

4. **A. drummondii** Gray (Fig. 227). Similar to No. 3; stem leaves toothed or entire; seeds in 2 rows in each cell.—Dry hills; local, from Iowa to Juneau counties along the Wisconsin River, Pierce and Fond du Lac counties to Brown and Door counties.

5. **A. missouriensis** Greene, var. **deamii** M. Hopkins. Stems slightly hispid at base, glabrous above, with 19-50 nodes to the first flower or branch.—A northern species of rocky places, from Washburn to Marinette counties, south to Columbia and Fond du Lac counties.

6. **A. canadensis** L. Sicklepod (Fig. 228). Leaves flat, 1.5-3.5 cm broad; pods 4 mm broad, hanging downward.—Rich woods and bluffs, mostly south of the Tension Zone, from Polk, Rusk, and Brown counties southward.

7. **A. hirsuta** (L.) Scop. Hairy Rock Cress. Stems slender, erect, sometimes branching; rosette leaves stellate-pubescent

Fig. 226. *Arabis lyrata,* × .25.
Fig. 227. *Arabis drummondii,* × .40.
× .40.

Fig. 228. *Arabis canadensis,* × .40.

on both surfaces; pods upright, seeds in 1 row in each cell.

Var. **pycnocarpa** (M. Hopkins) Rollins. Stem pubescence mostly of simple spreading stiff hairs; stem leaves often heavily pubescent.—Calcareous rocks, not common; Grant and Lafayette counties, east to Racine County, and north to Door County.

Var. **adpressipilis** (M. Hopkins) Rollins. Stem pubescence mostly of appressed forked hairs; stem leaves often nearly glabrous.—Woods, calcareous rocks, and pastures; Grant County north to Polk County, Grant County east to Walworth and Milwaukee counties, and north to Sauk and Shawano counties; also Price County.

8. **A. laevigata** (Muhl.) Poir. Leaves very long and tapering; pods long and narrow, spreading; seeds in 1 row in each cell.—Woods and shaded rocky hillsides; common from

Barron and Chippewa counties with range mostly south of Tension Zone, and along Lake Michigan to Manitowoc and Brown counties.

9. **A. divaricarpa** Nels. Plants erect, to 1 m tall; basal leaves with fine stellate pubescence on both sides, stem leaves glabrous and with auriculate bases.—Rocky banks; local, throughout the state except for the Driftless Area.

8. Lepidium   Peppergrass

Annual or perennial herbs; flowers small, on elongate racemes with spreading pedicels; petals white or greenish; fruit a small round silicle, flat, about 3 mm broad, each cell 1-seeded.

a. Leaves tapered to the base *b*
    b. Leaves simple, toothed *c*
        c. Petals shorter than the sepals or absent . . . 1. *L. densiflorum*
        c. Petals longer than the sepals . . . . 2. *L. virginicum*
    b. Leaves twice pinnately divided . . . . 3. *L. ruderale*
a. Leaves broadest at base *d*
    d. Leaves embracing the stem . . . . 4. *L. perfoliatum*
    d. Leaves with a lobe on each side of the stem . . . 5. *L. campestre*

1. **L. densiflorum** Schrad. (Fig. 229). Pods slightly narrower than long; seed with back of one cotyledon against the embryonic root.—Naturalized from Europe; a common weed, throughout the state.

Fig. 229. *Lepidium densiflorum*, × .25.

2. **L. virginicum** L. Pods as wide as long; seed with edges of both cotyledons against the embryonic root.—Waste places, southern Wisconsin, north to St. Croix and Marinette counties.

3. **L. ruderale** L. Upper leaves lobed nearly to the midrib.—Adventive from Europe; a weed in Milwaukee County.

4. **L. perfoliatum** L. Upper leaves nearly as broad as long; lower leaves much divided.—Adventive from Europe; a rare weed in southeastern Wisconsin, from Walworth to Sheboygan counties.

5. **L. campestre** (L.) R. Br. Plants minutely downy; fruits slightly cupped.—Naturalized from Europe; abundant in cultivated fields in southern Wisconsin, north locally to Langlade and Door counties.

9. Nasturtium   Water Cress

Perennial aquatic plants, smooth; leaves pinnately compound; petals white; fruit a silique.

N. officinale R. Br. (Fig. 230). Plants somewhat fleshy; leaflets often rounded, the edges not toothed, the terminal usually larger than the lateral ones.—A European species, cultivated as a salad plant, now naturalized and found commonly in springs and along streams, especially in southern Wisconsin and locally to Douglas and Ashland counties. Not related to the "Nasturtium" (*Tropaeolum*) of flower gardens.

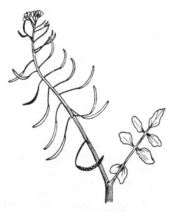

Fig. 230. *Nasturtium officinale,* × .50.

Fig. 231. *Rorippa sylvestris,* × .40.   Fig. 232. *Rorippa islandica,* × .40, silicle × .80.

10. Rorippa   Yellow Cress

Aquatic to terrestrial plants; leaves pinnate to pinnatifid or dentate; petals yellow; fruit a short silique or silicle, slender to globular.

a. Leaves simple, dentate . . . . .1. *R. austriaca*
a. Leaves pinnate or pinnatifid *b*
 b. Petals 4-6 mm long, exceeding the sepals . . . . . . 2. *R. sylvestris*
 b. Petals 1-2 mm long, equaling or shorter than the sepals . . . .
 . . . . .3. *R. islandica*

1. **R. austriaca** (Crantz) Bess. Austrian Cress. Perennials, to 1 m tall; leaves oblanceolate, tapering to an auriculate base, dentate to entire; flowers small, 3-5 mm long; fruit globose, 1-3 mm long; style and stigma persistent.—Native of Europe; Dane and Rock counties.

2. **R. sylvestris** (L.) Bess. Creeping Yellow Cress (Fig. 231). Leaflets elliptical, toothed, the terminal on stem leaves, about as large as the lateral; on leaves at the base, the terminal leaflet often larger.—Introduced from Europe; damp ground, Grant to Green Lake and Brown counties and eastward; also Lincoln and Pierce counties.

3. **R. islandica** (Oeder) Borbas. Marsh Cress (Fig. 232). Lower leaves pinnately cleft almost to the midrib; upper leaves less deeply cleft; stems with scattered spreading hairs; petals yellow; fruit almost globose, about 2 mm broad.

Var. **hispida** (Desv.) Butt. & Abbe. Stems hispid nearly throughout, with stiff spreading hairs.—Local, throughout the state in wet places.

Var. **fernaldiana** Butt. & Abbe. Stems glabrous; pods about twice as long as broad.—Common in wet places, throughout the state.

11. Draba

Small herbs; leaves mostly in a basal rosette; flowers on

Fig. 233. *Draba reptans,* × .40, silicle × .60.

small racemes; petals white or yellow; fruit narrowly oblong, flattened parallel to the partition; seeds in 2 rows in each compartment, marginless.

a. Petals round *b*
   b. Style none *c*
      c. Leaves not toothed .... 1. *D. reptans*
      c. Leaves few-toothed ....2. *D. nemorosa*
   b. Style well developed *d*
      d. Pods without hairs ....3. *D. arabisans*
      d. Pods with minute starlike hairs .... 4. *D. lanceolata*
a. Petals cleft .... 5. *D. verna*

1. **D. reptans** (Lam.) Fern. (Fig. 233). Leaves 1 dm or less long, not toothed, whitened by close matted hairs; peduncle slender, hairy, bearing a few leaves toward the base; flowers on short pedicels on a short common axis which elongates as the fruits mature; petals white.—Rocky or dry sandy soil; south of the Tension Zone, except for Manitowoc County.

Var. **micrantha** (Nutt.) Fern., with pods finely hairy, is rare from Dane County southeastward.

2. **D. nemorosa** L. Stems leafy to the base of the inflorescence; leaves 4-8 mm long, ovate; petals yellow.—Open sandy soil; Marinette and La Crosse counties; doubtfully native.

3. **D. arabisans** Michx. Stems much branched and prostrate below, with numerous crowded leaves; erect branches, mostly simple with scattered toothed leaves; petals white; fruit usually twisted.—Rocky woods and cliffs, Door County and at Oakfield in Fond du Lac County.

4. **D. lanceolata** Royle. Similar to No. 3.—Limestone cliffs; Fish Creek and perhaps elsewhere in Door County.

5. **D. verna** L. Whitlowgrass. Small annual or winter annual; leaves oblanceolate, hairy; petals white, cleft to half their length.—Naturalized from Europe; dry sandy soil; Douglas, Wood, Portage, and Shawano counties.

12. Arabidopsis   Mouse-ear Cress

Small annuals or winter annuals, with a basal rosette of petioled leaves and a slender taproot; petals white or purplish; fruit a silique.

A. thaliana (L.) Heynh. Plants 0.5-4 dm high; rosette leaves oblong or narrowly obovate, stem leaves linear sessile, remote; petals 3-4 mm long.—Dry fields and prairies, Dane County.

13. Alyssum   Madwort

Small pubescent annuals or perennials, with stellate hairs; leaves entire to toothed; flowers pale yellow to white; fruit a roundish silicle, flattened parallel to the partition.

A. alyssoides L. Small annuals, simple or branched from the base, hoary-pubescent throughout; silicles 2.5-3.5 mm long, with winged margins.—Native of Europe; locally abundant in sandy places from Iowa and Green counties northeastward to Door County.

Fig. 234. *Berteroa incana,* × .30.

14. Berteroa   Hoary Alyssum

Annual or perennial herbs, pubescent with stellate hairs; leaves simple, mostly entire; sepals ascending; petals deeply cleft, white (our species); fruit a silicle, elliptical, flattened; style persistent.

B. incana (L.) DC. (Fig. 234). Plants 3-6 dm high, branched or simple; gray green throughout, with dense stellate pubescence; many basal leaves in spring which may disappear in summer; petals 4-6 mm long; silicles 5-8 mm long.—Naturalized from Europe; a common weed, throughout the state.

15. Camelina   False Flax

Slender erect annuals, with simple stellate pubescence; leaves linear to lanceolate, sagittate-clasping; petals small, yellow; sepals erect; silicles obovoid to pear-shaped, pointed.

a. Pubescence at base of stem appressed or lacking, or with few long simple hairs; silicles 7-12 mm long, about 3-4 times as long as the style . . . . 1. *C. sativa*
a. Base of stem densely pubescent, the simple hairs projecting beyond the stellate hairs; silicles 5-7 mm long, about twice the length of the style . . . . 2. *C. microcarpa*

1. **C. sativa** (L.) Crantz. Stem and leaves glabrous to sparsely pubescent, the simple hairs not projecting beyond the stellate; silicles commonly 7-10 mm long, 5-7 mm wide, 3-4 times as long as the style.—Very local, but widely scattered throughout the state.

2. **C. microcarpa** Andrz. Stem and leaves rough-pubescent, with both simple and branched hairs, the short stellate hairs exceeded by simple hairs 1-2 mm long; silicles erect, 5.5-8 mm long, 4-5 mm wide, about twice as long as the style.—Dane to Outagamie and Door counties eastward, and in Washburn County.

16. Alliaria   Garlic Mustard

Tall biennials or perennials, with the odor of garlic; leaves broad, cordate with dentate margins; flowers white; fruit a linear silique.

A. petiolata (Bieb.) Cav. & Grande. Biennials, to 1 m high; lower leaves kidney-shaped, others triangular; petals 5-6 mm long; fruit 4-6 cm long.—A European weed of moist woods; Grant, Walworth, Waukesha, and Milwaukee counties.

Fig. 235. *Thlaspi arvense,* × .25.

17. Thlaspi　　Penny Cress

Annuals or perennials; leaves simple, with wavy margins and mostly with basal lobes; petals white (our species); fruit a silicle, 8-15 mm broad, much flattened at right angles to the partition, keeled or winged at the margin, notched at the summit; seeds 2-8 in each cell.

T. arvense L. (Fig. 235). Plants mostly glabrous; lower leaves without petioles, the upper clasping the stem, sometimes toothed, the leaves abundant in spring, falling in early summer, giving the plant a very different appearance; petals 2.5-4 mm long.—Naturalized from Europe; a common weed of early stages of succession and in cultivated ground; throughout the state.

18. Cardaria

Perennials; leaves simple, serrate-dentate, sessile or clasping; petals white; silicle ovoid to cordate, somewhat inflated; style persistent.

C. draba (L.) Desv. Hoary Cress. Short hairy plants, to 6 dm tall; petals 3-4 mm long; silicle glabrous, broader than long; cordate and notched at base.—A European weed; Walworth and Waukesha counties.

19. Erysimum　　Treacle Mustard

Annuals to perennials; pubescent plants, with closely

Fig. 236. *Erysimum cheiranthoides,* × .40.

appressed, stellate, or 2-parted hairs; leaves simple, linear to lanceolate, mostly entire; petals yellow to orange; fruits linear, 4-sided, the midribs of the carpels keeled; seeds in 1 row in each carpel.

a. Petals 15-23 mm long; fruits 4-10 cm long . . . . 1. *E. asperum*
a. Petals 3-8 mm long; fruits 1-4 cm long *b*
  b. Pedicels spreading; flowers bright yellow to orange; leaves lanceolate-linear to broadly lanceolate, green . . 2. *E. cheiranthoides*
  b. Pedicels strongly ascending; flowers lemon yellow; leaves linear, gray green . . . . 3. *E. inconspicuum*

1. **E. asperum** (Nutt.) DC. Western Wallflower. Biennials or short-lived perennials; erect, 2-4 dm tall, appressed-pubescent nearly throughout; petals bright yellow to orange; pedicels 5-7 mm long.—Rare, introduced from the West; Bayfield, Waupaca, and Grant counties.

2. **E. cheiranthoides** L. Wormseed Mustard (Fig. 236). Annuals; leaves covered with minute 3-branched hairs, but bright green; stems with closely appressed 2-branched hairs; pods 1-2 cm long, on spreading pedicels.—Sterile soil, throughout the state.

3. **E. inconspicuum** (S. Wats.) MacM. Perennials; similar to No. 2; hairs more dense, whitening the leaves; pods 2.5-6 cm long, on short pedicels. Can be positively distinguished from the preceding species only in the fruiting condition.—Open soil; local, throughout the state.

20. Brassica

Erect annuals; lower leaves often incised, pinnatifid, or

Fig. 237. *Brassica kaber,* var. *pinnatifida,* × .40.

Fig. 238. *Brassica juncea,* × .40.
Fig. 239. *Brassica nigra,* × .40.

lyrate; petals yellow, 6-15 mm long; siliques round to 4-sided, with a stout beak.—This genus includes Mustard, Turnip, Rape, and Cabbage, which sometimes persist after cultivation or escape from gardens.—The fruit characters are more reliable for identification of species than the vegetative ones used in the following key.

a. Leaves neither clasping or lobed at base *b*
    b. Leaves without petioles, or with short-winged petioles (Fig. 237) . . . . 1. *B. kaber*, var. *pinnatifida*
    b. Leaves petioled, the petiole not winged *c*
        c. Plants smooth or nearly so . . . . 2. *B. juncea*
        c. Plants with scattered hairs . . . . 3. *B. nigra*
a. At least the middle and upper leaves clasping and with lobes at base . . . . 4. *B. rapa*

1. **B. kaber** (DC.) L. C. Wheeler, var. **pinnatifida** (Stokes) L. C. Wheeler. Wild Mustard; Charlock (Fig. 237). Beak half as long as the rest of the capsule.—Naturalized from Europe; a common weed, throughout the state.

2. **B. juncea** (L.) Coss. Indian Mustard (Fig. 238). Pods 3.5 cm long, on slender spreading pedicels.—Naturalized from Europe; a weed; local, throughout the state.

3. **B. nigra** (L.) Koch. Black Mustard (Fig. 239). Capsules 1.5-1.8 cm long, appressed to the stem, with a short beak.—Naturalized from Europe; occasional as a weed.

4. **B. rapa** L. Field Mustard; Turnip. Erect robust plants; petals yellow, 6-11 mm long; siliques 4-8 cm long.

Var. **rapa**. Turnip. Root swollen and fleshy.—Persisting after cultivation as a local weed.

Var. **silvestris** (Lam.) Briggs. Field Mustard. Root neither swollen nor fleshy.—A rare European weed of waste places; Dane, Waupaca, and Forest counties.

Fig. 240. *Barbarea vulgaris,* × .40.

21. Neslia   Ball Mustard

Annuals or biennials, to 6 dm tall; leaves clasping; flowers on long racemes; petals yellow, 2-3 mm long; fruit a silicle on divergent pedicels.

N. paniculata (L.) Desv. Stems and leaves covered with stiff usually starlike hairs; pod subglobose, the surface marked with netlike ridges; style persistent as a beak; seeds usually but 1 in each pod.—Naturalized from Europe; uncommon in waste ground; Dane, Milwaukee to Door counties along Lake Michigan, and Douglas and Bayfield counties on Lake Superior.

22. Barbarea   Winter Cress; Yellow Rocket

Biennials or perennials, mostly glabrous; basal leaves pinnate with large terminal lobe, stem leaves pinnatifid to entire; flowers yellow; fruits linear siliques.

B. vulgaris R. Br. (Fig. 240). Petals 4-8 mm long; fruits 1.5-3 cm long, rounded or 4-angled, with a short beak, spreading at maturity.—Introduced from Europe; common weed, throughout southern Wisconsin, north locally to Bayfield and Iron counties.

23. Erucastrum

Annual or perennial herbs, pubescent with simple hairs; leaves pinnatifid; flowers yellow; stigma capitate; fruit a silique, 4-angled.

E. gallicum (Willd.) O. E. Schulz (Fig. 241). Annuals or biennials, to 8 dm tall; leaves rough-textured; petals 7-8 mm long; siliques 2-3 cm long, ascending on widely divergent pedicels.—Scattered in the state, in waste places, roadsides, and beaches, especially along Lake Michigan, and north to Douglas and Ashland counties.

Fig. 241. *Erucastrum gallicum,* × .20.

24. Bunias

Tall coarse herbs, with lower and basal leaves with toothed lobes; flowers yellow on glandular racemes; fruits ovoid, beaked.

B. orientalis L. Biennials, to 1 m tall; lower leaves dandelionlike, upper leaves gradually reduced.—Adventive from Eurasia; Green County.

25. Descurainia   Tansy Mustard

Annuals or biennials, with branched or glandular pubescence; leaves finely dissected, 1-3-times pinnate; petals yellow; fruit a linear or club-shaped silique.

a. Stems pubescent and glandular; siliques 6-10 mm long, club-shaped; leaves once or twice pinnate . . . .
. . . . 1. *D. pinnata,* var. *brachycarpa*
a. Stems lightly appressed pubescent to canescent, but never glandular; siliques linear, 12-21 mm long; leaves twice or three times pinnate . . . . 2. *D. sophia*

1. **D. pinnata** (Walt.) Britt., var. **brachycarpa** (Richards.) Fern. (Fig. 242). Plants very leafy; leaves much divided, the ultimate leaflets 2-5 mm long and often lobed; lower part of the stem with almost microscopic gland-tipped hairs; pods about 1 cm long, shorter than the slender spreading pedicels; seeds in 2 rows in each cell.—A weed along railroads, roadsides, and on limestone beaches, particularly along the Mississippi River and Lake Michigan; also southern and extreme northwestern Wisconsin.

Fig. 242. *Descurainia pinnata,* var. *brachycarpa,* × .60; silicle × 1.20.

Fig. 243. *Sisymbrium altissimum,* × .40.
Fig. 244. *Sisymbrium officinale,* × .40.

2. **D. sophia** (L.) Webb. Hairs on stem not gland-tipped; pods 2 cm long; seeds in 1 row in each cell.—A rare weed, with range similar to the preceding.

26. Sisymbrium

Annuals or winter annuals; leaves pinnately deeply lobed to compound; petals yellow; fruit a silique; stigma 2-lobed.

a. Leaflets of upper leaves linear, rarely more than 2 mm broad; fruits widely divergent, 6-10 cm long; petals pale yellow . . . .
. . . . 1. *S. altissimum*
a. Leaflets of upper leaves angularly toothed, the terminal segment ovate or elliptical; fruits closely appressed to the stem, 1-2 cm long; petals deep yellow . . . . 2. *S. officinale*

1. **S. altissimum** L. Tumble Mustard (Fig. 243). Leaves compound, the segments of the lower ones elliptical or narrowly triangular and coarsely toothed, segments of the upper ones ribbonlike or threadlike; hairs on stem scattered, long and spreading, or lacking; flowers pale yellow; pod 6-10 cm long.—Introduced from Europe; waste places and roadsides, throughout the state.

2. **S. officinale** (L.) Scop., var. **officinale.** Hedge Mustard (Fig. 244). Lower leaves with the terminal segment larger than the triangular and toothed lateral ones; leaves and fruits densely covered with simple hairs; flowers yellow; pod 1-1.5 cm long.—Naturalized from Europe; uncommon weed; Lafayette, Dane, Kenosha, and Waushara counties.

Var. **leiocarpum** DC. has few hairs on the leaves and none on the fruit, is also from Europe, and is common north to Polk, Lincoln, and Oconto counties, also Bayfield County.

CAPPARIDACEAE   Caper Family

Our species annuals, with alternate palmately compound leaves and terminal racemes; flowers regular or irregular; sepals 4; petals 4, commonly long-clawed; stamens 6-many, filaments elongate and exserted; ovary 2-carpellate; fruit an elongate capsule without the partion of the Brassicaceae.

Polanisia   Clammyweed

Viscid-pubescent annuals with petiolate trifoliate leaves; inflorescence terminal racemes of small white to pinkish flowers.

P. dodecandra (L.) DC. (Fig. 245). Stems 2-5 dm high; petioles as long as the leaf blades; petals 4-8 mm, the blade usually exceeding the claw; stamens little exserted; capsules 2-4 cm.—Dry sandy or gravelly soil; throughout southern Wisconsin, north to Douglas County in western Wisconsin, especially along the Mississippi River, and north to Sheboygan County in eastern Wisconsin.

RESEDACEAE   Mignonette Family

Our species herbs, with alternate leaves and small irregular flowers; fruit a capsule.

Reseda   Mignonette

Sepals and petals 4-8, the sepals almost equal, the petals unequal, the backs bearing projecting appendages; stamens 12 or more to one side of the flower and with a large nectariferous disk on the opposite side; inflorescence a dense terminal raceme; fruit a 3-carpellate capsule opening toward the top.

Fig. 245. *Polanisia dodecandra,*
× .35.

Fig. 246. *Sarracenia purpurea,* × .30.

R. lutea L. Taprooted biennials or perennials, to 8 cm high; leaves pinnatifid; flowers greenish yellow. Introduced from Europe; Dane and Rock counties.

SARRACENIACEAE   Pitcher Plant Family

Bog plants, with hollow pitcher- or trumpet-shaped leaves; flowers solitary, perfect; sepals 5; petals 5, early deciduous; stamens many; style 1.

Sarracenia   Pitcher Plant

Leaves hollow, with a wing on the upper side, the extended part beyond the opening usually covered with down-ward-pointing hairs, smooth inside; stigma very broad, covering the ovary and the stamens, extending into 5 rays to form an umbrella-shaped body with minute stigmas beneath at the angles; fruit a 5-valved capsule.

S. purpurea L. (Fig. 246). Leaves curved, ascending from the base; petals dark red purple.—Bogs, throughout the state except southwestern Wisconsin.

CRASSULACEAE   Orpine Family

Succulent herbs, with simple leaves; inflorescence usually cymose; flowers regular; sepals, petals, and pistils in equal numbers; stamens double the number; fruit a follicle.

Sedum   Stonecrop

Leaves fleshy, alternate, opposite, or whorled; sepals and petals 4-5, either distinct or barely united at base.

S. acre L. Mossy Stonecrop (Fig. 247). Low perennials, spreading on the ground; leaves small, fleshy, scalelike; flowers yellow.—A European species escaping from cultivation; dry soil, not common; scattered localities from Rock to Ashland counties.

Fig. 247. *Sedum acre,* × .65, flower × .80.

SAXIFRAGACEAE   Saxifrage Family

Herbs or shrubs; leaves alternate or opposite; flowers mostly perfect and regular; sepals and petals 4-5 (sometimes 7-10, or reduced or lacking); stamens usually twice as many as the petals; carpels usually 2, united below; fruit a capsule, follicle, or berry.

a. Herbs *b*
  b. Scapes upright, leafless or with a few leaves *c*
    c. Leaves lanceolate, with short broad petioles . . . . 1. *Saxifraga*
    c. Leaves ovate, with long slender petioles *d*
      d. Stamens 5 *e*
        e. Scape simple, leafless, with spreading white hairs. . . . .
        . . . . 2. *Heuchera*
        e. Scape branching above, with a few reduced leaves,
          glabrous . . . . 3. *Sullivantia*
      d. Stamens 10 *f*
        f. Petals entire . . . . 4. *Tiarella*
        f. Petals fringed . . . . 5. *Mitella*
  b. Stems creeping, leafy . . . . 6. *Chrysosplenium*
a. Shrubs *g*
  g. Leaves opposite . . . . 7. *Philadelphus*
  g. Leaves alternate . . . . 8. *Ribes*

1. Saxifraga   Saxifrage

Leaves all clustered in a basal rosette, rough-hairy, 1-2 dm long, 2-7 cm wide, blunt; scapes stout, hairy, 4-5 cm high; flowers on short branches, whitish.

a. Carpels united at least as far as the middle . . . . . 1. *S. pensylvanica*
a. Carpels united only at base or entirely separate . . . . . 2. *S. forbesii*

1. **S. pensylvanica** L. Swamp Saxifrage. The 2 carpels that comprise the fruit are erect for most of their length, the tips divergent, the styles erect to spreading; petals greenish white.—Bogs and wet meadows, sometimes wet woods, north to Sheboygan, Waushara, Burnett, and Iron counties, and in the Lake Superior region.

The typical material, ssp. **pensylvanica**, has narrowly lanceolate 1-nerved petals, styles more than 1 mm long, and leaves mainly lanceolate; a variant with large bracts, 7-10 cm long, in the inflorescence occurs in Wisconsin, f. **fultior** (Fern.) Burns; ssp. **interior** Burns has ovate to lanceolate-ovate 3-nerved petals, with styles less than 1 mm long, the leaves ovate to obovate. Within the ssp. **interior** are: var. **crassicarpa** (Johnson) Burns, with the follicles large, 4-5 mm long, 3-4 mm wide, and styles 0.5-1 mm in fruit; and var. **congesta** Burns, with carpels 2 mm long, 1-1.5 mm wide, and the styles 0.5 mm or less in fruit. Ssp. **pensylvanica** is a

Fig. 248. *Heuchera richardsonii,* × .20.

tetraploid with n=28 chromosomes, ssp. **interior** is an octo-ploid with n=56 chromosomes.

2. **S. forbesii** Vasey. Cliff Saxifrage. The 2 carpels divergent for most of their length, the styles spreading or recurved; petals white or greenish white.—Moist shaded sandstone cliffs in the Driftless Area, Lafayette and Green counties north to Juneau and Adams counties.

2. **Heuchera**   Alumroot

Perennial herbs; leaves mostly basal, long-petioled, round-cordate; flowering scape usually leafless; calyx lobes 5; petals 5; stamens 5; carpels 2; fruit a 2-beaked capsule.

H. richardsonii R. Br. (Fig. 248). Leaves clustered at base of the scape, the blades rounded, with a broad sinus, 5-9-lobed; lobes toothed; petioles long, hairy; scape 5-12 dm high; flowers in a loose panicle, whitish.—Wooded bluffs and prairies, north to Douglas and Lincoln counties.

Var. **hispidior** Rosend., Butt. & Lak. has glandular petals and barely exserted stamens; var. **grayana** Rosend., Butt. & Lak., with flowers 6-10 mm long, and var. **affinis** Rosend., Butt. & Lak., with flowers 5-7 mm long, both have glandular petals with minute bumps and exserted stamens.

3. **Sullivantia**

Perennial herbs; leaves mostly basal; flowering scape with a leaf near the base; calyx lobes 5; petals 5, white; stamens 5; carpels 2; fruit a 2-beaked capsule.

S. renifolia Rosend. Spreading herbs with bright green rounded and toothed leaves; flowers small and white, in an

Fig. 249. *Mitella diphylla,* × .30,   Fig. 250. *Mitella nuda,* × .30,
flower × .80.   flower × .80.

openly branched inflorescence.—Moist mossy cliffs in the
Driftless Area.

4. Tiarella   Foamflower

Perennial herbs with broad lobed basal leaves and a
leafless flowering scape of white flowers; sepals 5; petals 5,
clawed; stamens 10; carpels 2.

T. cordifolia L. Plants 1-3.5 dm high, stoloniferous; basal
leaves cordate-ovate, shallowly lobed; petals 3-5 mm
long.—Rich woods, rare; Door and Florence counties.

5. Mitella   Bishop's-cap

Perennial herbs; leaves mostly at the base of the scape,
with long hairy petioles; scapes slender, hairy, bearing a
raceme of white or yellowish green flowers; petals delicately
fringed.

a. Flowering scape with a pair of nearly sessile leaves . . 1. *M. diphylla*
a. Flowering scape without leaves . . . . 2. *M. nuda*

1. **M. diphylla** L. (Fig. 249). Petals white, 1.5-2 mm long,
the width of the flat body less than the length of the
fringes.—Damp woods, throughout the state.

2. **M. nuda** L. (Fig. 250). Petals yellowish green, 4 mm
long, the body as well as the fringes threadlike.—Northern
Wisconsin, southward to Walworth County in tamarack bogs.

6. Chrysosplenium   Golden Saxifrage

Small creeping perennial herbs; leaves opposite or alternate; flowers in the leaf axils; calyx lobes 4; petals 0;
stamens 4-8; carpels 2; fruit a 2-lobed capsule.

Fig. 251. *Chrysosplenium americanum,* × .40, flower × 1.50.

C. americanum Schwein. (Fig. 251). Stems slender; leaves opposite, the blades roundish; flowers inconspicuous, without petals, the 8-10 anthers purple or yellow.—Northern Wisconsin, south in cold canyons and springy streambanks to the Baraboo Hills.

7. **Philadelphus**   Mock Orange

Shrubs, with simple opposite leaves; flowers showy; calyx lobes and petals 4 (3-6); stamens 20-40; styles united to or beyond the middle; stigmas separate; fruit a capsule.

P. coronarius L. Flowers large, cream-colored, odorous.— Escaping from cultivation.

8. **Ribes**   Gooseberry; Currant

Shrubs, often spiny; leaves alternate, palmately lobed; sepals 5, longer than the petals; petals 5, yellowish, purplish, or greenish; stamens 5; fruit a berry.

a. Flowers solitary or in bunches of 2-4, often with spines at the base of each bunch (Figs. 252, 253, 255) *b*
  b. Calyx lobes shorter than the tube; ovary and berries usually prickly; petioles with simple or slightly fringed gland-tipped hairs . . . . .1. *R. cynosbati*
  b. Calyx lobes longer than the tube; ovary and berries not prickly; petioles with long branched hairs *c*
    c. Stamens about twice as long as the calyx lobes, conspicuous; bracts of the inflorescence fringed by minute stalked red or yellow glands . . . . 2. *R. missouriense*
    c. Stamens about equaling the calyx lobes; bracts fringed with minute hairs, and sometimes glands *d*
      d. Stamens about twice as long as the petals .3. *R. hirtellum*
      d. Stamens about as long as the petals 4. *R. oxyacanthoides*
a. Flowers in racemes (Fig. 258) *e*
  e. Calyx tube about as broad as long *f*
    f. Leaves with resinous dots or glands beneath *g*
      g. Calyx lobes equaling the tube *h*
        h. Bract at the base of each flower longer than its pedicel . . . . 5. *R. americanum*

    h. Bract shorter than the pedicel . . . . 6. *R. nigrum*
   g. Calyx lobes much longer than the tube . . . .
     . . . . 7. *R. hudsonianum*
  f. Leaves frequently with hairs, but never with resinous dots
    beneath *i*
    i. Stems densely covered with prickles . . . . . . 8. *R. lacustre*
    i. Stems without prickles *j*
       j. Ovary and fruit with stalked glands 9. *R. glandulosum*
       j. Ovary and fruit without glands *k*
         k. Pedicels and peduncles without stalked glands . . . .
          . . . . .10. *R. sativum*
         k. Pedicels and peduncles with stalked glands . . . .
          . . . . 11. *R. triste*
 e. Calyx-tube several times as long as broad . . . . .12. *R. odoratum*

1. **R. cynosbati** L. Prickly Gooseberry (Fig. 252). Stems usually with a sharp reddish spine at the base of each leaf; internodes often covered with slender spines; leaves round-ovate, velvety beneath; berries usually prickly.—Open woods and pastures; common, throughout the state.

2. **R. missouriense** Nutt. Missouri Gooseberry (Fig. 253). Spines long, stout, and red; flowers white or yellowish; sepals usually turned back or spreading, longer than the petals and much shorter than the conspicuous stamens.—Woods and thickets, north to Barron, Marathon, and Manitowoc counties.

3. **R. hirtellum** Michx. Smooth Goosebery (Fig. 254). Usually unarmed; leaf blades deeply 3-lobed, usually slightly pubescent beneath; flowers greenish or purplish.—Moist woods in northern Wisconsin, south to Polk and Wood counties, and rare in boggy places south to Waukesha County in eastern Wisconsin.

Var. **calcicola** Fern. Leaves velvety beneath.—Widespread, but mostly east of the Wisconsin River.

4. **R. oxyacanthoides** L. Similar to No. 3, but usually more spiny.—Rare; Barron, Douglas, and Door counties.

Fig. 252. *Ribes cynosbati*, × .30, flower × 1.
Fig. 253. *Ribes missouriense*, × .30, flower × 1.

Fig. 254. *Ribes hirtellum*, × .30, flower × 1.

5. **R. americanum** Mill. American Black Currant (Fig. 255). Stems unarmed; leaf blades sharply 3-5-lobed, a little longer than broad, with glands on the upper as well as the lower surface; bracts of the inflorescence green, hairy, lan-

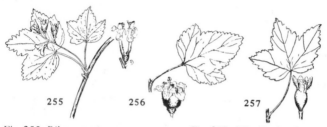

Fig. 255. *Ribes americanum,* × .30, flower × 1.
Fig. 256. *Ribes nigrum,* × .30, flower × 1.

Fig. 257. *Ribes hudsonianum,* × .30, flower × 1.

Fig. 258. *Ribes lacustre,* × .30, flower × 1.50.
Fig. 259. *Ribes glandulosum,* × .30, flower × 2.

Fig. 260. *Ribes sativum,* × .30, flower × 1.

Fig. 261. *Ribes triste,* × .30, flower × 1.

Fig. 262. *Ribes odoratum,* × .30, flower × .65.

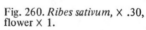

ceolate; calyx yellow and greenish, 8-10 mm long; ovary not glandular.—Open woods and fields, throughout the state.

6. **R. nigrum** L. Black Currant (Fig. 256). Similar to No. 5; calyx 5-6 mm long; ovary with sessile glands.—Cultivated, sometimes escaping.

7. **R. hudsonianum** Richards. Hudson Bay Currant (Fig. 257). Similar to No. 5, but upper surface of the leaves without glands; ovary, and often the petioles, with sessile glands; leaf blade a little broader than long.—Rare; Douglas to Florence counties.

8. **R. lacustre** (Pers.) Poir. Swamp Black Currant; Prickly Currant (Fig. 258). Leaf blades heart-shaped at base, deeply 3-5-lobed; ovary with long-stalked glands.—Cold woods and swamps, northeastern Wisconsin from Kewanee and Door counties to Forest, Florence, and Ashland counties.

9. **R. glandulosum** Grauer. Skunk Currant (Fig. 259). Stems reclining; leaf blades somewhat heart-shaped at base, 3-5-lobed; fruit ill-scented.—Swamps and cool woods, south to Dunn and Manitowoc counties.

10. **R. sativum** Syme. Garden Currant (Fig. 260). Leaf blades velvety, heart-shaped at base and 3-5-lobed; the middle lobe ovate, a little longer than broad; flowers yellow green.—Escaping from cultivation, and sometimes appearing as if native when growing in woods.

11. **R. triste** Pall. Swamp Red Currant (Fig. 261). Stems reclining; leaf blades densely hairy beneath, 3-lobed, the sides almost parallel; middle lobe triangular, broader than long.

Var. **albinervium** (Michx.) Fern. is more common, and has the leaf blades almost without hairs.—Wet woods, south to Dunn, Washington, and Milwaukee counties.

12. **R. odoratum** Wendl. Clove Currant (Fig. 262). Stems without spines; leaf blades usually 3-lobed, wedge-shaped or almost square at base; flowers trumpet-shaped, golden yellow, spicy-fragrant.—Commonly cultivated and often persisting, mostly southward.

PLATANACEAE   Plane Tree Family

Deciduous trees, with alternate simple palmately lobed leaves; monoecious, male and female flowers in separate dense heads on a pendulous stalk; fruit a spherical head of linear achenes subtended by long hairs.

Fig. 263. *Platanus occidentalis,* × .30.

Platanus   Sycamore

Generic characters as for family.

P. occidentalis L. (Fig. 263). Bark splitting off and exposing the smooth mottled white to buff inner bark; leaves maplelike, very woolly when young, with stellate pubescence.—Rare in river bottoms in southern Wisconsin, from Grant to Walworth counties, also Sauk and Dane counties; adventive in Ozaukee County.

ROSACEAE   Rose Family

A large family of trees, shrubs, or herbs; leaves alternate, with stipules; either the flower with a very small cuplike hypanthium composed of 5 fused sepals, 5 petals, (5) 10-many stamen bases separate from a superior ovary, or a hypanthium fused with the ovary so that the ovary appears inferior; pistils 1-many; fruits pomes, achenes, follicles, drupes, or berries.

a. Ovary superior, i.e., with open hypanthium or sepals borne at its base (see Fig. 565) *b*
  b. Trees or shrubs; flowers not yellow *c*
    c. Pistils solitary or several united *d*
      d. Leaves roundish, 3-lobed (Fig. 264) . . . . . .1. *Physocarpus*
      d. Leaves lanceolate or ovate, not lobed *e*
        e. Inflorescence a compound panicle (Fig. 265) . . . .
        . . . . .2. *Spiraea*
        e. Inflorescence a raceme or umbel (see Figs. 569*c*,

     569*d*) ....3. *Prunus*
  c. Pistils many, on a dome-shaped receptacle ....4. *Rubus*
 b. Herbs, or if shrubs, then yellow-flowered *f*
  f. Bractlets outside of and alternating with the calyx lobes *g*
   g. Styles jointed, the terminal part silky-hairy or fringed,
    persisting in fruit ....5. *Geum*
   g. Styles not silky-hairy, falling as the fruit ripens *h*
    h. Leaves 3-foliate, borne at the base of the plant *i*
     i. Petals white ....6. *Fragaria*
     i. Petals yellow .... 7. *Waldsteinia*
    h. Leaves 5-many-foliate, or 3-foliate on the stem ......
      ....8. *Potentilla*
  f. Calyx without bractlets ....4. *Rubus*
a. Ovary inferior, i.e., with sepals apparently borne at its summit (see
 Fig. 566) *j.*
 j. Flowers 5 cm or more broad; leaves compound ....9. *Rosa*
 j. Flowers 4 cm or less broad *k*
  k. Petals at least twice as long as broad ......10. *Amelanchier*
  k. Petals hardly longer than broad *l*
   l. Plants with thorns or spinescent branchlets *m*
    m. Leaves simple *n*
     n. Leaves usually lobed ....11. *Crataegus*
     n. Leaves merely dentate or serrate .... 12. *Pyrus*
    m. Leaves compound ....9. *Rosa*
   l. Plants unarmed *o*
    o. Leaves compound .... 13. *Sorbus*
    o. Leaves simple *p*
     p. Inflorescence compound; shrubs ....14. *Aronia*
     p. Inflorescence simple or in umbellike clusters; trees
      ....12. *Pyrus*

1. Physocarpus Ninebark

  Shrubs, with simple lobed leaves; inflorescence corymbose; hypanthium shallowly cup-shaped; sepals 5; petals 5; stamens 20-40; carpels 1-5, superior; fruit an inflated follicle splitting halfway to the base along both edges.

  P. opulifolius (L.) Maxim. (Fig. 264). Shrubs, 1-3 m high;

Fig. 264. *Physocarpus opulifolius,*
× .40.

Fig. 265. *Spiraea alba,* × .40.

bark loose and separating into thin layers; leaves 3-lobed, the lobes toothed; flowers white, on pedicels longer than the axis of the raceme; calyx lobes woolly; fruit of 2-5 dry inflated 2-seeded carpels.—Banks of streams and rocky places, throughout the state.

2. Spiraea   Meadowsweet

Shrubs, with simple leaves and white to purple or pink flowers in terminal or lateral inflorescences; hypanthium cup-shaped or top-shaped; sepals 5; petals 5, small; stamens 15-many; pistils 5; fruit a follicle.

S. alba Du Roi (Fig. 265). Low shrubs, with lanceolate pinnately veined toothed leaves; flowers white, in cylindrical or pyramidal panicles; fruit of several carpels, not inflated.— Low wet meadows and marshes, throughout the state.

3. Prunus   Cherry; Plum

Trees or woody shrubs; leaves alternate, finely toothed; flowers solitary or in umbels or racemes; hypanthium urn-shaped to cup-shaped to obconic; sepals 5; petals 5, white or pinkish; stamens 15-30; carpel 1, fully superior; fruit fleshy, with a stone enclosing the single seed.

a. Flowers coming with the leaves, in racemes *b*
  b. Petals fan-shaped, decidedly narrowed to the attachment (Fig. 266); calyx persisting on the fruit . . . . 1. *P. serotina*
  b. Petals nearly round, not narrowed to the attachment (Fig. 267); calyx not persisting on the fruit . . . . 2. *P. virginiana*
a. Flowers coming before the leaves, in umbels *c*
  c. Petals 4-6 mm long; branches without thorns *d*
    d. Leaf blades toothed nearly to the base . . . 3. *P. pensylvanica*
    d. Leaf blades not toothed below the middle . . . . .4. *P. pumila*
  c. Petals 6-16 mm long; some branches thorny *e*
    e. Calyx lobes glandular-margined, not woolly . . . . . 5. *P. nigra*
    e. Calyx lobes woolly, not glandular . . . . 6. *P. americana*

1. P. serotina Ehrh. Black Cherry (Fig. 266). Large trees, with aromatic inner bark; leaves with incurved teeth; midrib with rusty pubescence, thickest at base; fruit black.—Woods and fencerows, throughout the state.

2. P. virginiana L. Choke Cherry (Fig. 267). Shrubs or small trees, the inner bark with a disagreeable odor; leaves with sharp spreading teeth; glabrous, except in axils of veins along the midrib; fruit dark red.

Forma **deamii** G. N. Jones is pubescent on new branch-lets, rachis, petioles, and lower leaf surfaces.—Woods, thickets, and fencerows; both the typical and the pubescent forms occur throughout the state.

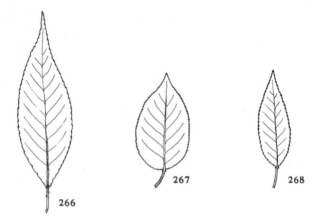

Fig. 266. *Prunus serotina,* × .45   Fig. 268. *Prunus pensylvanica,*
Fig. 267. *Prunus virginiana,* × .45.   × .45.

3. **P. pensylvanica** L. f. Pin Cherry (Fig. 268). Shrubs or small trees; leaves oval or lanceolate, rounded or broadly wedge-shaped at base; fruit light red, sour.—Woods and clearings, throughout the state.

4. **P. pumila** L. Sand Cherry (Fig. 269). Erect or prostrate small shrubs; leaf blades somewhat whitened beneath, 3-4 times as long as broad.—Sandy or rocky ground.

Broader-leaved forms are often separated as var. **susquehanae** (Willd.) Jaeg. (includes *P. susquehanae* Willd.). The typical form occurs throughout the state, the variety in scattered localities.

5. **P. nigra** Ait. Canada Plum (Fig. 270). Shrubs or small trees; some twigs short and leafless, serving as thorns; leaves with rounded teeth; calyx green or pink.—North to Douglas and Vilas counties.

6. **P. americana** Marsh., var. **americana**. Wild Plum (Fig. 271). Small trees, with spinescent branchlets as in No. 5; leaves sharply toothed; glabrous.

Var. **lanata** Sudw. has branchlets, petioles, and lower leaf surfaces pubescent.—Pastures, fencerows, and borders of woods; both the species and the variety are common in southern Wisconsin, and occur in scattered locations north to Lincoln County.

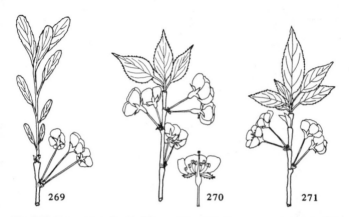

Fig. 269. *Prunus pumila,* × .60.   Fig. 271. *Prunus americana,* × .60.
Fig. 270. *Prunus nigra,* × .60.

4. Rubus   Raspberries and Blackberries

Mostly shrubs, the underground rhizomes bearing biennial stems (canes), the first-year canes vegetative and known as primocanes, the second-year canes fertile and known as floricanes; plants mostly prickly with thorns, bristles, or slender prickles; leaves mostly compound with 3 or 5 leaflets on the primocanes, their axillary buds in the floricanes forming short flowering branches with 3 leaflets; sepals 5; petals 5; stamens many; carpels many, borne on a dome-shaped receptacle; fruits fleshy drupelets, each with a single stone.

For the identification of the species of this genus it is best to make field observations of the entire plants rather than to rely on small specimens.

The blackberry subgenus known as Eubatus has many species, and is very difficult taxonomically. H. A. Davis, Albert M. Fuller, and Tyreeca Davis, authorities for this group, have prepared this treatment of the Wisconsin species especially for this manual.

a. Leaves simple, maplelike; plants unarmed . . . . 1. *R. parviflorus*
a. Leaves compound; plants armed with bristles or prickles except No. 2 *b*
    b. Plants low, herbaceous, soft-woody at base . . . . 2. *R. pubescens*
    b. Plants woody *c*
        c. Leaves woolly-whitened beneath, primocanes pinnately compound *d*

    d. Pedicels with glands and slender bristles; sepals bristly on
       the back . . . . 3. *R. idaeus*
    d. Pedicels with stout-hooked prickles but no glands; sepals
       without bristles . . . . 4. *R. occidentalis*
  c. Leaves sometimes velvety beneath, but not woolly-whitened,
    the primocanes 3-5-foliate . . . . Subgenus Eubatus

1. R. parviflorus Nutt. Thimbleberry (Fig. 272). Shrubs
without prickles, but the new growth densely glandular;
leaves maplelike; flowers large, white.—Rich woods, south to
Lincoln and Manitowoc counties.

2. R. pubescens Raf. Dwarf Red Blackberry (Fig. 273).
Stems 1-4 dm high, trailing or ascending; leaflets 3-5, thin,
coarsely toothed; petals small, white or pinkish.—Bogs,
throughout the state.

3. R. idaeus L., var. **idaeus** Red Raspberry (Fig. 274).
Stems upright, covered with needlelike prickles; fruit red,
easily detached from the receptacle.

Var. strigosus (Michx.) Maxim. Pedicels and rachis of inflorescence with glands and minute bristles.—Common
throughout the state.

Var. canadensis Richards. Stems of the first year's growth
minutely hairy beneath slender prickles.—Usually growing
with var. **strigosus** but most common in the northern
counties.

Var. aculeatissimus Regel & Tiling has the prickles

Fig. 272. *Rubus parviflorus,*
× .25.

Fig. 273. *Rubus pubescens,* × .25.

Fig. 274. *Rubus idaeus,* × .25.

broadened at base, and is approached by material from northern Wisconsin.

4. **R. occidentalis** L. Black Raspberry. Stems recurved, rooting at tip; leaves usually 3-foliate; fruit purple black, easily detached from the receptacle; whole plant somewhat powdery-whitened.—Common in openings in woods in southern Wisconsin, in scattered localities northward to Lincoln and Polk counties.

Key to the Sections of Eubati

a. Canes without broad-based prickles, but bearing stiff hairs, bristles, spicules, slender prickles, or a combination of them; leaves firm and glossy *b*
    b. Canes trailing or mounding, at least the terminal third of the mature well-developed canes prostrate, and tip-rooting under favorable conditions .... I. Section Hispidi
    b. Canes erect or arching; an occasional cane arching to the ground and tip-rooting, but not trailing for any considerable distance .... II. Section Setosi
a. Canes with broad-based prickles, even the "thornless" types with a few minute prickles nearly as broad at the base as long; no bristles or spicules present; leaves dull or glossy, not particularly firm *c*
    c. Canes trailing or mounding, at least the terminal third of the mature well-developed canes prostrate, and tip-rooting under favorable conditions .... III. Section Flagellares
    c. Canes erect or arching; an occasional cane arching to the ground and tip-rooting, but not trailing for any considerable distance *d*
        d. Inflorescence decidedly glandular . IV. Section Alleghenienses
        d. Inflorescence glandless, or rarely with a few glands on some pedicels ....V. Section Arguti

I. Section Hispidi

a. Armature of main primocane axis consisting of hairs and bristles, soft (or brittle when dry), not scratchy to the touch (late season

terminal pieces and branches often provided with little hooks
which may partly or entirely replace the hairs and bristles of the
main axis) *b*
- *b.* Main primocane armature consisting of stiff hairs only, often
dense enough on vigorous canes to nearly hide the surface;
more sparse on weak canes *c*
  - *c.* Primocane axis glandless or nearly so, sometimes with occasional gland-tipped hairs; primocane leaves 3-foliate, the
central leaflet mostly less than 5 cm long . . . . 1. *R. hispidus*
  - *c.* Primocane axis decidedly glandular; some primocane leaves
5-foliate, the central leaflet mostly more than 5 cm long . . .
. . . . 2. *R. exter*
- *b.* Primocane armature consisting, at least in part, of long slender
bristles, rigid, but brittle when dry, usually breaking off before
the finger is torn . . . . 3. *R. plus*
- *a.* Primocane armature consisting partly or entirely of prickles
scratchy to the touch; when dry, some strong enough to tear the
finger; primocane axis decidedly glandular; inflorescence racemiform, standing well above the foliage; primocane leaflets widest
near or slightly above the middle, acuminate . . . . . 4. *R. permixtus*

II. Section Setosi

- *a.* Primocane leaves not pubescent beneath to the touch *b*
  - *b.* Primocane armature not strong enough to tear the finger *c*
    - *c.* Inflorescence compound when well developed *d*
      - *d.* Inflorescence open and diffuse, much exceeding the
foliage *e*
        - *e.* Primocane axis densely armed with prickles and hairs;
pedicels armed with spicules and hairs, some gland-
tipped . . . . .5. *R. dissensus*
        - *e.* Primocane axis sparsely armed with slender prickles;
pedicels nude, or nearly so . . . . .6. *R. stipulatus*
      - *d.* Inflorescence more compact, about equaling, or buried
in, the foliage *f*
        - *f.* Primocane axis and pedicels glandular; pedicels with
many spicules; floral bracts 10-15 mm long . . . .
. . . . .7. *R. groutianus*
        - *f.* Primocane axis and pedicels glandless; pedicels with
few or no spicules; floral bracts smaller . . .8. *R. navus*
    - *c.* Inflorescence simple *g*
      - *g.* Simple floral leaves present, 3-5 times as long as wide;
primocane leaflets about twice as long as wide, caudate-
acuminate . . . . 9. *R. junceus*
      - *g.* Simple floral leaves, if present, wider; primocane leaflets
wider, not caudate-acuminate *h*
        - *h.* Inflorescence extending well beyond the foliage,
racemiform *i*
          - *i.* Central primocane leaflet ovate, gradually acum-
inate . . . . 10. *R. vermontanus*
          - *i.* Central primocane leaflet elliptical, short-acuminate
. . . . .6. *R. stipulatus*
        - *h.* Inflorescence extending little, if any, beyond the
foliage, cymiform or short-racemiform *j*
          - *j.* Primocane axis glandless, or nearly so *k*
            - *k.* Central primocane leaflet widest well above the
middle . . . . 11. *R. wheeleri*

k. Central primocane leaflet widest near or below
the middle *l*
    l. Primocane provided with hairs only, not
    scratchy . . . . 12. *R. uniformis*
    l. Primocane armed with slender scratchy
    prickles or spicules *m*
        m. Pedicels glandless . . . . 8. *R. navus*
        m. Pedicels glandular . . . . . . .13. *R. spectatus*
  j. Primocane axis decidedly glandular *n*
    n. Pedicels with numerous spicules 2-3 mm long . .
    . . . . 7. *R. groutianus*
    n. Pedicels with few or no prickles *o*
        o. Primocane leaflets ovate, the central one
        mostly more than half as wide as long . . . . .
        . . . . 14. *R. regionalis*
        o. Primocane leaflets obovate, the central one
        mostly less than half as wide as long . . . .
        . . . . .15. *R. gulosus*
b. Primocane prickles, when mature, strong enough to easily tear
the finger *p*
  p. Inflorescence elongate, exceeding the foliage, leafy only at
  the base . . . . 10. *R. vermontanus*
  p. Inflorescence about as wide as long, equaling, or buried in,
  the foliage; most pedicels subtended by leaves . . . .
  . . . . .16. *R. wisconsinensis*
a. Primocane leaves pubescent beneath to the touch *q*
  q. Primocane axis decidedly glandular . . . . 17. *R. glandicaulis*
  q. Primocane axis glandless, or nearly so *r*
    r. Pedicels glandular; primocane leaflets shouldered, abruptly
    caudate-acuminate .18. *R. perspicuus*
    r. Pedicels glandless; primocane leaflets not shouldered,
    gradually acuminate . . . . .19. *R. missouricus*

III. Section Flagellares

a. Plants glandless *b*
  b. Leaves not pubescent beneath to the touch *c*
    c. Inflorescence ascending *d*
      d. Inflorescence several-flowered *e*
        e. Well-developed primocane leaves mostly 5-foliate, the
        central one widest below the middle and broadly
        rounded to cordate at the base . . . . . . 20. *R. multifer*
        e. Well-developed primocane leaves mostly 3-foliate, the
        central one widest near the middle and narrowed at
        the base . . . . 21. *R. steelei*
      d. Inflorescence 1-flowered, or occasionally 2- or 3-flowered
      . . . . 22. *R. baileyanus*
    c. Inflorescence racemiform or cymiform *f*
      f. Canes trailing; inflorescence cymiform, about as wide as
      long, not surpassing the foliage . . . . . . 23. *R. plicatifolius*
      f. Canes arching, only the terminal portion trailing; in-
      florescence racemiform, longer than wide, surpassing the
      foliage . . . . .24. *R. multiformis*
  b. Leaves pubescent beneath to the touch; canes arching . . . .
  . . . . 25. *R. ithacanus*
a. Plants glandular, at least on the pedicels *g*
  g. Primocane leaflets wide and overlapping; inflorescence with

flowers on nearly erect pedicels . . . . 26. *R. exsularis*
g. Primocane leaflets narrow, not overlapping; inflorescence with flowers on spreading-ascending to wide-spreading pedicels . . . . .
. . . . 27. *R. schoolcraftianus*

IV. Section Alleghenienses

a. Inflorescence definitely racemiform, cylindrical, or narrowed upward *b*
  b. Racemes long, usually more than 3 times as long as wide, measured from the first pedicel; primocane leaflets not overlapping, the central one mostly less than two-thirds as wide as long . . . . 28. *R. allegheniensis*
  b. Racemes short and condensed, less than 3 times as long as wide; primocane leaflets wide and overlapping, the central one more than two-thirds as wide as long . . . . 29. *R. rosa*
a. Inflorescence scarcely racemiform, widening upward . . . .
. . . . 30. *R. alumnus*

V. Section Arguti

a. Primocane leaves pubescent beneath to the touch *b*
  b. Inforescence racemiform, extending well beyond the foliage, leafy only at base *c*
    c. Racemes long, more than twice as long as wide; canes lightly armed . . . . 31. *R. pergratus*
    c. Racemes short and condensed; canes well armed . . . .
. . . . 32. *R. bellobatus*
  b. Inflorescence cymiform, leafy, not surpassing the foliage *d*
    d. Primocane prickles many, mostly 2-5 mm apart; central primocane leaflet gradually long-acuminate . 33. *R. recurvans*
    d. Prickles fewer, mostly 1 cm or more apart; central primocane leaflet abruptly short-acuminate . . . 34. *R. frondosus*
a. Primocane leaves not pubescent beneath to the touch *e*
  e. Canes unarmed or nearly so; inflorescence long-racemiform, exceeding the foliage . . . . 35. *R. canadensis*
  e. Canes armed; inflorescence short-racemiform, about equaling the foliage *f*
    f Canes armed with many slender prickles, 2-5 mm apart; primocane leaflets ovate, long-acuminate . . 36. *R. elegantulus*
    f. Canes armed with fewer, heavier prickles, mostly 1 cm or more apart; primocane leaflets mostly obovate, abruptly short-acuminate . . . . 37. *R. quaesitus*

1. **R. hispidus** L. Dewberry. Evergreen trailer.—Sphagnum bogs and moist ground.

2. **R. exter** Bailey. Usually a long trailer, freely tip-rooting; primocane axis densely shaggy above, glandular.—Occurs over much of Wisconsin.

3. **R. plus** Bailey. Trailer.—Juneau, Lincoln, Price, and Wood counties.

4. **R. permixtus** Blanch. Trailer; inflorescence racemiform, standing well above the foliage.—Juneau County.

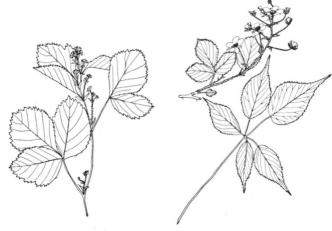

Fig. 275. *Rubus wheeleri,* × .30.   Fig. 276. *Rubus allegheniensis,* × .25.

5. **R. dissensus** Bailey. Canes erect or arching.—Iron County.

6. **R. stipulatus** Bailey. Erect.—Common over much of Wisconsin on moist ground.

7. **R. groutianus** Blanch. Erect.—Iron and Oconto counties.

8. **R. navus** Bailey. Erect.—Ashland County.

9. **R. junceus** Blanch. Erect.—Bayfield and Sawyer counties.

10. **R. vermontanus** Blanch. Canes usually erect.—Common in the northern counties.

11. **R. wheeleri** Bailey. (Fig. 275).—Bayfield, Juneau, Price, and Sawyer counties.

12. **R. uniformis** Bailey. Primocane armature consisting entirely of uniform soft bristles.—Northern Wisconsin.

13. **R. spectatus** Bailey.—Bogs and other moist places; across northern Wisconsin, and in Adams County.

14. **R. regionalis** Bailey.—Northern Wisconsin.

15. **R. gulosus** Bailey.—Shawano County.

16. **R. wisconsinensis** Bailey. Canes erect or recurving. While typical *R. wisconsinensis* is entirely glandless, a form

with decidedly glandular pedicels occurs in the northern part
of the range.—Throughout Wisconsin.

17. **R. glandicaulis** Blanch. Canes erect to recurving.—
Many counties in northern Wisconsin.

18. **R. perspicuus** Bailey.—Marinette and Wood counties.

19. **R. missouricus** Bailey.—Adams and Juneau counties.

20. **R. multifer** Bailey.—Abundant in Dunn, Pierce, and St.
Croix counties.

21. **R. steelei** Bailey.—Adams, Green, Iowa, Waupaca, and
Waushara counties.

22. **R. baileyanus** Britt.—Juneau County.

23. **R. plicatifolius** Blanch.—Scattered throughout the
state.

24. **R. multiformis** Blanch.—Clark County.

25. **R. ithacanus** Bailey.—Scattered localities in the state.

26. **R. exsularis** Bailey.—Milwaukee County.

27. **R. schoolcraftianus** Bailey.—Sawyer County.

28. **R. allegheniensis** Porter. Common Blackberry (Fig.
276). Throughout the state.

29. **R. rosa** Bailey. This species has a shorter raceme than
R. allegheniensis, and stouter pedicels more closely placed.—
Scattered localities in the state.

30. **R. alumnus** Bailey.—Southern, western, and central
Wisconsin.

31. **R. pergratus** Blanch.—Dodge, Door, Iowa, Milwaukee,
Waukesha, and Vilas counties.

32. **R. bellobatus** Bailey.—Scattered localities.

33. **R. recurvans** Blanch. Low-arching, glandless plants,
occasionally tip-rooting, but not trailing; inflorescence normally a short leafy few-flowered raceme, about equaling the
foliage; pedicels and peduncles woolly.—Adams, Ashland,
Juneau, and Waukesha counties.

34. **R. frondosus** Bigel. Canes erect to arching; central
primocane leaflet ovate, widely rounded to cordate at the
base.—Adams, Bayfield, Dodge, Douglas, Marinette, Marathon, Oconto, and Waukesha counties.

35. **R. canadensis** L.—Northern and eastern Wisconsin
south to Sheboygan County.

36. **R. elegantulus** Blanch. Canes armed with many
slender prickles 2-5 mm apart.—Bayfield and Iron counties.

37. **R.** quaesitus Bailey. Excellent fruit, drought resistant.—Bayfield and Waupaca counties.

5. Geum Avens

Perennial herbs; leaves pinnate, the largest often in a basal rosette; stipules large, kidney-shaped; petals white, yellow, or sometimes suffused with red; stamens 10-many; calyx with bractlets outside the lobes and alternating with them; achenes many, with long plumose persistent styles.

The character of the receptacle, important in the first 4 species, may be seen by removing some of the fruits and examining the structure on which they were borne.

a. Calyx lobes green, turned back (Fig. 280) *b*
  b. Petals white or greenish yellow; stipules 7-15 mm long *c*
    c. Calyx usually with alternating bractlets; head of carpels sessile in the calyx *d*
      d. Petals equaling the calyx lobes (Fig. 277); receptacle densely hairy . . . . .1. *G. canadense*
      d. Petals shorter than the calyx lobes (Fig. 278); receptacle nearly glabrous . 2. *G. laciniatum*, var. *trichocarpum*
    c. Calyx without alternating bractlets; head of carpels raised on a short stipe . . . . 3. *G. vernum*
  b. Petals bright yellow; stipules longer, deeply cut *e*
    e. Lower internode of style not glandular; receptacle densely hairy . . . . 4. *G. aleppicum*, var. *strictum*
    e. Lower internode of style with minute stalked glands; receptacle nearly glabrous . . 5. *G. macrophyllum*, var. *perincisum*
a. Calyx lobes purple, ascending (Fig. 281) *f*
  f. Bractlets about half as long as the calyx lobes (Fig. 281) . . . . . .
  . . . . 6. *G. rivale*
  f. Bractlets longer than the calyx lobes (Fig. 282) . 7. *G. triflorum*

1. **G.** canadense Jacq. (Fig. 277). Stems 6-11 dm high, forked above, slightly hairy or woolly; leaves pinnately 3-5-foliate.—Woods and thickets, throughout the state.

2. **G.** laciniatum Murr., var. **trichocarpum** Fern. (Fig. 278). Stems and petioles bristly-hairy; lowest leaves compound, or often simple with blades heart-shaped at base; stem leaves 3-cleft.—Local, throughout the state.

3. **G.** vernum (Raf.) T. & G. (Fig. 279). Plants 3-6 dm tall; basal leaves both long-petioled and simple, some smaller and pinnate; stem leaves pinnate or trifoliate, several-toothed or deeply cleft; sepals triangular, reflexed; petals pale yellow, 1-2 mm long; head of achenes on a stipe 1-2 mm long.—Disturbed woods in Dane, Green, and Rock counties.

4. **G.** aleppicum Jacq., var. **strictum** (Ait.) Fern. (Fig. 280). Stems 9-15 dm high, hairy; leaflets of stem leaves much like those of No. 2, but narrower; petals conspic-

Fig. 277. *Geum canadense,* × .40.   Fig. 279. *Geum vernum,* × .40.
Fig. 278. *Geum laciniatum,* var.
trichocarpum, × .40.

uous.—Moist openings in woods, throughout the state.

5. **G. macrophyllum** Willd., var. **perincisum** (Rydb.) Raup. Similar to No. 4.—Rare, Washburn County.

6. **G. rivale** L. Water Avens (Fig. 281). Stems with few leaves; leaflets of lower leaves roundish, toothed all around, the terminal one 3-lobed; style about 1 cm long in

Fig. 280. *Geum aleppicum,* var. *strictum,* × .40.

Fig. 281. *Geum rivale,* × .40.   Fig. 282. *Geum triflorum,* × .40.

fruit.—Moist meadows along Lake Michigan and scattered localities from Fond du Lac to La Crosse counties northward.

7. **G. triflorum** Pursh. Prairie Smoke (Fig. 282). Plants with soft hairs; leaves mostly at base of the stem; leaflets many, narrowed at base, coarsely toothed toward the apex only; sepals about 1 cm long, exceeding the white to purplish petals; style becoming 5 cm long in fruit.—Dry fields and prairies, north to Pierce, Waupaca, and Marinette counties.

6. Fragaria   Strawberry

Perennial herbs, with numerous elongate stolons; leaves all basal, compound with 3-toothed leaflets; hypanthium flat; sepals 5, with alternating bracts of equal size; petals 5, white, round; stamens many; carpels many, borne on the enlarged receptacle; styles lateral; fruits achenes on greatly enlarged juicy red receptacle.

a. Flowering scape shorter than the leaves; tip tooth of leaflet shorter than the next 2 teeth ....1. *F. virginiana*
a. Flowering scape taller than the leaves; tip tooth of leaflet equal to or longer than the next 2 teeth .... 2. *F. vesca*, var. *americana*

1. **F. virginiana** Duchesne (Fig. 283). Inflorescence flattish-topped, the primary branches subequal, usually overtopped by the leaves; flowers about 2 cm in diameter; sepals appressed to the young fruit; achenes in pits in the mature receptacle; petioles and scapes with appressed or spreading hairs; pedicels with appressed hairs.—Common in sunny fields, throughout the state.

2. **F. vesca** L., var. **americana** Porter. Woodland Strawberry (Fig. 284). Inflorescence more irregular, the primary branches unequal, usually overtopping leaves; flowers about 1.3 cm in diameter; sepals turned back from the young fruit; achenes superficial on the mature receptacle; fruit slenderly ovoid to ellipsoid.—Frequent throughout the state, mostly in woodland or mesic habitats.

7. **Waldsteinia**   Barren Strawberry

Low stoloniferous perennial herbs; leaves mostly basal, 3-foliate; hypanthium obconic; sepals 5, alternating with early deciduous bractlets; petals 5, yellow; stamens many; carpels 2-6, free, on a hairy receptacle; fruits achenes.

W. fragarioides (Michx.) Tratt. (Fig. 285). Plants 1-1.8 dm high; leaf blades almost round, the leaflets short-stalked and

Fig. 283. *Fragaria virginiana,* × .30.

Fig. 284. *Fragaria vesca,* var. *americana,* × .30.

Fig. 285. *Waldsteinia fragarioides,* × .30.

shallowly lobed at tip; flowers on the bracted scapes, yellow; achenes dry, 2-6 on a dome-shaped receptacle.—Dry sandy woods and barrens, from Marathon to Shawano counties northward.

8. **Potentilla** Cinquefoil; Five-finger

Herbs or shrubs, annual to perennial; leaves compound, stipulate; hypanthium shallow to cup-shaped; sepals (4) 5, alternating with as many bractlets; petals (4) 5; stamens 5-many, usually 20; carpels 10-many, borne on a prolonged receptacle; styles terminal, lateral or nearly basal, jointed to the ovary; fruits dry achenes on a dome-shaped receptacle.

a. Leaflets 3 *b*
  b. Petals yellow; leaflets toothed along the sides . . . .
    . . . . 1. *P. norvegica,* var. *hirsuta*
  b. Petals white; leaflets 3-toothed at tip . . . . 2. *P. tridentata*
a. Leaflets 5 or more, at least on lower leaves *c*
  c. Leaves palmately compound *d*
    d. Leaflets not woolly beneath, but often with long spreading hairs *e*
      e. Flowers many, in a much-branched inflorescence . . . .
        . . . . 3. *P. recta*
      e. Flowers few, each on a long peduncle in the axil of a leaf . . . . 4. *P. simplex*
    d. Leaflets woolly beneath *f*
      f. Leaflets with margins strongly inrolled; calyx and pedicels densely white-woolly . . . . 5. *P. argentea*
      f. Leaflets with flat margins; calyx and pedicels silky-hairy . . . . .6. *P. intermedia*
  c. Leaves pinnately compound *g*
    g. Stems woody . . . . 7. *P. fruticosa*
    g. Stems not woody *h*
      h. Stems leafy, erect or reclining at base but not rooting at each node; flowers several or many in a branched inflorescence *i*
        i. Petals purple; leaflets several times as long as wide . . .
          . . . . .8. *P. palustris*
        i. Petals cream-colored; leaflets seldom more than twice as long as wide . . . . 9. *P. arguta*
      h. Stems mostly flat on the ground, radiating from a rosette of leaves and rooting at each node; flowers borne singly or a few together at each node or in the rosette . . . .
        . . . . .10. *P. anserina*

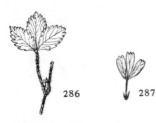

Fig. 286. *Potentilla norvegica,*
var. *hirsuta,* × .40.
Fig. 287. *Potentilla tridentata,*
× .40.

Fig. 288. *Potentilla recta,* × .40.

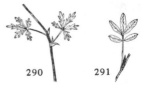

Fig. 290. *Potentilla argentea,*
× .40.
Fig. 291. *Potentilla fruticosa,*
× .40.

Fig. 289. *Potentilla simplex,*
× .40.

Fig. 292. *Potentilla palustris,*
× .40.

Fig. 293. *Potentilla arguta,* × .25.

Fig. 294. *Potentilla anserina*, × .30.

1. **P. norvegica** L., var. hirsuta (Michx.) T. & G. (Fig. 286). Stems erect, 2-9 dm high, with stiff spreading hairs; petals small, about as long as the sepals.—Dry soil, throughout the state.

2. **P. tridentata** Ait. Three-toothed Cinquefoil (Fig. 287). Stems 3-22 dm high, somewhat woody at base; leaflets 3, entire except for the 3 teeth at the apex; flowers white.—Dry sterile rocky or sandy soil, south to the Wisconsin Dells and rarely to Dane County.

3. **P. recta** L. (Fig. 288). Plants coarse, erect, hairy; leaflets 5-7, with many coarse blunt teeth.—A European plant recently introduced into this region; locally abundant in southern Wisconsin, northward in scattered localities to Ashland County.

4. **P. simplex** Michx. Old-field Cinquefoil (Fig. 289). Stems erect or reclining, often rooting at the nodes, and usually with spreading hairs; leaflets 3, the 2 lateral deeply cleft so that there appear to be 5 leaflets; flowers bright yellow.—Very common in pastures and on roadsides.

Var. calvescens Fern. Stems glabrous or with appressed hairs.—Adams, Eau Claire, and Douglas counties.

Var. argyrisma Fern. Leaflets silvery-silky.—Richland County.

5. **P. argentea** L. Silvery Cinquefoil (Fig. 290). Plants usually low, sometimes 5 dm high, much branched; sepals and lower surfaces of the leaflets whitened with a silky wool; petals a little longer than the sepals.—Dry soil, throughout the state.

6. **P. intermedia** L. Plants 3-7 dm high; leaflets coarsely dentate above, entire near the base; inflorescence much branched, flowers 8-10 mm wide; achenes with conspicuous longitudinal ridges.—Introduced from Europe; a rare weed in 5 scattered counties.

7. **P. fruticosa** L. Shrubby Cinquefoil (Fig. 291). Stems

much branched, woody, 1-8 dm high; leaves pinnately compound; leaflets 5-7, small, entire, whitened beneath; flowers bright yellow, conspicuous.—Occasional in springy meadows and borders of bogs southward and eastward, also on cliffs in Crawford and St. Croix counties.

8. **P. palustris** (L.) Scop. Marsh Five-finger (Fig. 292). Stems somewhat woody at base; leaves pinnately compound; bases of the petioles sheathing the stem.—Bogs, throughout the state except in the southwestern part.

9. **P. arguta** Pursh. Tall Cinquefoil (Fig. 293). Stems 3-10 dm high, stout, rough-hairy; basal leaves long-petioled, the petioles decreasing in length in the upper leaves; leaflets velvety beneath; flowers many, crowded, the petals cream-colored.—Dry open ground, throughout the state.

10. **P. anserina** L. Silverweed (Fig. 294). Spreading by long runners; leaflets whitened beneath with silky hairs.—Sandy shores, Lake Michigan and occasionally inland.

9. Rosa   Rose

Shrubs, characterized partly by the presence of various prickles, bristles, or thorns, sometimes a pair of enlarged thorns known as infrastipular prickles just below each node; leaves pinnate, the leaflets toothed and mostly blunt or rounded at apex; stipules joined for most of their length with the petiole, the free part triangular; calyx tube (or receptacle) urn-shaped, constricted at apex, becoming red and fleshy in fruit, enclosing the bony achenes; stamens very numerous; petals 5, usually pink or white.—A variable, probably freely hybridizing group, in which many species and varieties have been described, and many more doubtless will be described. The present treatment is necessarily very conservative, including only the more well-marked forms.

a. Styles united in a column, exserted above the receptacle *b*
  b. Leaflets large, 3-9 cm long, 3-5 per leaf; petals pink . . . .
      . . . . 1. *R. setigera*
  b. Leaflets small, 1.5-3 cm long, 5-11 per leaf; petals white . . . . . .
      . . . . .2. *R. multiflora*
a. Styles distinct from one another, stigmas forming a closed head *c*
  c. Stems usually arching, with strong, curved, and flattened thorns, 8 mm or longer; sepal lobes pinnate *d*
      d. Leaflets more or less glandular; pedicels and hypanthium glandular-hispid . . . . 3. *R. eglanteria*
      d. Leaflets not glandular; pedicels and hypanthium without glands . . . . .4. *R. canina*
  c. Stems mostly upright, thorns less than 8 mm long; sepal lobes mostly entire *e*

e. Infrastipular prickles present and differentiated from internodal prickles (Fig. 295) *f*
  f. Canes and underside of stipules densely tomentose; leaflets strongly veined, thick with prominent reticulation on underside . . . . **5.** *R. rugosa*
  f. Canes glabrous or essentially so *g*
    g. Leaflets oblong; stipules and sepals glandless . . . .
    . . . . **6.** *R. cinnamomea*
    g. Leaflets oblong-obovate; stipules and sepals glandular
    . . . . **7.** *R. woodsii*
e. Infrastipular prickles absent, or undifferentiated from internodal prickles *h*
  h. Prickles lacking, or mostly confined to the base of the stem . . . . **8.** *R. blanda*
  h. Prickles present *i*
    i. Pedicels and hypanthium glandular *j*
      j. Flowers solitary or corymbose; if solitary, the pedicel bracteate near base *k*
        k. Prickles stout, curved, broad-based; terminal leaflet with 14-20 fine teeth on each side above the middle (Fig. 297) . . . .**9.** *R. palustris*
        k. Prickles slender, straight, not very broad-based; terminal leaflet with 9-13 coarser teeth on each side above the middle (Fig. 298).**10.** *R. carolina*
      j. Flowers solitary; the pedicel not subtended by a bract . . . .**11.** *R. spinosissima*
    i. Pedicels and hypanthium nonglandular, prickly all the way to the floral branches *l*
      l. Leaflets 5-7; flowers single (rarely few) . . . .
      . . . . **12.** *R. acicularis*
      l. Leaflets 7-11, obovate; flowers 3-10 . . . .
      . . . . **13.** *R. arkansana*

1. R. setigera Michx. Prairie Rose. Stems climbing or arching, with stout flattened prickles; leaflets mostly 3, on flowering stems, glabrous to tomentose beneath; stipules very narrow; pedicels, hypanthium, and sepals stipitate-glandular; flowers numerous, 4-8 cm wide, pink to white; styles united in a column as long as the stamens.—Escaping from cultivation; rare, Dane and Milwaukee counties.

2. R. multiflora Thunb. Multiflora Rose. Stems climbing or arching; leaflets mostly 7-9; inflorescence many-flowered; flowers white, 2-4 cm wide; styles united in a column.— Native of East Asia, becoming naturalized after cultivation; from Dane and Rock counties eastward.

3. R. eglanteria L. Sweetbrier; Eglantine. Stems usually arching, armed with very broad based flattened prickles; leaflets 5-9, dark-dotted with glands beneath, aromatic; sepals mostly pinnately lobed; flowers pink, 3-5 cm wide.—Native of Europe; from Iowa and Lafayette counties eastward to Lake Michigan.

4. **R. canina** L. Dog Rose. Tall or arching shrubs; stems armed with stout decurved prickles; leaflets 5-7, glabrous; pedicels and hypanthium usually glabrous; flowers pink to white, 4 cm wide.—Native of Europe, escaping from cultivation; Columbia County.

5. **R. rugosa** Thunb. Rugose Rose (Fig. 295). Stems densely prickly, the infrastipular prickles larger, decurved; young stems and prickles conspicuously villous-tomentose; leaflets 7-9, wrinkled above, strongly net-veined on under side; flowers 7-12 cm wide, pink or white.—Native of East Asia, escaping from cultivation; Douglas and Dane counties.

6. **R. cinnamomea** L. Cinnamon Rose. Stems 1-2 m high, with prominent curved infrastipular prickles and a few between the nodes; leaflets 5-7, densely pubescent beneath; hypanthium and pedicels glabrous; flowers pink, 4-5 cm wide.—Native of Eurasia, sometimes escaping from cultivation; Lincoln, Sheboygan, and Walworth counties.

7. **R. woodsii** Lindl. Leaflets 5-9, obovate, coarsely serrate; flowers pink, about 3 cm broad.—A western species of prairies; Columbia and Lafayette counties.

8. **R. blanda** Ait. Meadow Rose (Fig. 296). Plants usually less than 1 m high; leaflets 5-9; leaf rachis sometimes smooth, often hairy or woolly, rarely with stalked glands; flowers 4-6 cm wide, usually 3 or more at the end of a branch 1 dm or more long, white or pink.—Usually in moist places, throughout the state.

Forma **carpohispida** (Schuette) Lewis is more or less armed with prickles.—Local in north-central Wisconsin.

9. **R. palustris** Marsh. Swamp Rose (Fig. 297). Stems to 2 m high, armed only with infrastipular prickles which are stout, conic, decurved, and flattened at base; leaflets usually 7, oblanceolate, finely serrate; flowers solitary or in corymbs 4-5.5 cm wide; pedicels and hypanthium stipitate-glandular.—Borders of swamps and streams in central and eastern Wisconsin.

Forma **inermis** (Regel) Lewis has stems without armature.—Rare.

10. **R. carolina** L. Pasture Rose (Fig. 298). Stems to 1 m high; many internodal prickles, mostly straight, slender, and round at base, alike in form but varying in size; leaflets 3-7, coarsely toothed; pedicels and hypanthium stipitate-glandular; sepals long-attenuate; flowers pink, 3.5-5.5 cm wide.—Borders of woods and roadsides, throughout Wisconsin.

Fig. 295. *Rosa rugosa,* × .20.

Fig. 296. *Rosa blanda,* × .40,
margin × 1.

Fig. 297. *Rosa palustris,* × .40.

Fig. 298. *Rosa carolina,* × .40.

Fig. 299. *Rosa acicularis,* ssp. *sayi,* × .40, margin × 1.

11. **R. spinosissima** L. Scotch Rose. Stems to 1 m tall, armed with dense slender straight prickles; leaflets 5-11; pedicels and hypanthium glabrous; flowers white, yellow, or pink, solitary, 2-4 cm wide, lacking bract subtending the pedicel.—Native of Eurasia, escaping from cultivation or persisting after cultivation; Eau Claire and Bayfield counties.

12. **R. acicularis** Lindl., ssp. **sayi** (Schw.) Lewis (Fig. 299). Stems to 1 m tall, armed with fine bristles to the apex of floral branches; leaflets 5-7, ovate; pedicels and hypanthium nonglandular; flowers 1-few, 4-6 cm wide, pink.—A

northern species in Wisconsin, occurring south to Crawford and Juneau counties.

13. R. arkansana Porter. Prairie Wild Rose. Stems densely prickly all the way to the floral branches; leaflets 7-11, obovate; petioles and leaf rachis pubescent; inflorescence corymbose; flowers 3-10, 4-6 cm wide; petals pale and often streaked with darker pink.—Rocky slopes, thickets, and dry prairies, southern Wisconsin and locally north to Bayfield County.

10. Amelanchier   Juneberry

Shrubs or small trees, with smooth gray bark; leaves simple, toothed; hypanthium obconic, bell-shaped or saucer-shaped; inflorescence mostly racemose; sepals 5; petals 5, white (rarely pinkish); stamens usually 20; ovary 5-celled, inferior but the upper surface exposed; styles 5, free or partly united; carpels 5, each divided by a partition so that there are 10 compartments in the fleshy edible fruit, each with 1 seed; fruit purplish black at maturity.—A variable group, the species probably freely hybridizing. The following treatment describes only the typical form of each species; a large proportion of the individuals in any region will combine to some extent the characters of several species.

a. Flowers in clusters of 2 or 3, 1 cluster terminal and the others in the axils of leaves; leaves glabrous . . . . 1. *A. bartramiana*
a. Flowers in racemes *b*
  b. Flowers showy, petals 11-25 mm long *c*
    c. Top of ovary glabrous; leaves finely serrate *d*
      d. Flowers appearing before leaves or leaves less than half grown; young leaves white-tomentose at flowering time; mature leaves firm, dark green above, paler beneath, pubescent along midvein . . . . .2. *A. arborea*
      d. Leaves half grown and glabrous at flowering time, often bronze-colored, glabrous and firm when mature . . . .
      . . . . 3. *A. laevis*
    c. Top of ovary tomentose; leaves coarsely serrate, the lower sides densely tomentose when young . . . . . . .4. *A. sanguinea*
  b. Flowers smaller, petals mostly less than 12 mm long *e*
    e. Flowers in dense erect short racemes, 1.5-4 cm long; leaves about half grown at flowering time, densely tomentose, becoming glabrous at maturity . . . . 5. *A. spicata*
    e. Flowers in nodding racemes, 4-7 cm long; leaves in bud stage or just unfolded at flowering time, sparsely pubescent to glabrous . . . . 6. *A. interior*

1. A. bartramiana (Tausch) Roem. (Fig. 300). Shrubs 0.5-3 m high; leaves rounded to acute at apex, finely toothed to below middle or near base, half grown at flowering time, veins irregular; flowers in terminal and axil-

Fig. 300. *Amelanchier bartramiana,* × .50.

Fig. 301. *Amelanchier arborea,* × .50.

Fig. 302. *Amelanchier laevis,* × .50.

Fig. 303. *Amelanchier interior,* × .50.

lary clusters of 2 or 3, petals 6-9 mm long; top of ovary tomentose; fruit 1-1.5 cm in diameter, edible but insipid. —Lake Superior region, Ashland and Vilas counties.

2. **A. arborea** (Michx. f.) Fern. (Fig. 301). Shrubs or trees, to 20 m tall; leaves white-tomentose when young, mature leaves firm, dark green, paler beneath and pilose at least along the midvein, 6-10 teeth per cm; flowers in racemes, appearing mostly before leaves; petals 12-25 mm long; top of ovary glabrous; fruit 6-10 mm in diameter, tasteless, falling early.—Throughout the state.

3. **A. laevis** Wieg. (Fig. 302). Shrubs or small trees, to 13 m tall; leaves usually purplish red or bronze-colored, half grown at flowering time, 6-8 teeth per cm, apex acute; flowers in racemes, petals 12-25 mm long; top of ovary glabrous; fruit 6-8 mm in diameter, sweet, succulent, and edible. —Common, throughout the state.

4. **A. sanguinea** (Pursh) DC. Shrubs or small trees, to 6 m tall; leaves half to full grown at flowering time, coarsely toothed, teeth 4-6 per cm, apex mucronate; petals 11-22 mm

long; top of ovary tomentose; fruit 6-8 mm in diameter, sweet, juicy, and edible.—Scattered localities, throughout the state.

5. **A. spicata** (Lam.) K. Koch. Dwarf colonial shrubs, to 2 m tall; young leaves densely tomentose at flowering time, mature leaves glabrous, finely serrate, teeth 5-8 per cm, apex mucronate or obtuse; flowers in dense short racemes, 1.5-4 cm long; petals 4-10 mm long; top of ovary tomentose; fruits 6-8 mm in diameter, glaucous, sweet, juicy, and edible. —Common, throughout the state.

6. **A. interior** Nielsen (Fig. 303). Straggly shrubs or small trees; leaves unfolded at flowering time, ovate, 5-6 teeth per cm, glabrous, apex acute; flowers in loose racemes, 4-7 cm long; petals 8-13 mm; top of ovary tomentose; fruit 6-8 mm in diameter, sweet, juicy, and edible.—Scattered localities, throughout the state.

11. Crataegus   Hawthorn

Usually thorny shrubs or small trees; leaves simple, toothed, usually lobed; inflorescence a simple or compound cyme; hypanthium cup-shaped; sepals 5; petals 5, white or pink; stamens 5-25; carpels 1-5; fruit a pome, red, yellow, blue, or black, the carpels bony, surrounded by the thin slightly fleshy receptacle. Approximately 100 species comprise this genus in Wisconsin. Frequent hybridization among the species makes the taxonomy very complex and difficult. Often both flowers and fruit are needed for proper identification. Therefore, a species treatment is not included in this manual, but some of the principal species are listed below. The genus occurs throughout the state.

C. calpodendron (Ehrh.) Medic.
C. chrysocarpa Ashe
C. grus-galli L.
C. disperma Ashe
C. dodgei Ashe
C. fontanesiana (Spach) Steud.
C. holmesiana Ashe
C. macracantha Lodd.
C. macrosperma Ashe
C. mollis (T. & G.) Scheele
C. oxyacantha L.
C. pedicellata Sarg.
C. pruinosa (Wendl.) K. Koch
C. punctata Jacq.
C. roanensis Ashe
C. succulenta Link.

12. Pyrus   Apple

Small trees or shrubs, sometimes thorny; leaves simple, toothed or lobed; inflorescences umbellike clusters on dwarf lateral branches; flowers large, showy; hypanthium globose to obovoid; sepals 5; petals 5, elliptical to obovate, short-clawed; stamens 15-50; ovary 5-celled, inferior; styles 2-5; fruit a pome, each papery or cartilaginously walled cell with 2 seeds.

a. Leaves finely serrate; anthers yellow; stems lacking thorns; sepals persistent in fruit . . . . 1. *P. malus*
a. Leaves coarsely serrate and usually shallowly lobed; anthers pink to salmon-colored; stems often with spinescent branches; sepals often deciduous in fruit . . . . 2. *P. ioensis*

1. **P. malus** L. Apple (Fig. 304). Small trees; leaves oblong-ovate, rounded to cordate at base, pubescent beneath; calyx-lobes white- to gray-tomentose; petals pinkish white.— Persisting after or escaping from cultivation; throughout the state.

2. **P. ioensis** (Wood) Bailey. Wild Crab Apple, Iowa Crab (Fig. 305). Small much-branched trees or shrubs, twigs often short and thornlike; leaf blades coarsely and doubly toothed and somewhat lobed, narrowed at base; flowers white or pinkish, very conspicuous and fragrant; fruit green when ripe, sticky.—Fields and open woods, north to Vernon and Dodge counties.

13. **Sorbus**   Mountain Ash

Small trees or shrubs; leaves pinnately compound (in our species), with 11-17 serrate leaflets; flowers white, numerous in rounded or flattened clusters; hypanthium obconic; sepals 5; petals 5; stamens 15-20; pistils 2-4, united below, half inferior; styles free; fruit a red pome.

a. Sepals and winter buds densely woolly; underside of leaves hairy . . . . 1. *S. aucuparia*
a. Sepals and winter buds with few scattered hairs or none; underside of leaves without hairs or very slightly hairy *b*

Fig. 304. *Pyrus malus,* × .40.

Fig. 305. *Pyrus ioensis,* × .40.

Fig. 306. *Sorbus aucuparia,* × .50.   Fig. 307. *Sorbus americana,* × .50.
                                         Fig. 308. *Sorbus decora,* × .50.

1. **S. aucuparia** (L.) Gaertn. European Mountain Ash; Rowan Tree (Fig. 306). Leaflets obtuse, 2-6 cm long, hairy beneath; hypanthium densely white-hairy; fruit about 10 mm in diameter.—A native of Europe, cultivated, sometimes escaping; southern Wisconsin and a few localities near Lake Michigan.

2. **S. americana** (Marsh.) DC. American Mountain Ash (Fig. 307). Small slender trees; leaflets lanceolate, 4-9 cm long, acuminate; flowers 5-6 mm broad, very many together in a flat-topped cluster; fruit 4-7 mm in diameter.—Woods and bluffs south to Barron, Dane, and Waukesha counties.

3. **S. decora** (Sarg.) Hyland. Mountain Ash (Fig. 308). Similar to No. 2; leaflets 11-17, oblong, acute to short acuminate; flowers 8-11 mm broad; inflorescence and leaf rachis sometimes hairy, particularly northward; fruit 6-10 mm diameter.—Woods in northern Wisconsin, occasionally coming south on sandstone bluffs in the Driftless Area.

14. Aronia   Chokeberry

Shrubs, with simple serrate leaves, the midvein on the upper side with a row of black glands; inflorescence a cluster of small flowers; hypanthium broadly obconic and extended beyond the inferior ovary; sepals 5; petals 5; stamens usually 20; ovary densely woolly on the summit; styles 5, united at base and long-persistent; fruit a small pome.

A. melanocarpa (Michx.) Ell. (Fig. 309). Shrubs 1-2 m high; leaves nearly smooth; petals white, roundish; fruit black, 6-8 mm thick.—Rocky places or bogs, throughout the state except Door County.

Fig. 309. *Aronia melanocarpa,* × .45.

CAESALPINIACEAE   Caesalpinia Family

Trees (our species); leaves alternate, stipulate, compound; flowers regular or irregular; hypanthium usually well developed; sepals 3-5; petals 3-5; stamens 3-10, distinct; ovary 1; fruit a legume.

a. Leaflets broadest below the middle, gradually tapering into a long point . . . . .1. *Gymnocladus*
a. Leaflets oval with parallel sides, blunt or with a short abrupt point . . . . .2. *Gleditsia*

1. Gymnocladus   Kentucky Coffee Tree

Leaves large, bipinnately compound; flowers in terminal panicles, mostly radially symmetrical; hypanthium tubular; sepals 5; petals 5; stamens 10, free, alternately long and short; fruit flat, oblong, woody, with few large seeds.

G. dioica (L.) K. Koch (Fig. 310). Tall trees, without thorns; leaves 6-9 dm long, once or twice pinnate; flowers nearly regular, about 2 cm long, short-hairy; pod oblong, 1.5-2.5 dm long, 3-4 cm broad, pulp inside, with flattish seeds.—Rather rare; southern Wisconsin, and north to Buffalo, Dane, and Outagamie counties.

2. Gleditsia   Honey Locust

Leaves, pinnate or bipinnate; flowers mostly unisexual in spikelike racemes, radially symmetrical; sepals and petals 3-5; stamens 3-10, exserted; fruit large, flat, oval to elongate.

G. triacanthos L. (Fig. 311). Trees, with long stout often 3-branched thorns; flowers small, nearly regular, in a long dense spike; fruit a broad flat pod 2-4.5 dm long.—River bottoms, north to Crawford and Dane counties; often planted, especially the thornless form.

Fig. 310. *Gymnocladus dioica,*   Fig. 311. *Gleditsia triacanthos,*
X .30.   X .45.

FABACEAE   Bean Family

Herbs, vines, shrubs, and trees; leaves alternate, stipulate, simple to mostly ternately or pinnately compound; sepals 3-5, often united into a tube which may be regular or irregular (divided into unequal lips) or toothed; flowers sometimes nearly regular, but usually bilaterally symmetrical, the upper petal (standard) largest and usually erect, overlapping the 2 lateral ones (wings) which are oblique and in turn outside the 2 lower which are united to form the keel; keel ordinarily enclosing the 10 (rarely 5) stamens and single pistil; only 1 petal, the standard, present in *Amorpha*; stamens distinct or usually 9 united and 1 free, or rarely all 10 united with fused filaments; fruit a legume.

a. Trees, shrubs, or plants at least partially woody *b*
  b. Stems whitened with copious hairs . . . . 1. *Tephrosia*
  b. Stems scarcely hairy *c*
    c. Flowers small, hardly pediceled, in a close cylindrical spike (Fig. 313) . . . . .2. *Amorpha*
    c. Flowers large, usually less than 20 in a loose raceme, on pedicels 5 mm or more long (Fig. 314) . . . . .3. *Robinia*

a. Herbs *d*
    d. Stamens not joined to one another . . . . 4. *Baptisia*
    d. Stamens joined in 1 or 2 groups *e*
        e. Leaves palmately foliate *f*
            f. Calyx 2-lobed . . . . 5. *Lupinus*
            f. Calyx 5-lobed . . . . 6. *Psoralea*
        e. Leaves not palmately foliate *g*
            g. Leaves 3-foliate, with no tendrils *h*
                h. Flowers in a close globular head; fruit a straight tiny pod . . . . 7. *Trifolium*
                h. Flowers in an elongate head or spike *i*
                    i. Spike many times as long as broad; pods wrinkled, straight . . . . 8. *Melilotus*
                    i. Spike not over 3 times as long as broad; pods coiled . . . . 9. *Medicago*
            g. Leaves pinnately compound, the terminal leaflet sometimes modified as a tendril *j*
                j. Leaves without tendrils *k*
                    k. Flowers in heads or umbels *l*
                        l. Flowers yellow to brick red *m*
                            m. Bracts subtending floral heads 3-cleft . . . . .10. *Anthyllis*
                            m. Bracts subtending floral heads entire . . . . .11. *Lotus*
                        l. Flowers white to pink . . . . 12. *Coronilla*
                    k. Flowers in racemes or elongate heads *n*
                        n. Each leaflet with many conspicuous lateral veins running from midrib to margin . . . . 1. *Tephrosia*
                      n. Each leaflet with a few obscure and irregular veins which seldom reach the margin *o*
                        o. Keel not tipped with a sharp point . . . . 13. *Astragalus*
                        o. Keel tipped with an abrupt sharp point . . . . 14. *Oxytropis*
            j. Leaves with tendrils *p*
                p. Lateral petals joined to the keel; style bearded at the tip . . . . 15. *Vicia*
                p. Lateral petals not joined to the keel; style bearded down the side . . . . 16. *Lathyrus*

1. Tephrosia   Hoary Pea; Goat's Rue

Partially woody perennial herbs, with long taproots; leaves pinnately compound; leaflets mucronate, with numerous straight parallel lateral veins; calyx 5-cleft; flowers zygomorphic; stamens 10, 9 united and 1 partially free; style bearded along inner side; pod linear, dehiscent.

T. virginiana (L.) Pers., var. **holosericea** (Nutt.) T. & G. (Fig. 312). Whole plants somewhat whitened with silky hairs; stems rather woody at base, 3-6 dm high; flowers 1-1.5 cm long, yellowish, marked with purple; pod linear, flat, several-seeded.—Dry sandy or rocky ground, north to Jackson and Juneau counties. Typical *T. virginiana,* with leaflets perfectly glabrous above, is rare in Wisconsin.

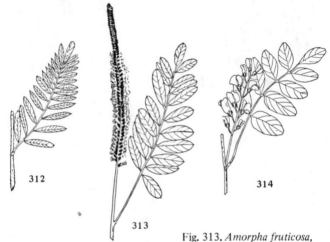

Fig. 312. *Tephrosia virginiana,*
var. *holosericea,* × .40.

Fig. 313. *Amorpha fruticosa,*
var. *angustifolia,* × .40.
Fig. 314. *Robinia pseudo-acacia,*
× .40.

2. Amorpha

Shrubs with pinnately compound leaves; calyx 5-toothed;
petals 1 (standard, wrapping around the stamens, others
wanting); stamens 10, united only at base.

A. fruticosa L., var. **angustifolia** Pursh False Indigo (Fig.
313). Tall shrubs; leaves with 9-25 leaflets, each about 2 cm
long; flowers in a long slender dense spike, each with but 1
petal, purple; pod oblong, longer than the persistent calyx,
roughened, 1-2-seeded.—River bottoms, up the Wisconsin
River to Sauk County, and up the Mississippi and St. Croix
rivers to St. Croix Falls.

3. Robinia   Locust

Trees and shrubs; leaves pinnately compound; stipules
bristlelike or modified into spines; calyx irregularly toothed;
flowers showy, in axillary racemes, zygomorphic; stamens
united, with 1 partially free; pod elongate, flat.

a. Trees; flowers white; twigs neither bristly nor glandular . . . .
. . . . 1. *R. pseudo-acacia*
a. Shrubs or small trees; flowers pink *b*

b. Twigs glandular; leaflets 13-25 . . . . .2. *R. viscosa*
b. Twigs bristly; leaflets 7-13 . . . . 3. *R. hispida*

1. **R. pseudo-acacia** L. Black Locust (Fig. 314). Trees, with stipules commonly modified into spines 1 dm long; branchlets scarcely hairy; flowers showy, white, fragrant, in drooping racemes in the leaf axils; fruit flat, about 1 dm long and 1 cm wide.—Often planted; many localities north to Adams and Waushara counties and very locally north to Bayfield and Marinette counties.

2. **R. viscosa** Vent. Clammy Locust. Large shrubs or small trees; the twigs and peduncles viscid with numerous glands; stipules bristly; flowers pink.—Occasionally planted, Washington County.

3. **R. hispida** L. Rose Acacia. Shrubs; densely to sparsely hispid, with glandular hairs 2-5 mm long; flowers rose or pink purple, 2.5-3 cm long.—Escaping from cultivation; Vilas and Milwaukee counties.

4. Baptisia   Wild Indigo

Large coarse herbs; leaves 3-foliate, 5-8 cm long; flowers zygomorphic, showy, in terminal racemes; calyx 2-lipped, 4-5 toothed; wings and keel nearly equal, sides of standard reflexed; stamens 10, distinct; pod stipitate in persistent calyx, rounded, cylindrical, or lens-shaped, with terminal beak.

a. Plants pubescent . . . . .1. *B. leucophaea*
a. Plants glabrous . . . . 2. *B. leucantha*

Fig. 315. *Baptisia leucophaea,* × .40.

Fig. 316. *Baptisia leucantha,* × .40.

1. **B. leucophaea** Nutt. (Fig. 315). Plants hairy; stipules large, persistent; flowers large, cream-colored.—Common on prairies north to Eau Claire and Juneau counties.

Var. **glabrescens** Larisey. Stem without hairs and leaves less hairy.—Less common, apparently, in sandy places.

2. **B. leucantha** T. & G. (Fig. 316). Plants without hairs; stipules soon dropping; flowers white.—Prairies, southern and western Wisconsin, north to Chippewa and Waupaca counties.

5. Lupinus   Lupine

Herbs, with palmately compound leaves; flowers zygomorphic, in long racemes; calyx deeply 2-lipped; sides of standard strongly reflexed, wings united toward the summit; keel convex on lower side, prolonged into a beaklike apex; stamens united; pod oblong, flattened, often constricted between the seeds.

a. Leaflets 7-11, oblanceolate, 1.5-5 cm long.........1. *L. perennis*
a. Leaflets 12-18, narrowly acute, 6-13 cm long ... 2. *L. polyphyllus*

1. **L. perennis** L. (Fig. 317). Herbs, to 7 dm high; leaflets 1.5-5 cm long; flowers blue purple. Less common than the following variety.

Var. **occidentalis** S. Wats. Long silky hairs on stems and petioles.—Dry sand, in southern two-thirds of the state and in Burnett and Washburn counties.

2. **L. polyphyllus** Lindl. Stout herbs, 0.6-1.2 m high; leaflets 6-13 cm long; flowers blue purple.—Vilas and Florence counties.

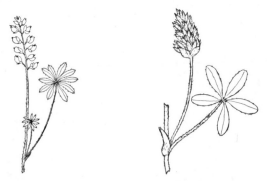

Fig. 317. *Lupinus perennis,* × .20.   Fig. 318. *Psoralea esculenta,* × .40.

6. Psoralea Pomme de Prairie

Perennial herbs, glandular; leaflets 3-5, palmately arranged (our species); flowers zygomorphic; calyx 5-lobed; stamens usually 9 united, 1 free; legume short, flattened.

P. esculenta Pursh (Fig. 318). Herbs, from a deep spindle-shaped taproot; stem and leaves very rough-hairy; flowers bluish or whitish, in dense spikelike racemes; pod thick, usually shorter than the persistent calyx, 1-seeded, with glandular dots.—Rare on prairies; southwestern Wisconsin and Pierce and St. Croix counties.

7. Trifolium Clover

Herbs, with 3 serrulate leaflets; flowers zygomorphic, in heads or headlike racemes; calyx and corolla persistent after anthesis; calyx 5-toothed; petals free except at base where more or less united with the stamen tube; stamens usually 9 united, 1 free; pods small, membranous, often shorter than the persistent calyx, 1-6-seeded.

a. Flowers without pedicels .... 1. *T. pratense*
a. Flowers on short pedicels *b*
  b. Flowers red or white; terminal leaflet not stalked *c*
    c. Stems creeping, rooting at the nodes .... 2. *T. repens*
    c. Stems erect or ascending, not rooting at the nodes ....
    .... 3. *T. hybridum*
  b. Flowers yellow *d*
    d. Terminal leaflet short-stalked .... 4. *T. campestre*
    d. Terminal leaflet sessile .... 5. *T. aureum*

1. T. pratense L. Red Clover (Fig. 319). Plants rather hairy; leaflets with a pale spot on the upper surface; flowers magenta to white.—Introduced from Europe, and escaped from cultivation to fields and roadsides, throughout the state.

2. T. repens L. White Clover. Plants without hairs; leaflets

Fig. 319. *Trifolium pratense,* × .40.

Fig. 320. *Trifolium hybridum,* × .40.
Fig. 321. *Trifolium campestre,* × .25.

nearly as broad as long, often notched at the apex; flowers white or pink; calyx lobes about equaling the tube.—Common everywhere; introduced from Europe.

3. **T. hybridum** L. Alsike Clover (Fig. 320). Similar to No. 2, but not rooting at the nodes; calyx lobes 1.5-2 times as long as the tube.—Common; introduced from Europe.

4. **T. campestre** Schreb. Low Hop Clover (Fig. 321). Stems low, 1-1.5 dm high; leaflets often notched at apex; stipules ovate, much shorter than the petiole, broad at base; heads small, 0.5-1.5 cm wide, globular, with 20-40 flowers; petals furrowed. (See *Medicago lupulina*, specimens of which may be confused with this species.)—Introduced from Europe; common in southern Wisconsin, less common northward.

5. **T. aureum** Poll. Hop Clover. Plants erect, 2-5 cm high; stipules elongate, attached to the petiole for half the length; all leaflets sessile; heads cylindrical ovoid, 1-2 cm wide; flowers yellow, becoming brown after anthesis.—Introduced from Europe; in scattered locations, north and west from Sauk County.

8. Melilotus   Sweet Clover

Annual or biennial herbs, fragrant when dried; leaves pinnately 3-foliate; flowers as in *Trifolium*, but the inflorescence a slender raceme; corolla deciduous and free from stamen tube; pods ovoid, wrinkled, longer than the calyx, 1-2-seeded.

a. Flowers white.................................... 1. *M. alba*
a. Flowers yellow .... 2. *M. officinalis*

1. **M. alba** Desr. White Sweet Clover. Tall plants; leaflets several times as long as broad.—Very common along roadsides, in fields and waste ground, and on railroad embankments; introduced from Europe.

2. **M. officinalis** (L.) Lam. Yellow Sweet Clover. Similar to No. 1.—Less common, but becoming abundant as it escapes from cultivation.

9. Medicago

Flowers similar to those of *Melilotus*; pods curved or coiled.

a. Flowers purple, 6-12 mm long..................... 1. *M. sativa*
a. Flowers yellow, 2-5 mm long.................... 2. *M. lupulina*

1. **M. sativa** L. Alfalfa; Lucerne. Plants smooth; leaflets several times as long as broad; flowers bluish purple.— Escaping from cultivation along roadsides and in fields; introduced from Europe; throughout the state.

Sometimes hybrids with **M. falcata** L., the Sickle Alfalfa, are found; their flowers range through cream to yellow, and the pods are only slightly coiled or sickle-shaped.

2. **M. lupulina** L. Black Medick (Fig. 322). Plants finely hairy; leaflets not over twice as long as broad; heads with 10-12 yellow flowers; petals not furrowed.—Roadsides, about dwellings, and borders of city streets; adventive from Europe; throughout the state.

10. Anthyllis Lady's-fingers

Herbs (our species), with pinnately compound leaves (except basal leaves often reduced, with a single terminal enlarged leaflet); inflorescence peduncled heads; calyx inflated; petals long-clawed; stamens united; ovary stipitate; fruit enclosed by the calyx.

A. vulneraria L. (Fig. 323). Plants 2-5 dm tall, branched from base; leaflets 5-11, terminal leaflet often larger than the lateral leaflets; floral heads subtended by cleft bracts; flowers about 1.5 cm long, yellow.—Native of Europe; escaping from cultivation in Dane County.

Fig. 322. *Medicago lupulina,*
× .40.

Fig. 323. *Anthyllis vulneraria,*
× .25.

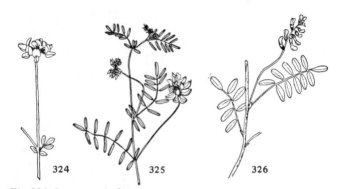

Fig. 324. *Lotus corniculatus,*
× .30.
Fig. 325. *Coronilla varia,* × .30.

Fig. 326. *Astragalus cooperi,*
× .40.

11. Lotus   Bird's-foot Trefoil

Herbs, with pinnately compound leaves; flowers solitary
or in umbels; calyx teeth nearly equal; petals clawed; stamens 10, 9 united, 1 free; pods linear.

L. corniculatus L. (Fig. 324). Perennials with stems prostrate or erect; leaflets 5; flowers in long peduncled headlike
umbels, yellow.—Native of Europe; naturalized on roadsides
and gravelly or sandy areas; Dane to Waukesha counties and
locally north to Bayfield and Vilas counties.

12. Coronilla   Crown Vetch

Glabrous herbs, with pinnately compound leaves; flowers
axillary on peduncled umbels; calyx 5-toothed with 2 lips;
petals clawed; pods transversely jointed.

C. varia L. (Fig. 325). Perennials; leaflets 11-21, oblong;
peduncles stout, longer than the subtending leaves; flowers
pink; pods 2-4 cm long.—Native of Europe; cultivated and
introduced for roadside planting; naturalized from Green and
Rock counties north to Waushara and Fond du Lac counties.

13. Astragalus   Milk Vetch

Perennial herbs, often slightly woody at base; leaves pinnately compound, with 11-25 leaflets; flowers in a spike,
zygomorphic; calyx 5-toothed; standard petal the largest;
stamens 10, 9 united, 1 free; pods mostly swollen.

a. Flowers purple; pod short-stalked in the calyx *b*
    b. Flowers 15-20 mm long.................1. *A. crassicarpus*
    b. Flowers 10-12 mm long.....................2. *A. alpinus*
a. Flowers white; pod not stalked in the calyx........3. *A. cooperi*

1. **A. crassicarpus** Nutt. Ground Plum. Stems many, from a taproot, 1-3 dm high; leaflets narrowly oblong, covered with stiff white hairs, as is the calyx.—Prairies, Pierce County.

2. **A. alpinus** L. Low, branching, sprawling; leaflets oval, over half as broad as long, with rather silky hairs; calyx with close jet-black hairs.—Gravelly shores of Pigeon Lake, near Drummond.

3. **A. cooperi** Gray (Fig. 326). Erect, 3-6 dm high; leaflets about 2 cm long, elliptical or oblong, without hairs; calyx with a few white hairs.—Counties bordering on Lake Michigan.

14. Oxytropis

Perennial herbs, nearly stemless; leaves pinnately compound; flowers zygomorphic, the keel petal tipped with a sharply projecting point; stamens 10, 9 united, 1 distinct; pods short-ovoid, partly 2-locular.

O. campestris L., var. **chartacea** (Fassett) Barneby. Leaves all at base of plant, silky with fine close white hairs; leaflets lanceolate; scapes with spreading silky or woolly hairs; calyx whitened with dense hairs; corolla blue, or sometimes white; pods leathery, silky-hairy, cylindrical and long-pointed.—Shores of Pigeon Lake, near Drummond, and lake shores east of Plainfield.

15. Vicia   Vetch; Tare

Trailing or climbing herbs; leaves pinnately compound with 1 or more of the terminal leaflets modified into a tendril; flowers zygomorphic, in axillary clusters or racemes; calyx 5-toothed; standard petal with a broad claw overlapping the wing petals; stamens 10, 9 united, 1 distinct; style pubescent all around the summit; pods flat, 2-several-seeded.

a. Leaflets with 10 or more pairs of veins *b*
    b. Racemes without peduncle................1. *V. angustifolia*
    b. Racemes peduncled.......................2. *V. americana*
a. Leaflets with 6 or fewer pairs of veins *c*
    c. Flowers barely 10 mm long, white with blue tip. ....
    .... 3. *V. caroliniana*

Fig. 327. *Vicia angustifolia,* × .40.

Fig. 328. *Vicia americana,* × .40.

Fig. 329. *Vicia caroliniana,* × .40.   Fig. 330. *Vicia villosa,* × .40.

   c. Flowers 10-15 mm long, blue, rarely white *d*
      d. Flowers less than 4 times as long as thick......4. *V. cracca*
      d. Flowers more than 5 times as long as thick....5. *V. villosa*

1. **V. angustifolia** Reichard (Fig. 327). Flowers solitary or in long pairs in the upper leaf axils, 1-1.8 cm long.—Occasional along railroads and in thickets; southern half of the state and Polk and Florence counties; adventive from Europe.

2. **V. americana** Muhl. (Fig. 328). Plants nearly or quite without hairs; leaflets elliptical or ovate, rounded at tip, wavy-veined; flowers purplish.—In woods and grassy places, throughout the state.

3. **V. caroliniana** Walt. (Fig. 329). Leaflets oblong, blunt; flowers whitish, tipped with blue.—Low rich thickets, north to Chippewa, Marathon, and Shawano counties.

4. **V. cracca** L. Leaflets with a short abrupt tip; flowers 1-1.2 cm long, in dense one-sided racemes.—Occasionally adventive in fields; eastern Wisconsin, and Taylor, Lincoln, and Sawyer counties.

Var. **tenuifolia** (Roth) G. Beck. Peduncles more elongate; blade of the standard petal twice as long as the claw.—Less common, Dane and Rock counties.

5. **V. villosa** Roth (Fig. 330). Hairy throughout; flowers white and violet.—Introduced from Eurasia; occasional in fields and roadsides, throughout the state.

16. Lathyrus   Wild Pea

Annual or perennial vinelike herbs; leaves pinnately compound with 1 or more leaflets modified into tendrils; stipules prolonged basally; flowers zygomorphic, in racemes; the standard petal broadly obovate to rotund; stamens 10, 9 united, 1 distinct; style usually bent at nearly a right angle, bearded along inner side; pods flat to round, several-seeded.

a. Stipules not more than 5 mm broad, or wanting *b*
  b. Principal leaves with 4-8 leaflets; peduncles 2-8-flowered . . . . . .
  . . . . 1. *L. palustris*
  b. Principal leaves with 8-12 leaflets; peduncles 6-18-flowered . . . .
  . . . . 2. *L. venosus*, var. *intonsus*
a. Stipules 1 cm or more broad *c*
  c. Each stipule shaped like half an arrowhead, with 1 basal lobe
  . . . . 3. *L. ochroleucus*
  c. Each stipule shaped like an arrowhead, with 2 basal lobes . . . . .
  . . . . 4. *L. japonicus*

1. **L. palustris** L. Vetchling (Fig. 331). Flowers pale purple, rather loosely arranged, with slender pedicels 3-6 mm long; calyx smooth.—Moist ground, southeastern Wisconsin and scattered localities northward.

The leaflets in this species vary greatly in hairiness, in width, and in length. Since the several named varieties intergrade, they are omitted here.

2. **L. venosus** Muhl., var. **intonsus** Butt. & St. John (Fig. 332). Flowers magenta, closely crowded on rather stout pedicels; calyx densely hairy.—Common, throughout the state.

3. **L. ochroleucus** Hook. (Fig. 333). Stems 3-9 dm high; flowers 1.5-1.8 cm long, yellowish white.—Common in rich open woods, throughout the state.

4. **L. japonicus** Willd. Beach Pea (Fig. 334). Flowers pink when young, purple in age, 1.8-2.5 cm long; leaves slightly fleshy.

Fig. 331. *Lathyrus palustris,*
× .40.

Fig. 332. *Lathyrus venosus,* var.
intonsus, × .40.

Fig. 333. *Lathyrus ochroleucus,*
× .40.

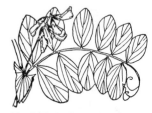

Fig. 334. *Lathyrus japonicus,*
× .40.

Var. **pellitus** Fern. Covered with fine hairs.—Rare; Door County.

Var. **glaber** (Ser.) Fern. Without hairs.—Common along beaches of Lake Michigan and Lake Superior.

GERANIACEAE   Geranium Family

Annual or perennial herbs, sometimes woody; leaves simple or compound, usually opposite and palmately cleft; flowers in a cymose or umbellike inflorescence, perfect, sometimes bilaterally symmetrical; sepals 5; petals 5; nectaries usually 5, conspicuous, alternating with the petals; stamens 5-15, sometimes reduced to staminodia; compound pistil of 3-5 lobes; fruit a capsule splitting into 5 portions, with the hygroscopic style attached to the ovarian beak, the portions splitting off when ripe and recurving elastically or spirally to eject the seeds.

Fig. 335. *Geranium maculatum,*
× .40.

Fig. 336. *Geranium bicknellii,*
× .40.

Geranium Wild Geranium; Cranesbill

Herbs, with palmately compound or deeply lobed leaf blades on long petioles; petals 5; stamens 10, 5 of them longer and with a gland at base; fruit a long slender cylindrical capsule.

a. Stem simple or little branched below, from a stout rootstock; petals 12-25 mm long . . . . 1. *G. maculatum*
a. Stem much branched below, spreading from a taproot; petals 10 mm or less in length *b*
  b. Petals about 10 mm long, twice the length of the sepals . . . . 2. *G. robertianum*
  b. Petals 3-6 mm long, 1-1.5 times the length of the sepals *c*
    c. Sepals with a long bristlelike point *d*
      d. Pedicels not more than twice as long as the sepals *e*
        e. Larger sepals 3-4.5 mm wide, 3-nerved . . . . 3. *G. carolinianum*
        e. Larger sepals 5-8 mm wide, 5-nerved . . . . 4. *G. sphaerospermum*
      d. Pedicels several times as long as the sepals 5. *G. bicknellii*
    c. Sepals without a long bristlelike point *f*
      f. Stylar beak very short to lacking; carpel bodies hairy . . . . 6. *G. pusillum*
      f. Stylar beak of fruit 2-5 mm long; carpel bodies glabrous and with transverse wrinkles . . . . 7. *G. molle*

1. **G. maculatum** L. Wild Cranesbill (Fig. 335). Perennials; stems simple below, above with several peduncles and leaves arising from one point; leaves about 1 dm wide, deeply 5-7-parted, the wedge-shaped divisions lobed and cut at the end; petals light purple, 10-25 mm long.—Common in woods and thickets, north to Douglas and Lincoln counties.

2. G. robertianum L. Herb Robert. Stems a little hairy; leaves compound, the 3-5 divisions pinnately divided, these divisions themselves pinnately lobed; petals pink or red purple.—Kewaunee and Door counties.

3. G. carolinianum L. Principal leaves round in outline but with 5-9 deeply cleft segments; flowers crowded; petals pale pink; fruit with a stout persistent beak about 2 mm long.—Rare; scattered localities in Wisconsin.

4. G. sphaerospermum Fern. Similar to No. 3, but with the differences noted in key above.—Rare; Douglas and Sheboygan counties.

5. G. bicknellii Britt. (Fig. 336). Leaf blades about 5-parted, the divisions forking and cut into somewhat ribbonlike segments; fruit with a slender persistent beak, 4-6 mm long.—In dry ground and clearings, mostly northward; south to Sauk County and along Lake Michigan to Racine County.

6. G. pusillum L. Leaf blades cut about two-thirds of the way to the base, the divisions with rounded lobes; petals little if at all exceeding the sepals.—Adventive from Europe; rare; Rock and Marquette counties.

7. G. molle L. Similar to No. 6, but with the stylar beak 2-5 mm long.—Native of Europe; Sheboygan County.

OXALIDACEAE   Wood Sorrel Family

Herbs, with compound leaves (our species); flowers perfect, single or in cymes, umbellike clusters, or racemes; sepals 5; petals 5; stamens 10, filaments usually fused, sometimes 5 lacking anthers; styles 5; fruit a capsule, rarely a berry.

Oxalis   Wood Sorrel

Small herbs, with sour juice; leaves long-petioled, with 3 heart-shaped leaflets, each of which is attached by its point; sepals 5; petals 5, delicate; stamens 10; styles 5; the stamens and styles of 2 or 3 relative sizes (heterostylous); fruit a small pod.

a. Leaves and flowering stalks all arising from base of the plant; flowers white or purple *b*
  b. Leaves and scapes densely pubescent; petals white with red or purple veins..............................1. *O. montana*
  b. Leaves and scapes glabrous; petals purple.......2. *O. violacea*
a. Leaves on the stem; flowers yellow *c*

    c. Plants creeping, rooting at most of the nodes. .3. *O. corniculata*
    c. Plants not rooting at the nodes *d*
        d. Plants erect; stipules lacking; flowers many, in an open
            cymose inflorescence...................... 4. *O. stricta*
        d. Plants caespitose; stipules present; inflorescence umbellate ..
        ....5. *O. dillenii*

1. **O. montana** Raf. Common Wood Sorrel (Fig. 337). Plants with thick creeping rootstocks; petals white with purple lines.—Cool damp woods in northern Wisconsin, south to Taylor to Oconto counties.

2. **O. violacea** L. Violet Wood Sorrel (Fig. 338). Plants from scaly bulbs and slender taproots (the latter sometimes thick and fleshy when diseased); petals violet.—Sandy and stony banks, north to Marquette to Polk counties.

3. **O. corniculata** L. Stems creeping and rooting at the nodes; subglabrous to pubescent, with appressed or spreading nonseptate hairs; stipules present, broad, often auriculate; inflorescence umbellate; petals yellow.—Rare; Grant, Dane, and Sheboygan counties.

4. **O. stricta** L. (Fig. 339). Stems erect, rhizomatous;

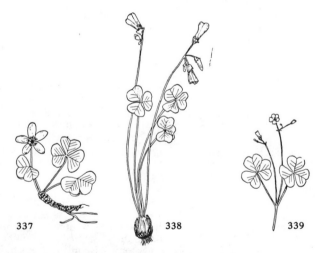

Fig. 337. *Oxalis montana,* × .40.   Fig. 339. *Oxalis stricta,* × .40.
Fig. 338. *Oxalis violacea,* × .40.

plants subglabrous to densely covered with shining, spreading septate hairs; stipules lacking.—Throughout the state.

5. **O. dillenii** Jacq. Plants caespitose, branching at base; stems densely covered with closely appressed white, non-septate hairs; stipules present, broad or narrow.—Throughout southern Wisconsin, locally northward to Bayfield and Florence counties.

LINACEAE   Flax Family

Herbs, with tough fibrous stems; leaves simple, entire, often with stipular glands at base; flowers perfect (our species); sepals 5; petals 5; stamens 5, united at base; fruit a capsule.

Linum   Flax

Annuals or perennials, with narrow sessile leaves; sepals 5, often ciliate; petals 5, often ephemeral; stamens 5; capsule ovoid to globose.

a. Flowers blue; capsules 6-10 mm in diameter *b*
   b. Annuals; margins of the inner sepals ciliate. .1. *L. usitatissimum*
   b. Perennials; margins of the inner sepals entire . . . . .2. *L. perenne*
a. Flowers yellow; capsules 4 mm or less in diameter *c*
   c. Styles united at base only; sepals persistent . . . . .3. *L. sulcatum*
   c. Styles united to above the middle; sepals early deciduous . . . . .
   . . . . 4. *L. rigidum*

1. **L. usitatissimum** L. Common Flax. Annuals, to 7.5 dm high; leaves narrow, 3-nerved; sepals 7-9 mm long, the inner ciliate on the hyaline margin; petals 10-15 mm long; fruit globose.—Escape from cultivation, naturalized on roadsides and waste places; locally in the southern half of Wisconsin, and in northwestern Wisconsin east to Price County.

2. **L. perenne** L. Perennials, 3-7 dm tall; stems usually clustered; leaves linear, 1-2 cm long, erect; sepals 5-7 mm long, mucronate, entire; petals 15-20 mm long; fruit sub-globose.—Prairies, rare; Sheboygan, Walworth, and Washburn counties.

3. **L. sulcatum** Riddell. Annuals, to 8 dm tall; leaves narrow, mostly alternate, with dark glands at base; sepals persistent, margins glandular-toothed; petals 5-10 mm long, obovate; styles united at base; fruit globose, acute at summit.—Prairies and dry hills in southeastern Wisconsin, Waukesha and Racine counties, north to Adams and Waushara counties in central Wisconsin, and to St. Croix and Eau Claire counties in western Wisconsin.

4. **L. rigidum** Pursh. Annuals, to 4 dm tall; leaves very narrow, entire or glandular-toothed, alternate; sepals glandular, toothed, the inner sepals with a hyaline margin, early deciduous; petals 9-17 mm long, broadly obovate; styles united to above the middle; capsules ovoid.—Rare; Pierce and Juneau counties.

RUTACEAE Rue Family

Trees or shrubs (our species), characterized by the presence of oil glands in most parts of the plants; leaves alternate, pinnately compound or trifoliate (sometimes simple), with transparent dots and a pungent oil; flowers perfect or imperfect; sepals 3-5; petals 3-5; stamens usually twice as many as the petals; a continuous nectar gland between the stamens and ovary; carpels 4-5; fruit a capsule, drupe, or berry.

a. Plants thorny; leaves pinnately compound; flowers in axillary clusters .... 1. *Xanthoxylum*
a. Plants thornless; leaves trifoliate, flowers in terminal panicles .... 2. *Ptelea*

1. Xanthoxylum Prickly Ash

Dioecious thorny shrubs or trees, with alternate pinnately compound leaves; flowers small, yellow, clustered; sepals none (our species); petals 4-5; stamens 4-5, alternate with the petals; styles at least partially united; fruit a somewhat fleshy follicle.

X. americanum Mill. (Fig. 340). Shrubs or small trees,

Fig. 340. *Xanthoxylum americanum,* ✕ .15.

Fig. 341. *Ptelea trifoliata,* ✕ .30.

with many broad-based prickles; flower clusters axillary, appearing before the leaves.—Disturbed woods, throughout the state but less common in the north.

Forma **impuniens** Fassett is without prickles.

2. Ptelea   Hop Tree

Small trees or shrubs, with alternate trifoliate leaves; flowers in a terminal panicle, greenish white or yellowish white, polygamous; sepals and petals 4-5 each; stamens as many as the petals and alternate with them; fruit flat, surrounded by a wing, 1 cm broad.

P. trifoliata L. (Fig. 341). Leaves with long petioles; leaflets ovate-pointed, to 15 cm long.—Native and cultivated; from Milwaukee County southward, also Dane, Rock, and Green counties.

POLYGALACEAE   Milkwort Family

Our species herbs; leaves simple; flowers zygomorphic, perfect, in racemes; sepals 5, the outer 3 often small and the inner 2 larger; petals 5 or less, partially united to each other and to the stamen tube; stamens 3-8, filaments usually united, anthers opening by a terminal pore; fruit a capsule.

Polygala   Polygala; Milkwort

Herbs, with alternate or whorled entire leaves; sepals 5, the outer 3 small and the inner 2 larger and often colored as the petals; petals 3, more or less fused, the lower one broader and appearing as a keel with a fringed crest; stamens 6-8, their filaments more or less united below with each other and with the petals.

a. Flowers 1-2 cm long, in the axils of the upper leaves............
.... 1. *P. paucifolia*
a. Flowers about 0.5 cm long, in a terminal spike *b*
  b. Flowers usually pink; lower petal with a crest
.... 2. *P. polygama*, var. *obtusata*
  b. Flowers white; petals not crested............... 3. *P. senega*

1. **P. paucifolia** Willd. Fringed Polygala (Fig. 342). Stems solitary, 7-10 cm high, from a rhizome; lower leaves small and scalelike, the upper ovate, stalked, crowded; flowers 1-3, 1-2 cm long, pink.—Woods, south to Wood and Outagamie counties and on dunes to Manitowoc County.

2. **P. polygama** Walt., var. **obtusata** Chodat. Stems densely tufted, from a slender taproot; underground stems with

Fig. 342. *Polygala paucifolia,* × .55.

Fig. 343. *Polygala senega,* × .55.

white fleshy flowers which are self-fertilizing without opening (cleistogamous).—Sandy regions, common except in the north-central area of the state.

3. **P. senega** L. Seneca Snakeroot (Fig. 343). Stems clustered, 1.5-3 dm high; leaves about 5 mm broad, 1-3 cm long; flowers white, in a terminal spike.—Prairies, north to Washburn and Portage counties.

EUPHORBIACEAE Spurge Family

Monoecious or dioecious herbs, shrubs, or trees, commonly with milky juice and often with parts specialized or reduced; leaves (our species) simple, alternate; inflorescence a cyme, often subtended by conspicuous bracts; calyx and corolla present or lacking; staminate flowers sometimes reduced to a single stamen; pistillate flowers consisting of a single pistil; fruit a capsule, splitting into 3 1-seeded nutlets.

Euphorbia Spurge

Plants 1-5 dm tall; leaves simple, alternate, not toothed; juice milky; flowers clustered, each cluster containing a

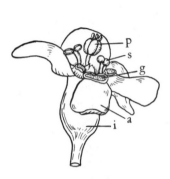

Fig. 344. Inflorescence of *Euphorbia:* *p*=pistillate flower; *s*=staminate flower; *g*=gland; *a*=appendage of gland; *i*=involucre (cyathium).

Fig. 345. *Euphorbia corollata,* × .40.

pistillate flower and many staminate flowers, each reduced to a single stamen, the whole cluster (cyathium) simulating a single flower, with a 4-5-lobed corollalike or calyxlike involucre, and 4 or 5 two-horned glands (Fig. 344); plants poisonous.

a. Involucre white . . . . 1. *E. corollata*
a. Involucre greenish or yellowish *b*
  b. Foliage leaves at least 3 times as long as wide *c*
    c. Leaves minutely serrulate on margin, heart-shaped, clasping at base; plants annual . . . . 2. *E. obtusata*
    c. Leaves entire-margined; plants perennial *d*
      d. Floral leaves 5 mm wide, ovate . . . . 3. *E. cyparissias*
      d. Floral leaves 10-15 mm wide, kidney-shaped . . . .
      . . . . 4. *E. podperae*
  b. Foliage leaves nearly or quite as wide as long . . . 5. *E. commutata*

1. **E. corollata** L. Flowering Spurge (Fig. 345). Perennials, usually with several stems, from a deep woody root; stems simple below, much branched in the inflorescence; leaves linear-oblong.—Open sandy soil, north to St. Croix and Columbia counties, and occasionally adventive northward.

2. **E. obtusata** Pursh. Plants annual; leaves oblong, finely serrate; floral leaves broadly ovate, green.—Rock County.

3. **E. cyparissias** L. Cypress Spurge (Fig. 346). Plants pe-

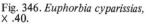

Fig. 346. *Euphorbia cyparissias,*
× .40.

Fig. 347. *Euphorbia podperae,*
× .40.

rennial, from rootstocks; leaves narrowly ribbonlike, very numerous, 0.5-2 cm long; floral leaves yellowish green.— Introduced from Europe; cemeteries and about abandoned houses, north to Rusk and Burnett counties.

4. **E. podperae** Croiz. (*E. esula*). Leafy Spurge (Fig. 347). Perennials, from rootstocks; leaves 2-6 cm long; floral leaves heart-shaped, yellow.—Adventive from Europe; becoming an abundant weed in fields and along roadsides; scattered localities north to Trempealeau and Oneida counties.

5. **E. commutata** Engelm. Leaves round obovate; floral leaves ovate, green, 7-10 mm broad.—Rock County.

LIMNANTHACEAE   False Mermaid Family

Annual herbs, with alternate, pinnately divided leaves; flowers solitary, axillary, perfect; petals and sepals 3-5; stamens twice the number of petals; carpels 2-5; style 1; fruit an achene.

Floerkea   False Mermaid

Delicate, ephemeral annuals; sepals and petals 3; petals shorter than the sepals, white.

F. proserpinacoides Willd. Stems 1-3 dm high, lax; leaves

pinnate; leaflets 3-5, lanceolate; flowers solitary in the leaf
axils, minute, on long pedicels.—Locally abundant in moist
shady maple woods, north to Clark and Rusk counties.

ANACARDIACEAE   Cashew Family

Trees, shrubs, or vines, with resinous or milky juice;
leaves alternate, compound; flowers perfect or unisexual;
sepals 5, united; petals 5, distinct (sometimes lacking); stamens 5; styles 3; fruit a drupe.

Rhus   Sumac; Poison Ivy

Erect or trailing shrubs or woody vines; leaves alternate,
compound; flowers greenish white or yellowish in lateral or
terminal panicles or heads; flowers polygamous or dioecious.

a. Leaflets 3 *b*
   b. Leaves and young branchlets glabrous; leaflets stalked; fruits
      white................................... 1. *Rhus radicans*
   b. Leaves and young branchlets soft-pubescent; leaflets nearly ses-
      sile; fruits red.......................... 2. *R. aromatica*
a. Leaflets 7-13 .... 3. *R. vernix*

1. **R. radicans** L. Poison Ivy; Poison Oak (Fig. 348). Leaflets 3, the terminal one on a long petiolule, often few-
toothed, red and shining when expanding; fruit white drupes.

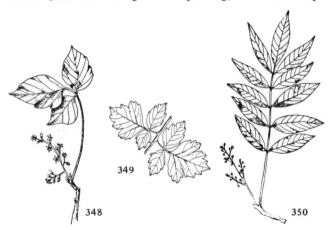

Fig. 348. *Rhus radicans,* × .40.   Fig. 350. *Rhus vernix,* × .40.
Fig. 349. *Rhus aromatica,* × .20.

This species has 3 distinct forms in Wisconsin: (1) an erect shrub to 1 m tall, sometimes branched; (2) a climbing woody vine with aerial rootlets; (3) a colonial rhizomatous short scarcely branched shrub. All forms are poisonous on all parts.—The shrubby forms occur throughout the state, the vine mostly in river-bottom vegetation of southern Wisconsin, Grant to Rock and Dane counties.

2. **R. aromatica** Ait. Fragrant Sumac (Fig. 349). Bushy shrubs, to ? m tall; leaflets sessile, margins lobed, the terminal leaflet larger than the lateral ones; flowers yellow, in spikelike clusters, opening before or with the leaves; fruits bright red, densely hairy.—Rare, on dry hillsides; Grant, Richland, Columbia, Dane, and Milwaukee counties.

3. **R. vernix** L. Poison Sumac (Fig. 350). Shrubs 2-3 m high; leaves pinnately compound; 7-13 leaflets, margins entire; fruit grayish white; plant poisonous on all parts.— Swamps and bogs of southern Wisconsin, north to Wood and Waupaca counties.

AQUIFOLIACEAE   Holly Family

Evergreen or deciduous trees and shrubs; leaves alternate, simple; flowers small, solitary or few in axillary cymes, perfect or unisexual; sepals 3-6, mostly united at base; petals 4-5, distinct or partially united; stamens 4-5, alternating with the petals and inserted at the base of the petals; pistil 1; fruit a berry.

a. Petals distinct; stamens free . . . . 1. *Nemopanthus*
a. Petals united at base; stamens borne on corolla tube . . . . . . 2. *Ilex*

1. Nemopanthus

Glabrous shrubs, with entire or serrulate leaves; flowers mostly perfect but usually with either the stamens or pistil much reduced; calyx none or minute and deciduous; petals 4-5, linear, distinct; stamens as many as the petals.

N. mucronata (L.) Trel. Mountain Holly (Fig. 351).

Fig. 351. *Nemopanthus mucronata,* × .40.

Deciduous shrubs, 0.3-3 m high; leaf blades elliptical to slightly obovate, round-tipped with a small abrupt point; flowers small, whitish, on elongate pedicels, 1-2 cm in flower and 3 cm in fruit; berries red.—Swampy areas, throughout the state except southwestern Wisconsin.

2. Ilex   Holly

Trees or shrubs, with alternate simple serrate leaves; flowers perfect or unisexual with either the stamens or pistil reduced; calyx 4-6-lobed; petals 4-8, united at base; stamens the same number as the petals and alternate with them, attached at the base of the corolla tube; style short; berries black or red.

I. verticillata (L.) Gray. Black Alder; Winterberry. Shrubs, to 5 m tall; leaves oblong-lanceolate, serrate; flowers in axillary clusters, almost sessile, or with pedicels to 5 mm; calyx lobes ciliate; berries bright red.—Rich woodlands, throughout the state.

CELASTRACEAE   Staff Tree Family

Small trees or shrubs (our species), often climbing or twining; leaves simple; flowers perfect or unisexual, the perfect flowers usually having either the stamens or pistil reduced and nonfunctional; sepals 4-5; petals 4-5; stamens as many as the petals and alternate with them, borne on the margin of a fleshy disk surrounding the ovary; fruit a capsule splitting along the median line of each cell of the ovary, the seeds with a colored appendage (aril).

a. Trailing or climbing shrubs; leaves alternate . . . . .1. *Celastrus*
a. Erect shrubs or small trees; leaves opposite . . . . .2. *Euonymus*

1. Celastrus   Bittersweet

Twining shrubs, with deciduous alternate serrate leaves; dioecious or with perfect flowers with reduced parts; sepals 5; petals 5; staminate flowers with stamens as long as the petals and a remnant of a pistil; the pistillate flowers with a well-developed ovary with a stout columnar style, a 3-lobed stigma, and vestigial stamens; the fruit 3-valved, covering 1 or 2 seeds within a fleshy red aril.

C. scandens L. (Fig. 352). Leaves finely toothed, pointed; inflorescence a terminal panicle, 3-8 cm long; flowers greenish; fruits orange, arils orange red.—Thickets and along fences, throughout southern Wisconsin and northwestern Wisconsin east to Price, Taylor, and Marathon counties.

Fig. 352. *Celastrus scandens,* × .40.

Fig. 353. *Euonymus atropurpureus,* × .45.

2. Euonymus   Wahoo

Shrubs or small trees (our species); flowers perfect; 4-5 sepals, petals, and stamens; stamens short; style lacking; stigma 3-5-lobed; fruit 3-5-lobed, the colored aril completely covering the seed.

E. atropurpureus Jacq. (Fig. 353). Leaves finely and sharply toothed; flowers several, on branched peduncles arising from the lower part of the current season's growth, aril bright red.—North and east to Dunn to Outagamie counties.

STAPHYLEACEAE   Bladdernut Family

Shrubs or trees, with mostly opposite leaves, pinnately compound or trifoliate; inflorescence a terminal panicle; flowers perfect; sepals 5; petals 5; ovary of 2-3 fused carpels; fruit an inflated capsule.

Staphylea   Bladdernut

Leaves stipulate, the stipules early deciduous; sepals and petals about the same length; the styles free below, united above, with 1 capitate stigma.

S. trifolia L. (Fig. 354). Twigs with striped bark; leaves

Fig. 354. *Staphylea trifolia,* ✕ .40.

trifoliate, the terminal leaflet stalked, each leaflet narrowed to a long point and finely serrate; flowers greenish white, bell-shaped, 8-10 mm long; capsule large, 3-parted, papery, and much inflated.—Woods and bluffs, north to Polk, Sawyer, and Brown counties.

ACERACEAE   Maple Family

Trees or shrubs, with opposite, palmately lobed, simple or pinnately compound leaves; flowers perfect or unisexual; calyx 5-lobed; petals 5, small or lacking; stamens 5-10, long, exserted in the staminate flowers, lacking in the pistillate

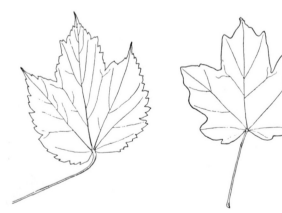

Fig. 355. *Acer spicatum,* ✕ .40.   Fig. 356. *Acer saccharum,* ✕ .40.

Fig. 357. *Acer platanoides,* × .20, flower × 1.50.

Fig. 358. *Acer saccharinum,* × .40.

Fig. 359. *Acer rubrum,* × .40.

Fig. 360. *Acer negundo,* × .40.

flowers; ovary of 2 carpels, each becoming winged, easily separable from each other at maturity.

Acer   Maple

Same characters as for family.

a. Stamens on a definite disk; leaves palmately lobed *b*
    b. Flowers in long racemes . . . . 1. *A. spicatum*
    b. Flowers in umbellike or paniclelike clusters *c*
        c. Flowers yellowish or greenish, long-pediceled *d*
            d. Inflorescence irregular and little branched; petals none . . .
                 . . . . 2. *A. saccharum*
            d. Inflorescence rather flat-topped and several times branched; petals well developed . . . . 3. *A. platanoides*
        c. Flowers red, almost without pedicels, appearing before the leaves *e*
            e. Petals none; leaves lobed more than halfway to the midrib (Fig. 358) . . . . 4. *A. saccharinum*
            e. Petals present; leaves lobed one-third of the way to the midrib (Fig. 359) . . . . 5. *A. rubrum*
a. Stamens not on a disk, leaves pinnately compound. . . 6. *A. negundo*

1. **A. spicatum** Lam. Mountain Maple (Fig. 355). Shrubs or small trees; leaves slightly heart-shaped at base of the blade, 3-5-lobed, toothed, downy beneath; fruit reddish.—Mostly northward, coming south on cool bluffs.

2. **A. saccharum** Marsh. Sugar Maple (Fig. 356). Trees, with scaly bark; buds slender, pointed; juice not milky; flowers hanging vertically on weak pedicels; leaf blades whitish and smooth beneath, 3-5-lobed with rounded sinuses, the lobes themselves somewhat lobed and wavy-margined.—Woods, throughout the state.

Var. **nigrum** Michx. f. Black Maple. Leaves green and downy beneath, mostly 3-lobed.—Scattered localities, with the Sugar Maple.

3. **A. platanoides** L. Norway Maple (Fig. 357). Large trees, with ridged bark; buds fat, blunt; juice milky; pedicels spreading to form umbellike clusters; leaf blades 5-lobed, the lobes cut into long-pointed coarse teeth or lobes.—Mostly cultivated as a shade tree; introduced from Europe.

4. **A. saccharinum** L. Silver Maple (Fig. 358). Usually large trees, with flaky bark; leaf blades much whitened beneath, 3-5-lobed, the lobes coarsely and sharply toothed, and the lobe bases narrowed; fruit 5 cm long.—Mostly along river bottoms; also common as a shade tree along city streets; throughout the state except extreme northwestern Wisconsin.

5. **A. rubrum** L. Red Maple (Fig. 359). Trees, similar to No. 4, but twigs without the rank odor and flavor of Silver

Maple; leaves with bases of lobes broad, whitened beneath; fruit 2 cm long.—Common, except southwestward.

6. **A. negundo** L. Box Elder (Fig. 360). Trees, with close bark, the twigs often blue waxy; leaflets 3-5, very veiny; flowers dioecious; petals none.—Mostly southwestward; frequently planted.

Var. **interius** (Britt.) Sarg. Twigs puberulent.—St. Croix County.

HIPPOCASTANACEAE   Buckeye Family

Trees or shrubs, with opposite, palmately compound leaves; sepals 5, partially united; petals 4-5, clawed; stamens (6) 7 (8), filaments elongate; ovary of 3 carpels, style 1; fruit a leathery 3-locular capsule with 1-3 large seeds.

Aesculus   Horse Chestnut; Buckeye

Characters as for family.

A. glabra Willd. Ohio Buckeye (Fig. 361). Small trees; opposite, palmate leaves with 5-7 leaflets; flowers showy, in a terminal panicle, yellowish; upper petals shorter than the stamens; fruit prickly.—Yahara River bottom near Fulton, Rock County, woods near Coon Valley, Vernon County, and southern La Crosse County.

Fig. 361. *Aesculus glabra,* × .20.

A. **octandra** Marsh. Yellow Buckeye, with dark yellow petals longer than the stamens, and **A. hippocastanum** L. Horse Chestnut, with white petals spotted with red, are frequently cultivated in Wisconsin. The latter is a European plant.

RHAMNACEAE   Buckthorn Family

Shrubs (our species), with simple leaves, alternate or opposite; flowers perfect or unisexual, with 4-5 sepals and petals; stamens as many as the petals and opposite them, alternate with the sepals; ovary 1; styles 1-4, partially fused; fruit a drupe.

a. Flowers scattered toward base of the current year's growth . . . . . .
. . . . 1. *Rhamnus*
a. Flowers in a compound panicle at the tip of current year's growth
. . . . 2. *Ceanothus*

1. Rhamnus   Buckthorn

Shrubs; leaves alternate or opposite; flowers greenish white, single or in umbels, in the axils of the leaves; petals small or wanting; calyx free from the ovary; fruit black, 2-4-seeded.

a. Leaves toothed *b*
    b. Calyx lobes and stamens 5; petals none . . . . 1. *R. alnifolia*
    b. Calyx lobes, petals, and stamens 4 *c*
        c. Leaves with 3-4 pairs of veins . . . . 2. *R. cathartica*
        c. Leaves with 6-7 pairs of veins . . . . 3. *R. lanceolata*
a. Leaves not toothed . . . . 4. *R. frangula*

1. **R. alnifolia** L'Hér. (Fig. 362). Leaves oval to elliptical, the shallow blunt teeth 4-8 per cm.—Swamps, throughout the state except the west-central part.

2. **R. cathartica** L. Common Buckthorn (Fig. 363). Branchlets rigid, often spinelike; leaf blades broadly ovate, the shallow somewhat incurved teeth 8-12 per cm; fruits poisonous; the plants alternate hosts to oat rust.—Occasionally escaping from cultivation; southern Wisconsin, north to La Crosse to Outagamie counties.

3. **R. lanceolata** Pursh (Fig. 364). Leaves narrowly ovate to lanceolate, the teeth 13-20 per cm, larger on the young growth.—Hillsides, Grant and Richland counties.

4. **R. frangula** L. Alder Buckthorn (Fig. 365). Leaves broadly oval, with 7-8 straight or slightly curved veins; flowers in sessile umbels in the axils of leaves.—Rare; southeastern and central Wisconsin; introduced from Europe.

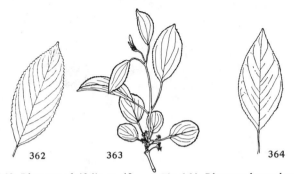

Fig. 362. *Rhamnus alnifolia,* × .40.
Fig. 363. *Rhamnus cathartica,*
× .35.

Fig. 364. *Rhamnus lanceolata,*
× .40.

Fig. 365. *Rhamnus frangula,*
× .30.

Fig. 366. *Ceanothus ovatus,*
× .50.

2. Ceanothus   New Jersey Tea

Shrubs, with alternate 3-nerved leaves; inflorescence terminal or in axillary panicles; petals long-clawed, blade partially enfolding the anthers; fruit a drupe, subtended by the persistent hypanthium.

C. ovatus Desf. (Fig. 366). Low shrubs; leaf blades finely toothed, the teeth ending in a dark tip; flowers white.—In sand, mostly northward.

Forma **pubescens** (S. Wats.) Soper has leaf blades hairy beneath.—Dry rocky soil; less common, with the typical variety.

Fig. 367. *Parthenocissus inserta,* × .25.

VITACEAE   Grape Family

Trailing or extensively climbing vines, with tendrils; leaves alternate, palmately lobed or divided; flowers in clusters, small and greenish; calyx very small and often lacking lobes; petals falling very early; fruit a berry with 4 seeds.

a. Leaves compound .... **1.** *Parthenocissus*
a. Leaves simple .... 2. *Vitis*

1. Parthenocissus   Woodbine; Virginia Creeper

Leaves compound, of 5 (often 3 on young growth) stalked, coarsely toothed leaflets; petioles as long as the leaflets; flowers many in a much-branched long-stalked inflorescence; sepals 5; petals 5; stamens 5; stigmas 2-lobed; fruit a small blue berry.

a. Inflorescence regularly forking; tendrils mostly coiling without well-developed adhesive disks .... .1. *P. inserta*
a. Inflorescence paniculate; tendrils with disklike adhesive tips ......
....2. *P. quinquefolia*

1. **P. inserta** (Kerner) K. Fritsch (Fig. 367). Inflorescence regularly forking, its main branches equal or nearly so; tendrils without well-developed adhesive disks.—Throughout the state, but more common southward.

2. **P. quinquefolia** (L.) Planch. Inflorescence a panicle, not regularly forking; tendrils often with well-developed adhesive

disks.—Northern; common in cultivation and occasionally escaping to roadsides in the Wisconsin River Valley north to Lincoln County.

2. Vitis   Grape

Flowers in a much-branched drooping panicle, fragrant; calyx reduced to a small disk at the base of the flower; petals separating only at the base and falling off without expanding; some plants with perfect flowers, others staminate with a vestigial ovary; petals and stamens 5; stigmas 2-lobed; fruit a pulpy berry; leaves somewhat heart-shaped at base, sometimes both shallowly and deeply lobed on the same plant.

a. Leaves glaucous, underside reddish-tomentose to reddish-floccose when mature; partition across pith at nodes to 4 mm thick in young woody canes . . . . 1. *V. aestivalis*
a. Leaves green on both sides; partition across pith at nodes 1 mm or less thick in young woody canes . . . . 2. *V. riparia*

1. **V. aestivalis** Michx. Summer Grape (Fig. 368). Teeth of

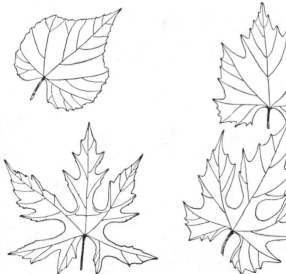

Fig. 368. *Vitis aestivalis,* × .25.   Fig. 369. *Vitis riparia,* × .25.

leaves broader than long, somewhat narrowed to a short abrupt point; leaves red-woolly beneath, especially when young.—Rare; Rock and Kenosha counties.

Var. **argentifolia** (Munson) Fern. Silverleaf Grape. Leaves whitened but not woolly beneath.—North to Clark and Fond du Lac counties.

2. **V. riparia** Michx. Frost Grape (Fig. 369). Teeth of leaves longer than broad, drawn out to a long point which is often slightly curved; leaves green beneath and without hairs.—Common.

Var. **syrticola** (Fern. & Wieg.) Fern. Dune Grape. Leaves hairy beneath.—Rare.

TILIACEAE   Linden Family

Trees, with simple serrate leaves; flowers mostly perfect; sepals and petals 4-5; petals usually conspicuous; stamens 10-many, distinct or united; style 1; fruit a capsule or nutlet.

Tilia   Linden; Basswood

Characters as for family, except the stamens in 5 groups, united at base to 5 broad petallike staminodia in our species.

T. americana L. (Fig. 370). Trees; leaf blades heart-shaped, without hairs except for minute tufts on the lower surface at the junctions of the conspicuous veins; flowers in a branching cluster in the axil of a leaflike bract about 1 dm

Fig. 370. *Tilia americana,* × .25.

long, which is partly fused with the peduncle; sepals 5; petals 5.—Rich woods, throughout the state.

MALVACEAE   Mallow Family

Herbs, with alternate long-petioled palmately veined or lobed leaves; flowers solitary or in cymose inflorescences, perfect; sepals 5; petals 5; stamens many, the filaments united into a column around the superior ovary and style; the style branches equal in number to the carpels; fruit at maturity separate carpels, dehiscent (a schizocarp), indehiscent, or a loculicidal capsule.

a. Petals notched at apex . . . . .1. *Malva*
a. Petals rounded at apex . . . . 2. *Napaea*

1. Malva   Mallow

Annual to perennial herbs; leaves broad, toothed or lobed; flowers single or in clusters; carpels united in a ring around a central axis from which they separate at maturity.

a. Leaves rounded (Fig. 371) *b*
  b. Sepals margined with hairs longer than the sepals; petals only a little exceeding the sepals . . . . .1. *M. rotundifolia*
  b. Sepals margined with short hairs; petals 3-4 times as long as the sepals . . . . .2. *M. neglecta*
a. Leaves cleft into ribbonlike segments . . . . 3. *M. moschata*

1. **M. rotundifolia** L. (Fig. 371). Stems reclining; leaf blades shallowly lobed, the lobes toothed; flowers pink, in clusters in the axils of the leaves; fruits marked with a raised network on the back.—A naturalized weed; not common; in most counties of southern Wisconsin and scattered localities northward to Douglas County.

2. **M. neglecta** Wallr. Common Mallow; Cheeses. Similar to No. 1; fruits finely velvety and without netlike markings on the back.—Naturalized from Europe; a common weed, throughout the state.

3. **M. moschata** L. Musk Mallow. Flowers rose-colored or white, in racemes, only the lowest in the axils of leaves.—In-

Fig. 371. *Malva rotundifolia,* × .25.

Fig. 372. *Napaea dioica,* × .25.

troduced from Europe, and rarely escaping from cultivation·
scattered localities in the state.

2. Napaea   Glade Mallow

Tall perennial dioecious herbs, with alternate palmately
lobed leaves; flowers in large terminal panicles, lacking extra
bracts around the calyx; staminate flowers with the stamens
in a column (monadelphous); pistillate flowers with abortive
sessile stamens in a ring around the ovary; carpels 10, the
fruit a beakless or very short-beaked dehiscent capsule.

N. dioica L. (Fig. 372). Plants 1-2 m tall; leaves
5-9-lobed, coarsely toothed or incised; petals white, 5-9 mm
long in the staminate flowers, smaller in the pistillate
flowers; carpels of fruit 5 mm long, ribbed on the back.—
Moist ground, Lafayette to Rock counties, north to Crawford to Richland counties.

CISTACEAE   Rockrose Family

Shrubs or herbs; leaves simple, entire; flowers single or in
a cymose inflorescence, perfect; sepals 5, irregular, 2 outer, 3
inner; petals 5 or lacking, often falling early; stamens many;
style 1; fruit a loculicidal capsule.

a. Leaves flat, narrow, green . . . . 1. *Helianthemum*
a. Leaves scalelike, woolly . . . . 2. *Hudsonia*

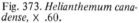

Fig. 373. *Helianthemum canadense,* × .60.

Fig. 374. *Hudsonia tomentosa,* × 1.40.

1. Helianthemum   Frostweed

Perennial rhizomatous herbs, with stellate hairs; leaves narrow; early flowers showy with 5 petals, later flowers cleistogamous with petals reduced or lacking; style with a large capitate stigma.

H. canadense (L.) Michx. (Fig. 373). Stems simple, later becoming branched; leaves about 2 cm long, rather narrow, ascending; first flower solitary, 2.5 cm broad, bright yellow, soon wilting; later flowers very numerous, small, clustered on small branches which rise from branches that overtop the first flower.—Dry prairies and sand barrens, throughout the state.

2. Hudsonia   False Heather

Prostrate bushy shrubs forming mats; leaves scalelike, linear; flowers solitary; sepals 3, resulting from the fusion of 2 outer with 3 inner sepals; petals 5, dropping early; stamens 10-30; style elongate; fruit a capsule enclosed in the calyx.

H. tomentosa Nutt. (Fig. 374). Plants 1-2 dm high, much branched; flowers short-pediceled, bright yellow.—In open sand along the Wisconsin River to Juneau and Adams counties, and in the northwestern corner of the state.

VIOLACEAE   Violet Family

Herbs (our species); leaves simple; flowers single, usually nodding or sometimes paniculate, perfect, bilaterally symmetrical; sepals 5; petals 5 or none; stamens 5; style 1; fruit a many-seeded loculicidal capsule.

a. Flowers blue, white, or yellow .... 1. *Viola*
a. Flowers greenish .... 2. *Hybanthus*

Viola   Violet

Plants small and delicate, seldom more than a few dm high; sepals usually auriculate; petals unequal, the 2 lateral ones often bearded at base, the lower one with a spur containing nectaries which project backward from the 2 lower stamens. Most species produce spring flowers with showy petals, followed by closed, self-pollinating (cleistogamous) flowers lacking petals for the remainder of the season. There is hybridization among the species. Norman H. Russell, who has made extensive studies of violets, prepared the following key to Wisconsin species (*V. palmata* added).

Spring Key to Wisconsin Violets

a. Plants caulescent, with leafy aerial stems *b*
   b. Stipules very large and leafy, palmately fimbriate at base ......
     ....1. *V. tricolor*
   b. Stipules smaller, not palmately fimbriate at base *c*
     c. Petals blue *d*
       d. Leaves ovate to triangular, the bases truncate to cordate, the apices blunt ....2. *V. adunca*
       d. Leaves orbicular, the bases cordate, the apices acute or acuminate *e*
         e. Petal spur short, one-quarter to one-third the length of the spur petal ....3. *V. conspersa*
         e. Petal spur long, one-half or more the length of the spur petal ....4. *V. rostrata*
     c. Petals white or yellow *f*
       f. Petals white or cream *g*
         g. Petals cream; leaves oval, acute; stipules large ....
         .... 5. *V. striata*
         g. Petals white with purple veins; leaves reniform to ovate, short to long-attenuate; stipules small *h*
           h. Leaves ovate, long-acute to acuminate; petals purple-lined on back; rootstock short, thick, without runners ....6. *V. canadensis*
           h. Leaves reniform to ovate, short-acute to long-acute; petals purple-backed; rootstock with long slender subterranean runners ....
           .... 6. *V. canadensis*, var. *rugulosa*
       f. Petals yellow *i*
         i. Plants with 1 flowering stem; 1 or 2 basal leaves, densely hairy .... 7. *V. pubescens*

      i. Plants with several flowering stems, several basal leaves; finely pubescent . . . .
. . . . .7. *V. pubescens,* var. *eriocarpa*

a. Plants acaulescent, with horizontal or vertical underground rhizomes *j*

  j. Petals white, flowers very small *k*

    k. Leaves reniform; plants without vegetative stolons; petals always glabrous . . . . 8. *V. renifolia*

    k. Leaves not reniform; plants with vegetative stolons, at least in late spring and summer; petals usually bearded *l*

      l. Leaves lanceolate, with cuneate bases . . . 9. *V. lanceolata*

      l. Leaves ovate or orbicular, with cordate bases *m*

        m. Leaf blades completely glabrous on both surfaces; petioles usually pubescent; lateral petals with slight beards . . . . .10. *V. macloskeyi,* ssp. *pallens*

        m. Leaf blades and petioles variously pubescent; lateral petals with heavy beard . . . . 11. *V. incognita*

  j. Petals blue or violet, rarely white; flowers large *n*

    n. Leaves deeply lobed or cut *o*

      o. All petals glabrous; flowers flattened in appearance; rhizomes fleshy, erect, nonstoloniferous . . . 12. *V. pedata*

      o. Lower lateral pair of petals bearded; flowers papilionaceous; rhizomes fleshy or woody, usually horizontal, somewhat stoloniferous *p*

        p. Leaves deeply lobed throughout, orbicular or reniform in shape *q*

          q. Plants glabrous . . . . 13. *V. pedatifida*

          q. Plants pubescent . . . . .14. *V. palmata*

        p. Leaves shallowly lobed at base, lanceolate in shape *r*

          r. Petioles shorter than the blades; basal teeth conspicuous in spring . . . . 15. *V. fimbriatula*

          r. Petioles longer than the blades; basal teeth inconspicuous in spring . . . . 16. *V. sagittata*

    n. Leaves entire, not lobed or deeply cut *s*

      s. Rhizomes narrow; basal petal spur 5-7 mm long . . . .
. . . . 17. *V. selkirkii*

      s. Rhizomes thick; basal petal spur about 2 mm long *t*

        t. Leaf blades glabrous on lower surfaces *u*

          u. Leaf blades glabrous on upper surfaces also *v*

            v. Leaves somewhat triangular in overall shape, the blunt apices definitely so, with few crenations; petals pale violet . . . . .18. *V. missouriensis*

            v. Leaves orbicular, with closely toothed, acute apices; petals blue violet, purple, or white with purple lines . . . . .19. *V. papilionacea*

          u. Leaf blades pubescent on upper surfaces only *w*

            w. Leaf blades cordate, with slightly attenuate apices; surface hairs long and stiff . . . .
. . . . 20. *V. affinis*

            w. Leaf blades reniform to cordate, with acute apices; surface hairs very small and inconspicuous *x*

              x. Spur petal glabrous; lateral petals bearded with clavate hairs . . . . 21. *V. cucullata*

              x. Spur petal bearded; lateral petals bearded with cylindrical hairs . . . 22. *V. nephrophylla*

        t. Leaf blades pubescent on both surfaces *y*

y. Blades more or less triangular in shape, longer than
   broad; sepals not ciliate ......23. *V. novae-angliae*
y. Blades reniform to orbicular, shorter than broad;
   sepals ciliate z
   z. Leaves small, with 8-15 teeth per side; early
      capsules small and globose. 24. *V. septentrionalis*
   z. Leaves larger, with 16-35 teeth per side; early
      capsules larger and more elongated...........
      ....25. *V. sororia*

1. **V. tricolor** L. Pansy (Fig. 375). Annuals, glabrous or
hairy, up to 45 cm high, branched from base; leaves crenate,
stipules large; flowers 1.5-2.5 cm wide, petals variously
colored.—Cultivated; introduced from Europe and rarely
escaping from cultivation.

2. **V. adunca** Sm. Sand Violet (Fig. 376). Stems not often
over 1 dm high, tufted; the whole plant almost micro-

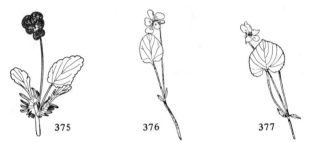

Fig. 375. *Viola tricolor,* × .25.   Fig. 377. *Viola conspersa,*
Fig. 376. *Viola adunca,* × .40.   × .40.

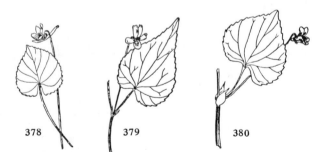

Fig. 378. *Viola rostrata,* × .40.   Fig. 380. *Viola pubescens.* var.
Fig. 379. *Viola canadensis,* × .40.   *eriocarpa,* × .40.

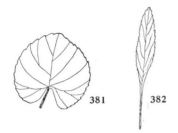

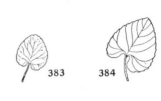

Fig. 381. *Viola renifolia,* × .40.
Fig. 382. *Viola lanceolata,* × .40.

Fig. 383. *Viola macloskeyi,* ssp. *pallens,* × .40.
Fig. 384. *Viola incognita,* × .40.

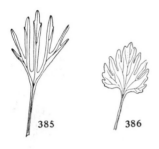

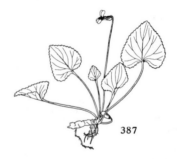

Fig. 385. *Viola pedata,* × .40.
Fig. 386. *Viola palmata,* × .40.

Fig. 387. *Viola cucullata,* × .25.

scopically woolly; leaf blades 1-2 cm wide; flowers small, blue.—Dry sand and ledges; throughout the state except for southern 2 tiers of counties.

3. **V. conspersa** Reichenb. American Dog Violet (Fig. 377). Stems several, 8-16 cm high; leaf blades roundish, 2-4.5 cm wide, without hairs; flowers small, blue.—Common, throughout the state.

4. **V. rostrata** Pursh. Long-spurred Violet (Fig. 378). Stems usually tufted, 1-1.5 dm high; leaf blades roundish heart-shaped, without hairs; flowers blue.—Rare; in counties bordering Lake Michigan.

5. **V. striata** Ait. Cream Violet. Plants caulescent; leaves uniform, with heart-shaped bases and acute apices, finely

serrate; sepals fringed; petals creamy white.—Rock County.

6. **V. canadensis** L., var. **canadensis**. Canada Violet (Fig. 379). Plants from a creeping rootstock; leaf blades heart-shaped, taper-pointed; stipules usually 4-5 times as long as broad, withering; corolla white with purple lines.—Rich woods, mostly northeastward, south to Rusk, Adams, and Fond du Lac counties.

Var. **rugulosa** (Greene) C. L. Hitchc. Plants with slender branching underground stolons; basal leaves broad, reniform, apices acute, slightly hairy; petals white but purple-tinged on the back side.—Scattered localities from Green and Rock counties to Bayfield to Florence counties.

7. **V. pubescens** Ait., var. **pubescens**. Downy Yellow Violet. Plants softly hairy; one or no leaves arising from the base of the stem; blades of stem leaves usually 7 cm or more wide, with 30-45 teeth; capsules woolly.—Woods; Dane and Waukesha counties, very scattered localities in north-central Wisconsin.

Var. **eriocarpa** (Schwein.) Russell (Fig. 380). Plants with several flowering stems; basal leaves 4 or more, with 25-30 teeth, finely pubescent to glabrate.—Common, throughout most of Wisconsin.

8. **V. renifolia** Gray. Kidney-leaved Violet (Fig. 381). Plants without stolons; leaf blades nearly round, 3-9 cm broad, shining, with stiff white appressed hairs on both surfaces; petals white.—Woods, northern Wisconsin; less common than the following variety.

Var. **brainerdii** (Greene) Fern. Hairs lacking on the upper leaf surface.—Northern tier of counties and along Lake Michigan, south to Sheboygan County.

9. **V. lanceolata** L. Lance-leaved Violet (Fig. 382). Leaf blades lanceolate, tapering to the petiole, glabrous; petals white, the lower 3 with purple veins.—Moist sandy shores, mostly northward, south along the Wisconsin River to Iowa County.

10. **V. macloskeyi** Lloyd, ssp. **pallens** (Banks) M. S. Baker (Fig. 383). Plants stemless, 1-3 cm tall; blades glabrous, petioles pubescent; petals white.—Wet situations, throughout the state.

11. **V. incognita** Brainerd. Sweet White Violet (Fig. 384). Plants stoloniferous; leaf blades 2-6 cm broad, pointed, dull, with delicate appressed white hairs; petals white.—Damp rich woods, throughout the state except for the southern tier of counties.

12. **V. pedata** L. Bird's-foot Violet (Fig. 385). Leaf blades 3-divided, the divisions cleft into ribbonlike segments; flowers flattened, pansylike; upper petals dark purple, the lower ones lilac purple.—Very rare, usually with the following variety.

Var. **lineariloba** DC. Petals colored alike, deep blue or rarely white.—Abundant in dry soil, especially in sand barrens.

13. **V. pedatifida** G. Don. Leaves divided much as in No. 12; petals violet, flower not pansylike.—Dry fields and prairies; southern Wisconsin, north to Fond du Lac County in the east, and north to Burnett County in western Wisconsin.

14. **V. palmata** L. (Fig. 386). Plants stemless, pubescent; rhizomes creeping or oblique; leaves palmately divided into 5-11 segments, the middle segment broadest; flowers 2-3 cm wide, usually deep violet.—Open woods; Dane and Iowa counties.

15. **V. fimbriatula** Sm. Plants stemless; leaves sagittate, blades up to twice the length of the petioles, basal teeth conspicuous in spring but later leaves obscurely toothed at the base, heavily pubescent; flowers large; petals blue or violet.—Very rare; Jackson County.

16. **V. sagittata** Ait. Arrow-leaved Violet. Plants mostly small, the flowers usually slightly overtopping the leaves; leaf blades on long petioles, more than twice as long as broad, slightly to very hairy; basal leaves with inconspicuous teeth in spring but later leaves with coarse elongated teeth; petals blue.—In sand and wet rocky woods, throughout Wisconsin.

17. **V. selkirkii** Pursh. Plants small and delicate; rhizomes slender; leaf blades heart-shaped, the lobes approaching each other or overlapping; petals pale blue, spur 5-7 mm long.—Along our northern borders, south to Clark and Sheboygan counties, and in cold canyons at Devil's Lake and the Dells of the Wisconsin River.

18. **V. missouriensis** Greene. Plants stemless, entirely glabrous; leaves with basal lobes, bluntly attenuate, the apical portion with 1-3 teeth; petals blue or violet.—Grant to Pierce counties along the Mississippi River, eastward to Rock and Sheboygan counties.

19. **V. papilionacea** Pursh. Plants with stout rhizomes with several crowns; leaves cordate-ovate; flowers violet with white centers.—Southern Wisconsin and north to St. Croix to Shawano counties.

Forma **albiflora** Grover. Confederate Violet. Petals pale or white with purple lines.—A popular cultivated variety, which has become weedy and naturalized in many habitats.

20. **V. affinis** Le Conte. Plants stemless; leaves heart-shaped, toothed; the peduncles and petioles about equal in length; upper surface of basal lobes of leaves with scattered stiff white hairs; flowers blue.—Deciduous woodlands from Vernon, Adams, and Portage counties eastward, and in scattered counties in northern Wisconsin.

21. **V. cucullata** Ait. Long-stemmed Marsh Violet (Fig. 387). Plants smooth except for a few short stiff hairs on upper surface of basal lobes; flowers on long stems above the leaves; petals purple violet, with darker centers; spurred petal beardless, shorter than the bearded lateral ones; lobed auricles of the sepals pointing straight back; leaves in general small, heart-shaped, pointed or blunt at the tips, teeth varying in size and number.—Widely ranging plant of cold springs, streams, and marshes, sometimes spreading to dryer ground; throughout the state.

22. **V. nephrophylla** Greene. Plants glabrous; early leaves round or reniform, often bluish beneath; later leaves heart-shaped, cordate, bluntly pointed, with low teeth; petals blue, hairy, the 2 upper ones only slightly so; sepals blunt, rounded at the base; auricles without cilia.—Cold mossy bogs and borders of streams and lakes; rare in northern half of the state, frequent southeastward.

23. **V. novae-angliae** House. Plants stemless; petioles villous; lower leaf surfaces minutely pubescent, crenulate-serrate to an acuminate tip; flowers violet purple.—Scattered localities in northern Wisconsin.

24. **V. septentrionalis** Greene. Plants stemless; leaves heart-shaped, pubescent, but only sparsely so on underside, a fringe of hairs around the margin; sepals closely ciliate; petals blue.—Scattered localities in southern Wisconsin, north to Vernon to Sheboygan counties.

25. **V. sororia** Willd. Hairy Blue Violet. Leaves heart-shaped, long-hairy beneath and on the stems; sepals rounded, ciliate toward the base; lateral petals bearded, spurred petal smooth or only slightly hairy; petals blue.—The common, widely ranging blue violet of rich woods, meadows, and river bottoms; throughout the state.

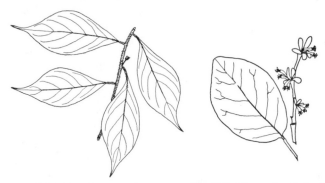

Fig. 388. *Hybanthus concolor,* ×.25.   Fig. 389. *Dirca palustris,* × .40.

2. Hybanthus   Green Violet

Perennial herbs, with leafy stems; leaves alternate; inflorescence axillary clusters of tiny greenish white flowers; lower petal spurred; anthers connate into a sheath surrounding the pistil; style hooked at tip.

H. concolor (T. F. Forst.) Spreng. (Fig. 388). Stems leafy to summit, blades pointed at both ends, with small green flowers in their axils.—Rich wooded slopes; Grant County.

THYMELAEACEAE   Mezereum Family

Shrubs (our species); leaves simple, alternate; flowers perfect; sepals 4-5, usually fused, or minute; petals 4-5, or lacking in our species; hypanthium bell-shaped; stamens twice the number of petals or sepals; style 1; fruit a drupe.

Dirca   Leatherwood; Wicopy

Shrubs 1-2 m high; wood brittle, but bark exceedingly tough and flexible; flower buds subtended by hairy bud scales; flowers pale yellow, appearing before the leaves in axillary clusters; hypanthium and calyx forming floral tube with 8 exserted stamens; petals lacking.

D. palustris L. (Fig. 389). Leaves short-petioled, obovate, entire, 5-8 cm long.—Rich wet woods; throughout the state, but more common northward.

ELAEAGNACEAE   Oleaster Family

Shrubs, characterized by the silvery appearance of the leaves caused by closely appressed hairs and scales; leaves simple, entire; flowers small, solitary or clustered; sepals corollalike, 4; stamens 8; style 1; fruit drupelike, an achene enclosed by a fleshy hypanthium.

Shepherdia

Plants dioecious; leaves opposite, at least the underside covered by silvery to rusty scales; flowers axillary.

a. Leaves and young twigs silvery . . . . 1. *S. argentea*
a. Leaves green, nearly glabrous above, under surface densely pubescent; young twigs and buds rusty red . . . . 2. *S. canadensis*

1. **S. argentea** Nutt. Buffaloberry. Shrubs or small trees, the stiff branches often thorny; leaves silvery on both sides, oblong; flowers appearing before the leaves; fruit red, edible. —Occasionally cultivated and rarely escaping; Grant, Langlade, and Juneau counties.

2. **S. canadensis** (L.) Nutt. Soapberry (Fig. 390). Shrubs 1-2 m high; leaves ovate to narrowly lanceolate, 3-5 cm long; flowers coming before the leaves; fruit nauseous.—On rocks and sandy shores; Lake Superior and Lake Michigan regions.

Fig. 390. *Shepherdia canadensis,*
× .50.

Fig. 391. *Oenothera serrulata,*
× .50.

ONAGRACEAE Evening Primrose Family

Herbs; leaves simple; flowers perfect, sepals 4, usually becoming reflexed at anthesis; petals 4, usually convolute in bud; stamens 8; ovary inferior; style 1; stigma 2-4-lobed; fruit a capsule.

Oenothera Evening Primrose

Annual, biennial, or perennial herbs; leaves alternate; inflorescence a terminal spike or single axillary flowers; hypanthium slenderly tubular, prolonged above the ovary, deciduous; sepals often with appendages; petals yellow (our species).

a. Ovary clavate, with well-defined wings; stamens alternately unequal ....1. *O. perennis*
a. Ovary cylindrical to 4-angled; stamens equal *b*
  b. Hypanthium tube prolonged more than 1 cm beyond the ovary; leaves mostly entire .... 2. *O. parviflora*
  b. Hypanthium tube prolonged less than 1 cm beyond the ovary; leaves sharply serrulate .... 3. *O. serrulata*

1. **O. perennis** L. Sundrops. Stems solitary or several in a cluster, 1.5-3.5 dm tall; leaves alternate, 1-4 cm long, entire, rounded at tip; petals 4-9 mm long, bright yellow; fruit club-shaped, winged, and with a narrowed base.—Dry ground, throughout the state except along Lake Michigan.

2. **O. parviflora** L. Evening Primrose. Biennials; stems simple, 1.4-13 dm high; leaves narrowly lanceolate, 2-15 cm long; inflorescence a terminal raceme; hypanthium tube 20-35 mm long; petals 6-18 mm long; capsule 1.5-4 cm long.—Throughout the state.

3. **O. serrulata** Nutt. Tooth-leaved Evening Primrose (Fig. 391). Perennials, 2-5 dm tall, simple to much branched, often woody; leaves linear-lanceolate, 1-6 cm long, sharply serrulate; flowers solitary, sessile in the axils of the upper leaves; hypanthium tube 4-8 mm long; petals 5-10 mm long; capsule 1-3 cm long.—Native of dry prairies in western Wisconsin along Mississippi River bluffs from Buffalo to St. Croix counties, and adventive along railroads and in gravel pits in scattered localities in the eastern half of Wisconsin.

ARALIACEAE Ginseng Family

Herbs, or somewhat woody plants; leaves compound, with petioles often sheathing the base; inflorescences heads or umbels; flowers small; sepals minute; petals and stamens 5 (our species); styles 2-5; fruit berrylike.

a. Leaves alternate or basal.........................1. *Aralia*
a. Leaves whorled ....2. *Panax*

1. Aralia

Leaves 3-parted, the divisions pinnately divided, ultimate leaf segments toothed; flowers in umbels; styles 5.

a. Umbels very numerous, in a raceme.............1. *A. racemosa*
a. Umbels 2-12, each on a long stalk *b*
   b. Ultimate leaf segments not over 2 cm wide.........2. *hispida*
   b. Ultimate leaf segments at least 5 cm wide.....3. *A. nudicaulis*

1. **A. racemosa** L. Spikenard (Fig. 392). Plants 1-2 m tall; ultimate leaf segments heart-shaped; flowers greenish, about 2 mm broad.—Rich woods, throughout the state.

2. **A. hispida** Vent. Bristly Sarsaparilla (Fig. 393). Stems 4-9 dm high, bristly on the woody base, openly branched above, each branch terminated by an umbel; flowers white, 2 mm broad.—Common northward in woods and clearings, rare southward, and not occurring in the southern tier of counties.

3. **A. nudicaulis** L. Wild Sarsaparilla (Fig. 394). Stems mostly underground; leaf long-petioled, overtopping the scape; branches of the scape rising all from one point; flowers greenish, 3 mm broad.—Rich woods and bluffs; common, throughout the state.

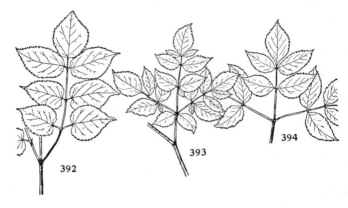

Fig. 392. *Aralia racemosa,* × .10.   Fig. 394. *Aralia nudicaulis,* × .10.
Fig. 393. *Aralia hispida,* × .10.

Fig. 395. *Panax quinquefolius,* × .20. Fig. 396. *Panax trifolius,* × .35.

2. Panax

Perennial herbs, stems with a whorl of 3 palmately compound leaves; inflorescence a single terminal umbel; styles 2-3.

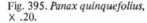

a. Leaflets with long petiolules . . . . **1.** *P. quinquefolius*
a. Leaflets sessile . . . . **2.** *P. trifolius*

1. **P. quinquefolius** L. Ginseng (Fig. 395). Roots large and spindle-shaped; stem simple, 3 dm high; leaves, mostly 3, large, palmately divided into 5 leaflets with long-pointed tips; flowers greenish, about 2 mm broad; fruit a cluster of scarlet berries.—Rich woods, originally throughout most of the state, now becoming rare.

2. **P. trifolius** L. Dwarf Ginseng; Groundnut (Fig. 396). Tuber globular; stem 1-2 dm high; the 3-5 leaflets usually much smaller than the above, and with blunt or rounded tips; fruit a cluster of small greenish pods (carpels).—Woods, northern Wisconsin, south to Richland and Sauk counties, and along the Lake Michigan shore.

APIACEAE (UMBELLIFERAE)   Parsley Family

Aromatic herbs; leaves basal or alternate, usually compound, the petioles sheathing the stem; inflorescence mostly compound umbels; flowers small; sepals and petals 5; stamens 5; styles 2; fruit 2 seedlike dry carpels which split at maturity. The genera are based primarily on fruit characters.

a. Petals white (rarely purplish; greenish in *Angelica*) *b*
  b. Leaves twice or more compound *c*

    c. Sheaths seldom over 2 cm long *d*
      d. Bracts of the umbel rounded or obtuse at tip *e*
        e. Stipules without hairs . . . . 1. *Erigenia*
        e. Stipules fringed with hairs . . . . 2. *Chaerophyllum*
      d. Bracts of the umbel acuminate at tip, or lacking *f*
        f. Divisions of leaves flat, toothed (Fig. 398) . . . .
        . . . . .3. *Osmorhiza*
        f. Divisions of leaves narrowly ribbonlike (Fig. 399) . . . .
        . . . . 4. *Carum*
    c. Sheaths 5-10 cm long *g*
      g. Leaflets with teeth or lobes 1 cm or more wide, these
        sharply toothed; stalks of leaflets with coarse flattened
        hairs . . . . 5. *Heracleum*
      g. Leaflets obscurely double-toothed, their stalks with
        delicate very short hairs or none . . . . 6. *Angelica*
  b. Leaves (except when submerged) once compound *h*
    h. Leaves pinnately 5-many-foliate . . . . 7. *Sium*
    h. Leaves palmately 3-foliate . . . . 8. *Cryptotaenia*
a. Petals yellow, purple, or greenish *i*
  i. Stem leaves pinnate, the leaflets not divided . . . . . . 9. *Pastinaca*
  i. Stem leaves divided into 3 parts above the sheath, each part
    simple or again compound *j*
    j. Ultimate divisions of leaves not toothed . . . . . . .10. *Taenidia*
    j. Ultimate divisions of leaves toothed *k*
      k. Each tooth ending in a fine bristle . . . . 11. *Sanicula*
      k. Teeth not ending in a bristle *l*
        l. Divisions of leaves somewhat ribbonlike, the sides
        roughly parallel . . . . 12. *Polytaenia*
        l. Divisions of leaves ovate, the sides curved *m*
        m. All flowers on short stalks . . . . .13. *Thaspium*
        m. Central flower of each cluster without a stalk . . . . .
        . . . . 14. *Zizia*

1. Erigenia   Harbinger-of-Spring

Perennial herbs; inflorescence a terminal umbel of comparatively large sessile flowers; fruit almost round, laterally flattened.

E. bulbosa (Michx.) Nutt. Stems 1-2.3 dm high, from a solid corm; leaves palmately much-divided.—Rare; Milwaukee, Racine, and Kenosha counties.

2. Chaerophyllum

Annuals; leaves ternately decompound; umbels with few flowers; fruits narrowly oblong, with equal ribs, notched at base.

C. procumbens (L.) Crantz. Stems slender, often spreading, 1-5 dm high, somewhat hairy; leaflets ovate, stalked, pinnately divided, divisions stalked, wedge-shaped at base, often 2-3-lobed; umbels with few flowers; fruit narrowly oblong, 5-10 mm long.—Wooded bottoms of the Sugar River, Rock County, and Platte River, Grant County.

Fig. 397. *Osmorhiza claytonii,*
× .25.

Fig. 398. *Osmorhiza longistylis,*
× .25.

3. Osmorhiza   Sweet Cicely

Stems 3-12 dm high; upper leaves practically sessile, divided into 3 leaflets, these rather long-stalked and pinnately 3-5-divided; fruit long and narrow, tapered to both ends, often curved, ribbed, with ascending hairs.

a. Umbels with bracts at base of branches *b*
  b. Anthers 0.2-0.3 mm long; style shorter than the petals, 0.8-1.5 mm long in fruit.........................1. *O. claytonii*
  b. Anthers about 0.5 mm long; style longer than the petals, becoming 3-4 mm long in fruit .... 2. *O. longistylis*
a. Umbels without bracts...................... 3. *O. chilensis*

1. **O. claytonii** (Michx.) C. B. Clarke (Fig. 397). Stems and leaves with soft white hairs; body of fruit 1-1.3 cm long. Woods, throughout the state.

2. **O. longistylis** (Torr.) DC. (Fig. 398). Stems glabrous, or rarely with spreading hairs; body of fruit 1.2-1.5 cm long.— Woods, throughout the state; less abundant than No. 1.

3. **O. chilensis** H. & A. Umbels with only 3-7 rays. − In Bayfield County near Lake Superior and Door County near Green Bay.

4. Carum   Caraway

Biennials or perennials, with long taproots; umbels compound; terminal and lateral pedicels unequal; fruits oblong, ribbed.

C. carvi L. (Fig. 399). Leaves pinnate, the leaflets once or twice pinnate.—Roadsides; scattered localities, throughout the state; naturalized from Europe.

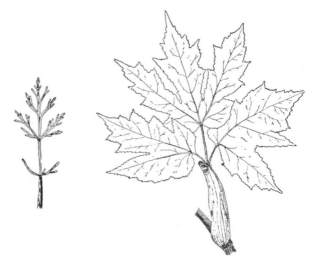

Fig. 399. *Carum carvi,* × .40.   Fig. 400. *Heracleum lanatum,* × .20.

5. Heracleum   Cow Parsnip

Tall stout perennials or biennials; leaves large, once ternate or once pinnate; umbels broad; flowers white or purplish; fruit flattened, obovate, broadly winged.

H. lanatum Michx. (Fig. 400). Plants 1-2.8 m high, coarse, woolly; leaves 3-parted, the leaflets usually 1 dm or more broad, deeply 3-cleft.—Roadsides and openings in woods, throughout the state.

6. Angelica   Angelica

Stout perennials; basal leaves long-petioled, pinnately decompound, upper leaves gradually reduced to bladeless petioles; umbels very large, compound, rounded; flowers white or greenish white; fruit flattened dorsally, ribs prominent, the lateral ones winged.

A. atropurpurea L. (Fig. 401). Plants stout, often 2 m high; stems streaked with purple and green, with a powdery bloom; leaves 3-parted, the divisions stalked and pinnately divided, without hairs; umbels spherical, long-stalked, many, aggregated into a large spherical umbel.—Wet open ground, from Lafayette, Dane, Columbia, and Shawano counties southeastward.

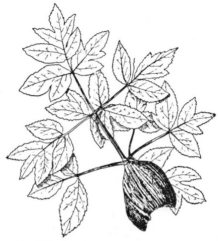

Fig. 401. *Angelica atropurpurea,* × .20.

Var. **occidentalis** Fassett. Leaflets very finely hairy beneath.—From Lafayette, Dane, and Milwaukee counties northwestward.

7. **Sium** Water Parsnip

Tall glabrous perennials; fruit ovate, with prominent corky ribs.

S. suave Walt. (Fig. 402). Plants usually erect, 0.8-2 m

Fig. 402. *Sium suave,* × .40.

high; leaves pinnate, the leaflets pointed, sharply toothed; submerged leaves, when present, usually finely dissected; fruit 2.5-3 mm long.—Marshy land, throughout the state.

8. **Cryptotaenia**   Honewort

Branching perennial glabrous herbs; inflorescences numerous loose irregular umbels both axillary and terminal; fruit linear-oblong, slightly flattened, with shallow broad ribs.

C. canadensis (L.) DC. (Fig. 403). Plants 3-9 dm high, not hairy; leaves long-petioled, blades 3-parted, leaflets ovate, tapered at base, doubly toothed, the 2 lateral ones often deeply 2-cleft; pedicels very unequal; fruit 4-6 mm long, often curved.—Rich woods, throughout the state.

9. **Pastinaca**   Parsnip

Glabrous biennials; lower leaves with long petioles, the upper on shorter wholly sheathing petioles; umbels large, compound; flowers yellow; fruit ovate, flattened, the edges with broad wings nerved to the outer margins.

P. sativa L. (Fig. 404). Plants stout, often tall; leaves pinnate, the leaflets coarsely toothed, sometimes deeply cleft; umbel branches 2-6 mm long.—Introduced from Europe; naturalized, throughout most of Wisconsin.

Fig. 403. *Cryptotaenia cana-densis,* × .25.

Fig. 404. *Pastinaca sativa,* × .20.

Fig. 405. *Taenidia integerrima,* × .30.

10. Taenidia   Yellow Pimpernel

Glabrous perennials; leaves ternately compound; umbels compound, the inner flowers staminate, short-pediceled, the outer flowers perfect and long-pediceled; petals yellow; fruit short, oblong, flattened, only slightly ribbed.

T. integerrima (L.) Drude (Fig. 405). Plants 5-10 dm high; leaves long-petioled, 3-divided, the leaflets often twice 3-divided (making theoretically 27 ultimate divisions); rays of the primary umbel 1.5-6 cm long; secondary umbels, in flower, less than 1 cm in diameter; fruit 4 mm long.—Dry woods and open ground; not common but in many localities throughout southern Wisconsin, and scattered localities northward to Burnett to Forest counties.

11. Sanicula   Black Snakeroot

Biennial or perennial plants 3-10 dm high; leaves mostly basal and long-petioled; leaves of stem sometimes 1-3, with petioles short or none; blades palmately 3-5-foliate, the 2 lateral leaflets often deeply 2-cleft, all sharply and somewhat doubly toothed, more or less narrowed to the base; umbels several times few-branched; pistillate flowers with very short pedicels or none; staminate flowers intermixed with the pistillate or in different heads, rather slender-pediceled; fruits with copious hooked bristles.

a. Styles longer than the bristles, conspicuous, recurved; some heads
   entirely staminate *b*
   b. Staminate calyx 1.5-2 mm long after stamens fall . . . .
   . . . . 1. *S. marilandica*
   b. Staminate calyx about 1 mm long after stamens fall . . . .
   . . . . 2. *S. gregaria*
a. Styles shorter than the bristles, obscure; all heads with some pistillate flowers *c*
   c. Pedicels of staminate flowers shorter than the entire pistillate
      flower . . . . 3. *S. canadensis*, var. *grandis*
   c. Pedicels of staminate flowers longer than the entire pistillate
      flower . . . . 4. *S. trifoliata*

   1. S. marilandica L. (Fig. 406). Leaves usually 5-parted
(appearing 7-parted); calyx lobes of staminate flowers very
narrow and acute; fruit 5-7 mm long, not stipitate.—Rich
woods, throughout the state.

Fig. 406. *Sanicula marilandica,*
× .25, flower × 1.50.

Fig. 407. *Sanicula gregaria,*
× .25, flower × 1.50.

Fig. 408. *Sanicula canadensis,* var.
grandis, × .25, flower × 1.50.

Fig. 409. *Sanicula trifoliata,*
× .25, flower × 1.50.

2. **S.** **gregaria** Bickn. (Fig. 407). Leaves usually 3-parted (appearing 5-parted); staminate flowers with calyx lobes not over twice as long as broad, obtuse or acute; fruit 3-4 mm long, short-stipitate.—Rich woods, throughout the state.

3. **S.** **canadensis** L., var. **grandis** Fern. (Fig. 408). Leaf blades 3-parted (appearing 5-parted); terminal leaflet 2-3 times as long as broad; fruit nearly globose, without conspicuous persistent calyx.—Not common; southwestward, mostly along the Mississippi and lower Wisconsin rivers.

4. **S.** **trifoliata** Bickn. (Fig. 409). Terminal leaflet 1.5-2 times as long as broad; fruit tapering at both ends, somewhat stipitate, with a conspicuous persistent beaklike calyx.— Grant, Wood, and Eau Claire counties.

12. Polytaenia   Prairie Parsley

Stout perennials; umbels compound, terminal and axillary; flowers yellow; fruits flattened dorsally, the lateral ribs with a thick corky wing.

P. **nuttallii** DC. Stems 5-10 dm high; leaves twice pinnate; fruit obovate to oval, flattened.—Not common; open ground, north to Monroe and Columbia counties.

13. Thaspium   Meadow Parsnip

Perennials, very similar to *Zizia*; flowers all pedicellate; petals yellow or purple; fruit oblong, prominently winged.

a. Basal leaves simple or ternate . . . . .1. *T. trifoliatum*
a. Basal leaves twice or more ternate . . . . .2. *T. barbinode*

1. **T.** **trifoliatum** (L.) Gray, var. **flavum** Blake. Leaflets with firm white glabrous margins.—Rather uncommon along the Rock, Sugar, and Pecatonica rivers. Easily confused, when fruit is not present, with the more abundant *Zizia aurea*; both have the flowers in little umbels which terminate long peduncles, but in *Thaspium* all the flowers are pediceled, while in *Zizia* each little umbel has one central flower sessile.

2. **T.** **barbinode** (Michx.) Nutt. Leaflets with thin green minutely ciliate margins.—Crawford County.

14. Zizia   Golden Alexanders

Mostly glabrous branched perennials; umbellets with very unequal pedicels, the central flower usually sessile; petals yellow; fruit oblong, ribbed, glabrous, laterally compressed.

a. Basal leaves compound . . . . 1. *Z. aurea*
a. Basal leaves simple . . . . 2. *Z. aptera*

Fig. 410. *Zizia aurea,* × .40.   Fig. 411. *Zizia aptera,* × .40.

1. **Z. aurea** (L.) Koch (Fig. 410). Lower leaves long-petioled, compound; leaflets of upper leaves again 3-5-divided, with stalked divisions; rays 2-5 cm long; fruit oblong, 4 mm long.—Woods and fields, throughout the state.

2. **Z. aptera** (Gray) Fern. (Fig. 411). Lower leaves not divided, the blades heart-shaped; upper leaves 3-5-parted, the divisions toothed or deeply cleft.—Prairies and barrens, and along railroad tracks; in all parts of state except northeastern Wisconsin.

Var. **occidentalis** Fern. Leaflets of stem leaves irregularly jagged-toothed.—Adventive from the West; along railroad tracks, Milwaukee County.

CORNACEAE   Dogwood Family

Trees, shrubs, or rarely herbs; leaves simple, entire; calyx lobes minute; petals 4 or 5; stamens as many as the petals, borne on a disk; style 1; fruit 1-2-seeded, fleshy.

a. Sepals 4, petals 4, stamens 4......................1. *Cornus*
a. Sepals, 5, petals 5, stamens 5-12 ....2. *Nyssa*

1. Cornus   Dogwood

Trees, shrubs, or rarely herbs; leaves opposite (except one species alternate and one species appearing whorled); inflorescence a cyme, or a close head surrounded by corollalike bracts; calyx lobes, petals, and stamens each 4; style 1; fruit a drupe.

a. Herblike plants with petallike bracts below the close cluster of flowers; fruit red...........................1. *C. canadensis*
a. Shrubs; flowers in open, flat-topped or pyramidal clusters, without large petallike bracts *b*
  b. Leaves alternate, crowded toward the tips of the branches; ripe fruit blue black ....2. *C. alternifolia*
  b. Leaves opposite, scattered along the stems; ripe fruit blue or white *c*
    c. Pubescence more or less spreading, often woolly; leaves broadly ovate, generally with 7-9 pairs of veins (Fig. 414), white-woolly below; branchlets yellow green, more or less streaked with purplish dots; fruit blue ....3. *C. rugosa*
    c. Pubescence of closely appressed, straight and silky hairs, or hairs lacking; leaves ovate or ovate-lanceolate, generally with 3-6 pairs of veins (Fig. 417); branchlets gray, brown, or reddish *d*
      d. Sepals 1-1.5 mm long; style with club-shaped tip; fruit pale blue; pith of twigs brown . . . . 4. *C. amomum*
      d. Sepals 0.7 mm long or less; style cylindrical; fruit white; pith of twigs white *e*
        e. Twigs gray, slender; young twigs lacking hairs; inflorescence pyramidal, about as tall as broad; fruiting pedicels bright red ....5. *C. racemosa*
        e. Twigs red, rarely green, stout; young twigs with appressed white pubescence, becoming hairless; inflorescence flat-topped; pedicels of fruit green ....
        .... 6. *C. stolonifera*

1. C. canadensis L. Bunchberry (Fig. 412). Stems not branched, 9-22 cm high, from a slender creeping rootstock; leaves crowded in an apparent whorl of 6 (rarely 1 or more pairs also on the stem); bracts white, petallike, usually 4; petals small, long-triangular.—Common northward; local southward along the Lake Michigan shore, and on sandstone bluffs to Dane County.

2. C. alternifolia L. f. Alternate-leaved Dogwood; Pagoda Dogwood (Fig. 413). Shrubs or small trees 2.5-8 m high; young twigs greenish; leaf blades oval, 4-10 cm long and 2-6 cm wide, short-pointed at apex; petals 3-3.5 mm long.— Common in rich woods, throughout the state.

3. C. rugosa Lam. Round-leaved Dogwood (Fig. 414).

Fig. 412. *Cornus canadensis,* × .15.   Fig. 414. *Cornus rugosa,* × .20.
Fig. 413. *Cornus alternifolia,* × .15.

Fig. 415. *Cornus amomum,* × .15.   Fig. 417. *Cornus stolonifera,*
Fig. 416. *Cornus racemosa,* × .15.   × .15.

Shrubs 2-3 m high; twigs green, tinged with red; leaf blades reaching 12 cm in length and 10 cm in width; petals 3-4 mm long.—Thickets and open woods, throughout the state.

4. C. amomum Mill. Silky Dogwood (Fig. 415). Shrubs 1.5-3.5 m high; twigs greenish when young, soon becoming brown; leaves slightly roughened beneath with minute hairs; petals 4-5 mm long.—Wet places, north to Pierce, Wood, and Waushara counties.

5. C. racemosa Lam. Gray Dogwood; Panicled Dogwood (Fig. 416). Shrubs 1-2.5 m high; twigs gray, slender; leaf blades narrower than those of *C. stolonifera*, tapering at base and apex; petals 3-4 mm long.—In sunny places in rich woods, throughout the state.

6. C. stolonifera Michx. Red-osier Dogwood (Fig. 417). Shrubs 1.5-3 m high, strongly stoloniferous and forming large thickets in open swamps; twigs stout, red (especially in winter), sometimes green, or rarely gray; leaves with fine close hairs beneath; petals about 3.5 mm long.—Common, usually in moist places, but occasionally on dry slopes, throughout the state.

2. Nyssa   Sour Gum; Black Gum

Dioecious trees; leaves alternate; flowers greenish; sepals 5; petals small, 5 or none; stamens 5-12, on a disk; style 1; fruit a drupe.

N. sylvatica Marsh. Branches horizontal; leaf blades not toothed, broadest above the middle, narrowed to a short point; fruit bluish black, 1-1.2 cm long.—Berryville, Kenosha County.

PYROLACEAE   Shinleaf Family

Herbaceous or partly woody plants, from perennial slender rootstocks; leaves mostly simple, sometimes evergreen; sepals 5; petals 5, separate; stamens 10, anthers opening by basal pores but bent back so that the pores appear terminal; fruit a capsule of 4-5 carpels.

a. Leaves evergreen, scattered on the stem, tapered to a very short petiole, sharply toothed, about 3 times as long as wide . . . . .1. *Chimaphila*
a. Leaves all basal, definitely petioled, rarely more than twice as long as wide b
   b. Flowers solitary . . . . 2. *Moneses*
   b. Flowers several in a raceme . . . . .3. *Pyrola*

1. Chimaphila   Prince's Pine; Pipsissewa

Low slightly woody plants, from horizontal rootstocks; leaves leathery and evergreen, shiny above; petals 5; stamens 10, the filaments enlarged and hairy in the middle; style almost immersed in the depressed summit of the ovary, the stigma broad; fruit a capsule.

C. umbellata (L.) Bart., var. cisatlantica Blake (Fig. 418). Leaves 4-5 cm long, sharply toothed above the middle; flowers 2-8, pink.—Common in sandy woodlands northward; rare southward to Iowa, Rock, and Walworth counties.

2. Moneses   One-flowered Pyrola

Tiny delicate perennials; leaves appearing basal; flower solitary on a long peduncle, white, nodding; anthers 2-horned; stigma large, 5-lobed.

Fig. 418. *Chimaphila umbellata*, var. *cisatlantica*, × .25.

Fig. 419. *Moneses uniflora*, × .15.

M. uniflora (L.) Gray (Fig. 419). Plants rarely 1 dm high; leaf blades almost round, scallop-toothed, narrowed at base to a long petiole; flower 1-2 cm wide.—Mostly in evergreen woods, south to Rusk, Shawano, Sheboygan, and very rarely to Grant counties.

3. Pyrola   Shinleaf

Low herbs, without hairs, from rootstocks; leaf blades mostly rounded; flowers nodding, in a raceme, on a naked or few-bracted scape; stigma 5-lobed.

a. Bracts intermingled with the leaves at the base of the stem, usually 1 cm or more long, rounded or blunt at tip *b*
  b. Calyx lobes longer than broad *c*
    c. Calyx lobes with rounded sides, blunt, twice as long as wide (Fig. 420) . . . . 1. *P. rotundifolia,* var. *americana*
    c. Calyx lobes triangular, with straight sides, sharp-pointed, about one and one-half times as long as wide (Fig. 421) . . . .
. . . . 2. *P. asarifolia*
  b. Calyx lobes about as broad as long (Fig. 422) *d*
    d. Style curved, slightly club-shaped below the small stigma . . .
. . . . 3. *P. elliptica*
    d. Style straight, ending in a flat shield-shaped stigma . . . .
. . . . .4. *P. minor*
a. Basal bracts 2-4 mm long and long-pointed, or absent *e*
  e. Flowers on all sides of the stem; style curved (Fig. 423) . . . . . .
. . . . 5. *P. virens*
  e. Flowers all turned in the same direction; style straight (Fig. 424) . . . . 6. *P. secunda*

1. **P. rotundifolia** L., var. **americana** (Sweet) Fern. (Fig. 420). Scape 1-3.5 cm high, with 1-5 bracts; leaf blades shining, not as long as the petiole; petals white or rarely pink, 6.5-10.5 mm long.—Mostly northward, coming south to the Dells of the Wisconsin River, and rarely to Lafayette County.

2. **P. asarifolia** Michx. Pink Pyrola (Fig. 421). Similar to No. 1, but petals pink, about 5 mm long; leaf blades as broad as long, heart-shaped or tapered at the base.—Woods and swamps, south to Barron, Adams, Milwaukee, and Racine counties.

Includes var. **purpurea** (Bunge) Fern.

3. **P. elliptica** Nutt. Shinleaf (Fig. 422). Scape without bracts or with 1-3 very narrow bracts; leaf blades thin, not shining, 3-7 cm long, longer than the petiole; flowers white.—Common in rich woods, throughout the state.

4. **P. minor** L. Scape without bracts; leaf blades round, 2-4 cm long, rounded at base.—Lake Superior shore; rare.

5. **P. virens** Schweigger (Fig. 423). Scape 0.5-3 dm high,

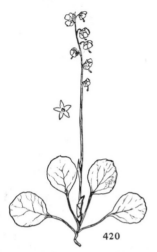

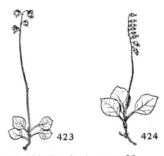

Fig. 421. *Pyrola asarifolia,* calyx × .80.
Fig. 422. *Pyrola elliptica,* calyx × .80.

Fig. 420. *Pyrola rotundifolia,* var. *americana,* × .20, calyx × .80.

Fig. 423. *Pyrola virens,* × .20.
Fig. 424. *Pyrola secunda,* × .20.

naked or with 1-2 minute hairlike bracts; leaf blades shorter than the petiole, not toothed, sometimes wavy-margined; flowers greenish. The typical form as a rosette of 4-11 leaves with rounded blades 1.5-3.3 cm wide and anthers not over 3 mm long.—Douglas to Forest counties, and south along the Lake Michigan shore to Racine.

Less common are the following 2 variants: forma **paucifolia** Fern., leafless or with 1-7 leaves with blades 0.7-2.5 cm wide; var. **convoluta** (Bart.) Fern., blades 2-4.5 cm wide and anthers 3-4 mm long.

6. **P. secunda** L. One-sided Pyrola (Fig. 424). Scape 1-2.5 dm high, with usually 2-4 bracts; leaf blades narrowed at tip, 1.5-6 cm long, with scalloped margins and usually minute teeth; flowers small, greenish.—Mostly northward and eastward, local southwestward.

Var. **obtusata** Turcz. Leaf blades rounded at tip, 0.8-3 cm long.—South to Rusk, Taylor, and Shawano counties.

ERICACEAE   Heath Family

Mostly small shrubs, prostrate or erect; flowers regular or nearly so; sepals 5; petals 5, united, at least at base, into a

bell-shaped corolla (except in *Rhododendron*); stamens 8-10,
anthers opening by terminal pores (except in *Epigaea*); fruit
a capsule or berry of 4-5 carpels.

a. Stems mostly erect, smooth or hairy but not with rusty hairs
   (except in *Ledum*) *b*
  b. Calyx borne at base of the ovary; leaves evergreen *c*
    c. Shrubs *d*
      d. Flowers violet *e*
        e. Leaves alternate, with minute rusty scales . . . .
        . . . . .1. *Rhododendron*
        e. Leaves opposite, without scales . . . . .2. *Kalmia*
      d. Flowers white or pinkish *f*
        f. Stems erect *g*
          g. Leaves with rusty wool beneath . . . . .3. *Ledum*
          g. Leaves not woolly *h*
            h. Leaves white-hairy beneath . . . . . 4. *Andromeda*
            h. Leaves with tiny round scales beneath . . . .
            . . . . .5. *Chamaedaphne*
        f. Stems prostrate . . . . .6. *Arctostaphylos*
    c. Herbs . . . . 7. *Gaultheria*
  b. Calyx borne at summit of the ovary; leaves not evergreen
    (except in cranberries) *i*
    i. Leaves resinous-dotted beneath . . . . 8. *Gaylussacia*
    i. Leaves not resinous-dotted beneath . . . . .9. *Vaccinium*
a. Stems creeping, with scattered rusty hairs *j*
  j. Leaf blades oval, cordate at base, 3-8 cm long . . . . .10. *Epigaea*
  j. Leaf blades pointed at both ends, 0.5-3 cm long . . **7**. *Gaultheria*

1. Rhododendron

Shrubs; leaves deciduous or evergreen, alternate; flowers
showy; calyx minute; petals 5, unequal; stamens 10; fruit a
capsule.

R. lapponicum (L.) Wahlenb. Lapland Rosebay. Much-
branched dwarf evergreen shrubs; leaves 2 cm or less long,
dotted above and below with glands, each of which is on a
minute scale; petals 1-2 cm broad, purple.—Trailing down the
face of cliffs, Wisconsin Dells, where it is known as Rock
Rose. An arctic-alpine plant; this is the only known locality
west of the Adirondacks and south of Hudson Bay.

2. Kalmia   Pale Laurel

Low evergreen shrubs, with opposite leaves; calyx deeply
5-lobed; corolla shallowly 5-lobed; the 10 short anthers first
fitting into pockets in the corolla, springing inward at
anthesis when touched; fruit a globose capsule.

K. polifolia Wang. (Fig. 425). Shrubs 1-6 dm high;
branches somewhat 2-edged; leaves scarcely petioled, nearly
veinless except for the conspicuous midrib, with inrolled
margins, white below; flowers on long slender pedicels, 1-2

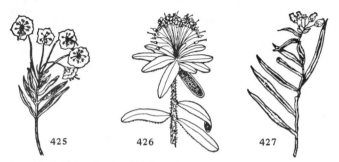

Fig. 425. *Kalmia polifolia,* × .50.   Fig. 427. *Andromeda glaucophylla,*
Fig. 426. *Ledum groenlandicum,*   × .60.
× .50.

cm wide, rose purple.—Sphagnum bogs; south to Monroe, Waushara, and Manitowoc counties.

3. Ledum   Labrador Tea

Freely branching shrubs, with thick, evergreen, sessile leaves; inflorescence a crowded terminal raceme; petals separate, spreading; style elongate.

L. groenlandicum Oeder (Fig. 426). Shrubs, to 1 m high, leaves without teeth, the margins inrolled, densely white to reddish brown woolly beneath, aromatic when bruised; flowers clustered, white.—Mostly northern, south in bogs to Wood and Sheboygan counties; on sandstone ledges at the Wisconsin Dells and in Richland County.

4. Andromeda   Bog Rosemary

Low evergreen shrubs; leaves alternate, narrow; calyx deeply 5-lobed, spreading; corolla cylindrical, somewhat globose, 5-toothed; anthers tipped with a slender ascending awn; fruit a depressed globose capsule.

A. glaucophylla Link (Fig. 427). Shrubs 5-50 cm high; stems round in cross section; leaves alternate, 2-4 cm long, 2-4 mm wide, strongly bluish whitened beneath and nearly veinless except for the conspicuous midrib, the margins strongly inrolled; flowers 3-5 mm wide, pink or white.—Bogs, south to Barron County in the west, and to Dane and Rock counties in the east.

5. Chamaedaphne   Leatherleaf

Low evergreen shrubs, forming dense thickets; leaves

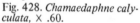

Fig. 428. *Chamaedaphne caly-*   Fig. 429. *Arctostaphylos uva-ursi,*
culata, × .60.   × .45.

alternate; inflorescence a leafy raceme of single axillary
flowers; bracts 2, subtending the calyx lobes; corolla cylindrical; anthers 10, oblong, prolonged into 2 erect tubes with
terminal pores; style elongate; fruit a depressed capsule.

C. **calyculata** (L.) Moench (Fig. 428). Low shrubs; flowers
white, small, in a single row in the axils of foliage leaves
along the last dm at the tips of the branches.—Bogs; locally
abundant except in the Driftless Area.

6. Arctostaphylos   Bearberry

Prostrate shrubs (our species); leaves evergreen, alternate;
sepals 5; calyx saucer-shaped; corolla bell-shaped, with shallow lobes, pink to white; stamens 10; anthers with 2 reflexed awns; style 1; fruit a fleshy drupe with 5 stones.

A. **uva-ursi** (L.) Spreng. (Fig. 429). Stems woody, matted
and trailing; fruit a red dry berrylike drupe.—On sand and
sandstone bluffs, mostly northward; reaching southern
Wisconsin in appropriate habitats along the Wisconsin River
to Grant County and also along Lake Michigan.

7. Gaultheria   Wintergreen

Low, erect or creeping, herbaceous or evergreen shrubs;
leaves alternate; calyx saucer-shaped, deeply divided; corolla
tubular to bell-shaped, with shallow lobes; stamens 8 or 10;
anthers tipped with 2 erect appendages; style 1; fruit a capsule enclosed in the fleshy calyx, berrylike.

a. Stems erect-ascending; sepals and petals 5 each . . 1. *G. procumbens*
a. Stems prostrate; sepals and petals 4 each . . . . 2. *G. hispidula*

1. **G. procumbens** L. Wintergreen (Fig. 430). Stems erect,
from woody rootstocks; leaves leathery, shining, aromatic,
the youngest ones red; flowers white; fruit red, edible.—Pine
woods and bogs; common northward, rare southward.

Fig. 430. *Gaultheria procumbens,* × .60.

Fig. 431. *Gaultheria hispidula,* × .60.

2. **G. hispidula** (L.) Bigel. Creeping Snowberry (Fig. 431). Creeping evergreen; leaves 0.5-1 cm long, on very short petioles, the blades pointed at both ends and beset beneath with scattered rusty hairs; flowers very small, solitary in the axils of leaves; fruit white.—Bogs; south to Chippewa County in the west, and rarely to Walworth and Milwaukee counties in the east.

8. Gaylussacia   Huckleberry

Low colonial shrubs, with tiny resinous globules on leaves and flowers; leaves deciduous (our species), alternate; inflorescence a raceme, sticky; calyx 5-lobed; corolla bell-shaped, with shallow lobes; stamens 10; fruit a drupe with 10 stones.

G. baccata (Wang.) K. Koch (Fig. 432). Flowers reddish, sticky; leaves sticky when young; drupe black.—Rocky woodlands, sandstone bluffs, and bogs in the southern half of the state; very scattered localities northward to Ashland County.

9. Vaccinium   Blueberry; Cranberry

Erect or trailing shrubs; leaves evergreen or deciduous, alternate; calyx and corolla 4- or 5-lobed (petals rarely separate); stamens 8 or 10, opening by apical pores; style 1;

Fig. 432. *Gaylussacia baccata,* × .40.

fruit a many-seeded berry, with 5 carpels in the blueberries
and 4 in the cranberries.

a. Lobes of the corolla 5, shorter than the tube; leaves not evergreen
   b
   b. Flowers clustered; anthers without bristles c
      c. Leaves finely toothed . . . . .1. *V. angustifolium*
      c. Leaves not toothed . . . . 2. *V. myrtilloides*
   b. Flower solitary; anthers with 2 bristles . . . . . .3. *V. caespitosum*
a. Lobes of the corolla 4; leaves evergreen d
   d. Corolla bell-shaped, with short lobes (Fig. 435) . . . .
      . . . . 4. *V. vitis-idaea*, var. *minus*
   d. Lobes of the corolla much longer than the tube (Fig. 436) e
      e. Leaves with margins strongly inrolled, becoming triangular
         in outline (Fig. 436) . . . . .5. *V. oxycoccos*
      e. Leaves with margins but slightly inrolled, elliptical in outline
         (Fig. 437) . . . . 6. *V. macrocarpon*

1. **V. angustifolium** Ait. Early Blueberry (Fig. 433). Stems
2-6 dm high; twigs green, warty, with 2 narrow lines of hairs
running down from each node; teeth of leaves finely bristle-
tipped; berry sweet, blue, with a bloom.—Common northward, south to Dane and Racine counties in bogs, and on
sand and sandstone ledges.

Var. nigrum (Wood) Dole. Berries black, without a
bloom; leaves somewhat thicker and somewhat bluish white
beneath.

2. **V. myrtilloides** Michx. Velvet-leaved Blueberry (Fig.
434). Shrubs 2-6 dm high; leaves and twigs soft-downy; fruit
ripening later than in the preceding, not so sweet.—More
frequent than the preceding northward but not common
southward.

3. **V. caespitosum** Michx. Dwarf Bilberry. Tufted shrubs
0.5-3 dm high; leaves smooth and shining, toothed; flowers
solitary, nodding.—On dry ledges and sand, rare; Barron,
Clarke, and Portage counties, and Wisconsin Dells.

4. **V. vitis-idaea** L., var. minus Lodd. Mountain Cranberry
(Fig. 435). Dwarf shrubs, rarely more than 10 cm high;
leaves leathery, 5-18 mm long, 3-9 mm wide, dark green
above, pale beneath and smooth except for scattered black
bristly hairs; flowers pink, in a small cluster; berry dark red,
hard.—Bogs and rocky areas in northwestern Wisconsin; rare.

5. **V. oxycoccos** L. Small Cranberry (Fig. 436). Stems
very slender; leaf blades 3-8 mm long, 1-3 mm wide, much
whitened beneath; pedicels 1-4, from a rachis 1-4 mm long
which terminates the stem; berry 6-8 mm in diameter.—Bogs,
south to Barron, Lincoln, and Manitowoc counties.

Var. ovalifolium Michx. Coarser, intermediate between
this and the next species; leaf blades 6.5-15 mm long, 3-6.5

Fig. 433. *Vaccinium angusti-folium,* × .40.
Fig. 434. *Vaccinium myrtilloides,* × .40.
Fig. 435. *Vaccinium vitis-idaea,* var. *minus,* × .40.

Fig. 436. *Vaccinium oxycoccos,* × .40.
Fig. 437. *Vaccinium macrocarpon,* × .40.

mm wide, less strongly inrolled; pedicels 2-10, from a rachis 5-10 mm long; berry 8-10 mm in diameter.—Occasional in bogs, throughout the state except in the Driftless Area.

6. **V. macrocarpon** Ait. American Cranberry (Fig. 437). Leaf blades blunt or rounded at tip, 6-17 mm long and 2-8 mm wide, only slightly whitened beneath; pedicels 1-10, from a rachis 1-3 cm long, beyond which a leafy branch continues; berry 1-2 cm long.—In bogs, where often cultivated, south to Polk, Jackson, Columbia, Waukesha, and Racine counties.

10. **Epigaea** Trailing Arbutus; Mayflower

Prostrate, trailing, only slightly shrubby plants; leaves evergreen, alternate; inflorescences short terminal and axillary spikes; flowers perfect or unisexual; sepals 5, subtended by 2 bractlets; corolla spreading above; anthers opening longitudinally; fruit a depressed capsule.

E. repens L. (Fig. 438). Plants bristly with rusty hairs;

Fig. 438. *Epigaea repens,* × .40.

leaves 2-10 cm long, rounded or heart-shaped at base; flowers in dense clusters, pink and white, very fragrant.—Dry rocky or sandy woods, south to the Wisconsin Dells, the Baraboo Hills, and rarely on bluffs to Dane and Green counties.

PRIMULACEAE   Primrose Family

Annual or perennial herbs; leaves simple, all basal, whorled or opposite; flowers perfect; calyx and corolla each mostly 5-lobed; stamens opposite the corolla lobes; style 1; fruit a capsule, with the few to many seeds on a central knob arising from the base.

a. Corolla with erect or spreading segments *b*
  b. Leaves all near the ground in a basal rosette *c*
    c. Corolla longer than the calyx ....1. *Primula*
    c. Corolla shorter than the calyx ....2. *Androsace*
  b. Leaves on the stem *d*
    d. Leaves along the entire stem .... 3. *Lysimachia*
    d. Leaves in a single whorl near the summit of the stem ......
    .... 4. *Trientalis*
a. Corolla with segments strongly turned back (Fig. 445) ....
.... 5. *Dodecatheon*

1. Primula   Primrose

Small perennials, with basal leaves; inflorescence an umbel; corolla narrowly tubular below, flaring above; stamens 5, included, inserted on the corolla tube.

P. mistassinica Michx. (Fig. 439). Plants 0.5-2 dm high; leaves 1-4 cm long; corolla pink or blue (rarely white), with a slender tube and spreading lobes.—Damp ledges of the Wis-

Fig. 439. *Primula mistassinica,*
× .40.

Fig. 440. *Androsace occidentalis,*
× .60.

consin and St. Croix rivers, and cliffs of the Kickapoo River; Bayfield Peninsula; Door and Marinette counties.

2. Androsace

Tiny annuals (our species), with basal leaves; inflorescence a bracted umbel on 1 or few scapes; calyx persistent, corolla salverform, each 5-lobed.

A. occidentalis Pursh (Fig. 440). Plants less than 1 dm high; leaves in a close rosette about 1 cm in diameter; flowers usually numerous in a dense cluster.—Bare hills; local in Sauk, Dane, Richland, Columbia, Rock, Manitowoc, and Pepin counties.

3. Lysimachia   Loosestrife

Perennial herbs, with opposite or whorled leaves; calyx and corolla each 5-lobed, corolla saucer-shaped, with short pointed lobes.

a. Flowers less than 1 cm in diameter; leaf blades pointed, longer than broad *b*
   b. Flowers in the axils of the whorled leaves .... 1. *L. quadrifolia*
   b. Flowers in dense axillary spikes ....2. *L. thyrsiflora*
a. Flowers more than 1 cm in diameter; leaf blades rounded, about as broad as long .... 3. *L. nummularia*

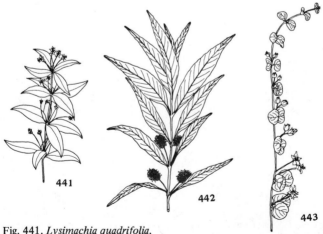

Fig. 441. *Lysimachia quadrifolia,*
× .30.
Fig. 442. *Lysimachia thyrsiflora,*
× .30.

Fig. 443. *Lysimachia nummularia,*
× .25.

1. **L. quadrifolia** L. Whorled Loosestrife (Fig. 441). Stems erect, 3-8 dm high; leaves in whorls of 4-5; corolla yellow, dark-dotted or streaked.—Dry rocky ground, throughout the state.

2. **L. thyrsiflora** L. Tufted Loosestrife (Fig. 442). Leaves opposite, the lower scalelike; corolla yellow, purplish-dotted. —Swampy ground, occasional throughout the state.

3. **L. nummularia** L. Moneywort (Fig. 443). Stems creeping; corolla bright yellow, not dotted or streaked.— Introduced from Europe; an occasional weed in shady river bottoms; southern Wisconsin, north to Brown and Pierce counties.

4. Trientalis   Starflower

Low stoloniferous perennials; leaves in a single whorl; calyx and corolla deeply divided into 7 lobes, the corolla saucer-shaped; stamens usually 7, borne at the base of the corolla.

T. borealis Raf. (Fig. 444). Stems 1-2 dm high, with a few scales below the single whorl of leaves; flowers solitary or few, white, starlike.—Deep woods, mostly northward; south to Sauk, Kenosha, and rarely Grant counties.

5. Dodecatheon   Shooting Star

Perennials; leaves all near the ground in a basal rosette, the blades without hairs, narrowed to a winged petiole; flowers in an umbel, nodding at flowering time, with a bract at the base of each pedicel; anthers 5, extending forward in a compact group and the bases joined, while the 5 corolla lobes bend backward.

a. Plants stout, 2.5-6 dm high; flowers white to lilac . . .1. *D. meadia*
a. Plants slender, 2-3.5 dm high; flowers deep rose purple . . . .
. . . . 2. *D. radicatum*

1. **D. meadia** L. Shooting Star (Fig. 445). Plants stout, 2.5-6 dm high; umbel with 6-30 pale lilac to white flowers; calyx lobes in the unfolding flower at least half as long as the corolla; capsule stout, dark brown, woody.—Prairies, open woods, and along railroads, north to Brown, Sauk, and St. Croix counties.

2. **D. radicatum** Greene (*D. amethystinum* Fassett). Jeweled Shooting Star. Plants slender, 2-3.5 dm high; umbel with 2-11 (rarely -18) deep rose purple flowers; calyx lobes in the unfolding flower not more than two-thirds as long as

Fig. 444. *Trientalis borealis,*
× .60.

Fig. 445. *Dodecatheon meadia,*
× .15.

the corolla; capsule slender, pale brown, papery.—Driftless
Area on bluffs along the Mississippi and La Crosse rivers.

OLEACEAE   Olive Family

Trees or shrubs; leaves opposite, simple or compound;
inflorescence a raceme or panicle; sepals and petals mostly 4
(sometimes lacking), united; stamens 2; style 1; stigma 2-
lobed; fruit a drupe, capsule, or samara.

a. Leaves pinnately compound . . . . .1. *Fraxinus*
a. Leaves simple . . . . 2. *Syringa*

Fig. 446. *Fraxinus americana,*   Fig. 447. *Fraxinus pennsylvanica,*
× .25.   var. *subintegerrima,* × .30.

1. Fraxinus   Ash

Trees; stamens and pistils in different flowers, on the same or different trees; calyx present or absent; fruit dry, flattened, winged samara.

a. Lateral leaflets narrowed to at least a short stalk *b*
    b. Twigs round in cross section *c*
        c. Leaflets pale beneath, rounded at base; leaf scars deeply
            concave on upper side . . . . 1. *F. americana*
        c. Leaflets green on both sides, tapered to the base; leaf scars
            nearly straight on upper side . . . . 2. *F. pennsylvanica*
    b. Twigs square in cross section . . . . 3. *F. quadrangulata*
a. Lateral leaflets with no stalk . . . . 4. *F. nigra*

1. **F. americana** L. White Ash (Fig. 446). Leaflets 5-9, with stalks 5-10 mm long, somewhat rounded at base; calyx present; body of fruit round in cross section, the wing extending only a short distance down its side.—In rich moist soil, north to Douglas, Lincoln, and Brown counties.

2. **F. pennsylvanica** Marsh. Leaflets 5-9, on stalks 2-5 mm long, narrowed at base; calyx present; boyd of fruit round in cross section, the wings extending down at least two-thirds of its length.

Var. **pennsylvanica.** Red Ash. Rachis of leaves and young branches velvety.—Throughout the state, in rich moist soils, mainly river bottoms.

Fig. 448. *Fraxinus nigra,* × .20.

Var. subintegerrima (Vahl) Fern. Green Ash (Fig. 447). Rachis and young branches glabrous.—Very common, throughout the state.

3. **F. quadrangulata** Michx. Blue Ash. Leaflets 7-11, with slender stalks; calyx absent or much reduced; fruit oblong, blunt.—Big Bend, Waukesha County.

4. **F. nigra** Marsh. Black Ash (Fig. 448). Leaflets 7-11, each with a tuft of brown hairs at its base; calyx absent; fruit with a flat body.—Banks of streams and in swamps, throughout the state.

2. Syringa   Lilac

Shrubs or small trees; leaves simple, entire; inflorescence a dense panicle, flowers fragrant; calyx small; corolla 4-lobed; stamens 2; fruit a capsule, seeds narrowly winged.

S. vulgaris L. Shrubs, to 6 m; leaves heart-shaped; flowers lilac purple or white.—Rarely escaping from cultivation or, more frequently, a remnant of cultivation.

GENTIANACEAE   Gentian Family

Herbs, with usually simple opposite leaves (our single spring-flowering species is an exception); flowers regular; both sepals and petals fused, with 4-5 lobes; stamens as

Fig. 449. *Menyanthes trifoliata,* × .40.

many as the corolla lobes and alternate with them, attached to the corolla; style 1; fruit a capsule.

Menyanthes   Buckbean

Perennial herbs; leaves compound, with 3 leaflets; inflorescence a raceme; corolla 5-lobed, white to pink; stamens 5; stigma 2-lobed; fruit a capsule.

M. trifoliata L. (Fig. 449). Petioles with sheathing bases, leaflets sessile; corolla 1.5-3 cm long, white, with many wavy hairs on its upper surface.—Sphagnum bogs, throughout Wisconsin except in the Driftless Area.

APOCYNACEAE   Dogbane Family

Herbs or twining woody vines; leaves simple, alternate or opposite; flowers perfect, regular; calyx 5-lobed; corolla 5-lobed, convolute in bud; stamens 5, borne on the corolla and alternate with the lobes; ovaries 2, united to a common style with a large stigma; fruit 2-many-seeded follicles per flower.

Vinca

Perennial herbs, with opposite leaves; flowers solitary, in

1 axil of a pair of leaves; corolla salverform, blue (rarely white).

V. minor L. Periwinkle. Plants creeping and forming mats; flower branches erect; leaves entire, evergreen; corolla 2 cm wide.—An escape from cultivation; upland woods and near old cemeteries in scattered localities in the state.

ASCLEPIADACEAE   Milkweed Family

Herbs or vines, with milky juice and mostly opposite or whorled simple leaves; flowers fragrant, 5-parted; anthers fused with the pistil to form a compound massive structure (gynostegium) which covers the 2 separate ovaries; the pollen in waxy masses (pollinia), these paired and joined to an adhesive disk which functions in pollination by insects; fruit a follicle containing seeds with hairs for wind dispersal.

a. Stems erect; corolla lobes cup-shaped, reflexed . . . . .1. *Asclepias*
a. Stems twining; corolla lobes triangular, spreading . . .2. *Cynanchum*

1. Asclepias   Milkweed

Perennial herbs, with alternate, opposite, or whorled leaves; inflorescences terminal or axillary umbels; sepals and petals reflexed; the stamens forming 5 cuplike hoods (nectaries) within each of which is usually a horn that curves inward over the massive gynostegium; 5 sets of pollinia produced around the top of this structure, the black adhesive disks appearing on the sides; receptive areas of the stigmatic disk within 5 slots on the sides.

a. Flowers brightly colored *b*
   b. Flowers orange; leaves alternate . . . . .1. *A. tuberosa*
   b. Flowers pink, red, or purple; leaves opposite *c*
      c. Leaves smooth beneath *d*
         d. Leaf tapered gradually to the tip; branch veins ascending
            (Fig. 451) . . . . .2. *A. incarnata*
         d. Leaf abruptly narrowed to a sharp point at the broadly
            rounded tip; branch veins spreading (as in Fig. 453) . . . .
            . . . . .3. *A. sullivantii*
      c. Leaves minutely velvety-hairy beneath *e*
         e. Flowers bright red or purple; umbels erect, hemispherical, at or close to tip of stem and above the leaves
            when in bloom . . . . .4. *A. purpurascens*
         e. Flowers dull pink; umbels forming complete spheres,
            borne laterally and exceeded by the upper leaves when in
            bloom . . . . .5. *A. syriaca*
a. Flowers greenish, yellowish, whitish, or rarely pinkish *f*
   f. Umbel on a terminal peduncle longer than uppermost leaf *g*
      g. Leaves rounded toward tip except for small sharp point . . . .
      . . . . 6. *A. amplexicaulis*
      g. Leaves tapered to an acute tip . . . . 7. *A. meadii*

f. Peduncles shorter than the leaves and often axillary *h*
    h. Leaves profusely hairy or downy beneath; stems scattered,
        from rootstocks *i*
        i. Stems with minute curved hairs or none *j*
            j. Leaves 1-2 dm long, elliptical-oblong . . . . 5. *A. syriaca*
            j. Leaves much shorter, ovate . . . . 8. *A. ovalifolia*
        i. Stems with spreading hairs 1 mm or more long . . . .
        . . . . 9. *A. lanuginosa*
    h. Leaves with few or no hairs; stems 1-few, all from the same
        point, without long rootstocks *k*
        k. Leaves broadly ovate, thin, smooth . . . . . 10. *A. exaltata*
        k. Leaves lanceolate, thick, rough . . . . 11. *A. viridiflora*

1. **A. tuberosa** L. Butterfly Weed; Pleurisyroot (Fig. 450).
Leaves linear to oblong, hairy, stems 3-6 dm high, often
branched at the tip, where the flat-topped umbels are borne.
—Prairies and sand barrens, north to Burnett, Wood, and
Marinette counties.

2. **A. incarnata** L. Swamp Milkweed (Fig. 451). Stems
5-15 dm high, without rootstocks, smooth except for the
pedicels and 2 downy lines along the upper stem; leaves
lanceolate; umbels flat-topped, borne above the leaves on
branches; flowers varying from pink to purplish red; hoods
2-3 mm long.—Low wet places; throughout the state.

3. **A. sullivantii** Engelm. Leaves oblong, very broadly
rounded, with very abrupt short tips, glabrous; flowers
purple to whitish; hoods 5-6 mm long.—Low prairies, southward; rare.

4. **A. purpurascens** L. Purple Milkweed (Fig. 452). Leaves
broad, but the upper ones often narrower and tapering more
gradually to the tip than do those of No. 5 (Fig. 453),
which it resembles; hoods 5-6 mm long.—Dry to damp
woods, thickets, and openings, mostly in the southern
counties.

5. **A. syriaca** L. Common Milkweed; Silkweed (Fig. 453).
Coarse downy plants; stems 0.5-1.5 m high, from deep-
seated, branching rootstocks; leaves thick, broad, acute to
apiculate; flowers old rose to whitish; corolla lobes 6-9 mm
long; hoods 3-4 mm long; umbels axillary, many-flowered,
numerous along upper half of the stem.—Common along
roadsides and in fields, throughout the state.

6. **A. amplexicaulis** Sm. (Fig. 454). Stem often solitary;
plants smooth; leaves blue green, heart-shaped at base, the
lower rather triangular, the upper more oblong and rounded
at the tip; umbel large, spherical, solitary, greenish.—Dry
sand, north to Eau Claire and Waushara counties.

Fig. 450. *Asclepias tuberosa*, × .50.

Fig. 451. *Asclepias incarnata*, × .40.

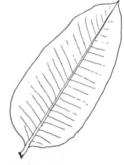

Fig. 452. *Asclepias purpurascens*, × .40.

Fig. 453. *Asclepias syriaca*, × .40.

Fig. 454. *Asclepias amplexicaulis*, × .40.

7. **A. meadii** Torr. Similar to No. 6 but with sharply triangular, acute leaves only 3-8 cm long.—Grant County.

8. **A. ovalifolia** Dcne. Stems 1.5-6 dm high; leaves usually tapering gradually to acute tips; flowers white, tinged with yellow, green, or pink.—Prairies and open woods; in the southwestern half of the state, the northwestern counties, and in Shawano and Marinette counties.

9. **A. lanuginosa** Nutt. Plants covered with soft spreading hairs; stems only 1-3 dm high, often curved; leaves numerous, lanceolate, either alternate or opposite; umbel solitary, erect, terminal; flowers greenish white, 6-8 mm long.—Sandy or gravelly prairies and hilltops, north to Waukesha and Portage counties.

10. **A exaltata** L. Poke Milkweed. Plants 5-15 dm tall, smooth or nearly so; leaves tapered to a long point, petioles about 1 cm long; umbels mostly axillary; pedicels long, drooping; flowers whitish with some green or pink.—Rich woods, throughout the state.

11. **A. viridiflora** Raf. Green Milkweed. Leaves 4-15 mm wide, mostly opposite; umbels several, axillary, sessile or on short down-curved peduncles.—Dry sandy prairies and open woods, north to Dunn, Portage, and Fond du Lac counties.

Fig. 455. *Cynanchum nigrum,* × .30.

2. Cynanchum   Black Swallowwort

Perennial vines, with opposite oval leaves, rounded or heart-shaped at base; inflorescences axillary umbellate clusters; flowers 5-parted, the petals enclosing a cuplike, fleshy, slightly toothed corona, which in turn surrounds the gynostegium and is slightly longer than it; stamens 5, united with pistil to form the gynostegium, forming 5 sets of pollinia; ovaries 2; fruit a follicle.

C. nigrum (L.) Pers. (Fig. 455). Scarcely hairy, twining herbs; leaves broad, rounded to heart-shaped at base; flowers small, dark purple, in peduncled, axillary, umbellate clusters. —Introduced from Europe; Grant, Walworth, and Waukesha counties.

CONVOLVULACEAE   Morning Glory Family

Herbs, mostly climbing or trailing; leaves alternate; calyx of 5 sepals; corolla 5-plaited or 5-lobed, pink or white, tubular and flaring, of delicate texture, twisted in bud; stamens 5; stigmas 2; fruit a capsule.

Convolvulus   Bindweed

Herbs or vines, with alternate leaves; calyx 5-lobed; corolla funnel-shaped, with broad points; stamens inserted on the corolla; fruit a nearly glabrous capsule containing 4 seeds.

a. Pedicels with 2 broad bracts enclosing the calyx (Fig. 457); corolla 3-5 cm long *b*
    b. Erect plants . . . . .1. *C. spithamaeus*
    b. Trailing vines . . . . 2. *C. sepium*
a. Pedicels without bracts enclosing the calyx (Fig. 458); corolla 1.5-2 cm long . . . . 3. *C. arvensis*

1. C. spithamaeus L. Low Bindweed. Plants 1.5-3 dm high, downy; stems mostly simple, erect or nearly so; petioles short; leaf blades rounded or truncate at base; corolla white.—Open sandy or rocky ground, locally abundant.

Var. stans (Michx.) Fogelberg (Fig. 456). Leaves heart-shaped at base, densely velvety.—Throughout the state.

2. C. sepium L. Hedge Bindweed (Fig. 457). Stems trailing, twining, or climbing; leaf blades triangular-arrow-shaped, on long petioles; corolla white or rose-colored.— Throughout the state.

Var. repens (L.) Gray. A narrow-leaved extreme of beaches.

Fig. 456. *Convolvulus spithamaeus,* var. *stans,* × .40.

Fig. 457. *Convolvulus sepium,*   Fig. 458. *Convolvulus arvensis,*
× .40.   × .40.

3. **C. arvensis** L. Field Bindweed (Fig. 458). Similar to
No. 2; nearly smooth; flowers and leaves smaller; lobes of
the leaf blades acute.—Naturalized from Eurasia; waste places
in southern Wisconsin.

Forma **cordifolius** Lasch (heart-leaved) has the broad
blade cordate, with rounded basal lobes; forma **auriculatus**
Desr. (eared) has linear-oblong to lanceolate blades with the
auricles acute. The forms are less common than the typical
plant.

POLEMONIACEAE   Phlox Family

Herbs; leaves simple or compound, alternate or opposite;
flowers perfect, both sepals and petals fused, each with 5
lobes; stamens 5, borne on the corolla, alternate with the
lobes; style 1, with 3 stigmas; fruit a capsule.

a. Leaves opposite, not divided . . . . .1. *Phlox*
a. Leaves alternate, pinnately compound . . . . 2. *Polemonium*

1. Phlox   Phlox

Perennial herbs; leaves opposite, entire; inflorescence a
terminal and axillary cyme; corolla tube long and narrow,
abruptly spreading; stamens inserted at different levels on
the corolla tube.

a. Leaves mostly over 2 cm long, without axillary tufts *b*
   b. Leaves at least 10 times as long as broad . . . . 1. *P. pilosa*
   b. Leaves not over 5 times as long as broad . . . .
   . . . . 2. *P. divaricata,* var. *laphamii*

a. Leaves 0.5-1.5 cm long, with axillary tufts *c*
   c. Notch in corolla lobes about 1 mm deep . . . . 3. *P. subulata*
   c. Notch in corolla lobes about 3 mm deep . . . . 4. *P. bifida*

1. **P. pilosa** L. (Fig. 459). Stems erect; leaves narrow; corolla pink, purple, or white, the tube woolly.

Ssp. fulgida Wherry. Prairie Phlox. With fine, lustrous pubescence in the inflorescence.—Dry sandy soil; common except in north-central Wisconsin.

Ssp. **pilosa**. With gland-tipped hairs.—Rare; Iowa, Grant, Shawano, and Marathon counties.

2. **P. divaricata** L., var. laphamii Wood. Wild Blue Phlox (Fig. 460). Stems reclining at base; leaves oblong-ovate; flowers usually blue.—In woods, north to Polk, Sawyer, and Kewaunee counties.

Var. **canadensis** (Sweet) Wherry. Corolla lobes notched.— Rare, Racine County.

3. **P. subulata** L. Moss Pink (Fig. 461). Plants low, forming mats.—Occasionally escaping from cultivation.

4. **P. bifida** Beck. Leaves stiff, linear to narrowly lanceolate; corolla lobes notched one-fourth to one-half their length.—Native in Rock County and escaping from cultivation northward.

2. **Polemonium**   Jacob's Ladder

Perennial herbs; leaves alternate, pinnately compound or deeply cleft; calyx and corolla bell-shaped; corolla blue or white.

P. reptans L. (Fig. 462). Stems 2-4 dm high, branching;

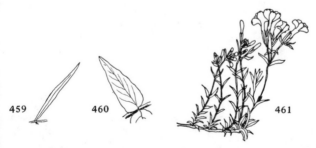

Fig. 459. *Phlox pilosa,* × .40.   Fig. 461. *Phlox subulata,* × .50.
Fig. 460. *Phlox divaricata,* × .40.

Fig. 462. *Polemonium reptans,* × .30.

calyx becoming inflated in fruit.—Woods, north to Barron, Marathon, and Brown counties.

HYDROPHYLLACEAE   Waterleaf Family

Herbs, usually hairy; leaves mostly alternate, entire to compound; flowers perfect; sepals 5, fused; petals 5, fused; corolla lobed to about the middle; stamens 5, borne on the corolla tube and alternating with the lobes; style 1, 2-cleft, or styles 2; fruit a capsule.

a. Flowers in terminal branched clusters . . . . .1. *Hydrophyllum*
a. Flowers long-pediceled in the leaf axils . . . . 2. *Ellisia*

1. Hydrophyllum   Waterleaf

Calyx lobes narrowly ribbonlike, bristly-hairy; corolla longer than the calyx, with the stamens more or less extending beyond its mouth; leaves variously patterned with white.

a. Stem leaves deeply pinnately divided; calyx lacking appendages . . .
. . . . .1. *H. virginianum*
a. Stem leaves shallowly palmately lobed; calyx with a reflexed lobe
at each notch between the sepals . . . . .2. *H. appendiculatum*

1. **H. virginianum** L. (Fig. 463). Plants with mostly appressed hairs, 2-7 dm high; leaves all pinnate, the leaflets sharply and coarsely toothed; flowers bluish purple, pink, or white.—Rich woods, north to Douglas, Lincoln, and Shawano counties.

Fig. 463. *Hydrophyllum virginianum,* × .40.

Fig. 464. *Hydrophyllum appendiculatum,* × .40.

2. **H. appendiculatum** Michx. (Fig. 464). Plants hairy; lower leaves pinnately divided; upper leaves shallowly lobed, the lobes toothed (resembling a small grapevine leaf); flowers bluish purple, pink, or white.—Rich woods, less common than No. 1; north to Pierce, Dane, Brown, and Calumet counties.

2. Ellisia

Annual herbs; at least the basal leaves opposite, the blades pinnately divided; flowers axillary or opposite the leaf axils, a few also in a terminal inflorescence; calyx deeply lobed, enlarging at fruiting stage; corolla bell-shaped, white to lavender; filaments included; style cleft, shorter than the corolla.

E. nyctelea L. Rough-hairy, much branched, 1-4 dm high; leaves pinnately lobed nearly to the midrib, the lobes coarsely 1-3-toothed; corolla scarcely longer than the calyx, whitish.—Occasional as a weed in southern Wisconsin; adventive from farther south.

BORAGINACEAE Borage Family

Herbs, usually rough-hairy, with simple alternate leaves; flowers perfect, usually on one side of a branch which is rolled up from the tip and straightens as the blossoms open; sepals 5, free or fused at the base; petals 5, fused, often with petaloid appendages opposite the lobes and almost closing the throat; stamens 5, borne on the corolla tube alternate with the lobes; ovary deeply 4-lobed, the single style arising from the middle, 2-branched at the tip; fruit of 4 1-seeded nutlets.

Genera are based largely on fruit characters. The following key is based on superficial characters, and plants should be carefully checked, when possible, with the description of mature fruit given under each genus.

a. Corolla 1 cm or more in length or breadth *b*
  b. Plants hairy *c*
    c. Upper leaves with a wing extending down the stem . . . .
      . . . . .1. *Symphytum*
    c. Leaves without such wings *d*
      d. Flowers reddish purple . . . . .2. *Cynoglossum*
      d. Flowers yellow or orange . . . .·. . . . . .3. *Lithospermum*
  b. Plants without hairs . . . . .4. *Mertensia*
a. Corolla 5 mm or less in length or breadth *e*
  e. Throat of the corolla closed by 5 scales *f*
    f. Flowers not in the axils of bracts . . . . .2. *Cynoglossum*
    f. Flowers, or at least some of them, in the axils of bracts *g*
      g. Hairs on the stem slender, curved, somewhat ascending *h*
        h. Pedicels short, erect in fruit . . . . .5. *Lappula*
        h. Pedicels recurved or deflected in fruit . . . . 6. *Hackelia*
      g. Hairs on the stem stout, bristlelike, straight, spreading . . .
        . . . . 7. *Lycopsis*
  e. Throat of the corolla without scales, open or with crests *i*
    i. Flowers, or at least the uppermost, not in the axils of bracts
      . . . . 8. *Myosotis*
    i. Flowers all in the axils of bracts or leaves . 3. *Lithospermum*

1. Symphytum   Comfrey

Large broad-leaved perennial herbs; inflorescences one-sided, with nodding flowers; stamens included within the corolla tube; nutlets nearly smooth, erect, with a hollow scar where attached, the scar finely toothed on its margin.

S. officinale L. (Fig. 465). Plants with white stiff short hairs; lower leaves broad, abruptly narrowed at base; upper leaves narrowed at both ends; corolla yellowish or pinkish

Fig. 465. *Symphytum officinale*, × .30.

Fig. 466. *Cynoglossum officinale,* × .30.

white to bluish or roseate purple, with 5 short spreading teeth, the throat closed by 5 scales.—Introduced from Europe; in very scattered localities in Wisconsin.

2. Cynoglossum

Corolla about equaling the calyx, with 5 rounded lobes, the throat closed by 5 scales; stamens included; nutlets attached by their sides, widely spreading in one plane, covered with barbed prickles.

a. Stems leafy to the summit, bearing axillary inflorescences near the tip . . . . 1. *C. officinale*
a. Upper part of stems leafless, terminating in a once- or twice-forked bractless inflorescence .2. *C. boreale*

1. **C. officinale** L. Hound's-tongue (Fig. 466). Stout plants, much branched above, copiously leafy and covered with soft hairs; corolla reddish purple.—Pastures, throughout Wisconsin; naturalized from Europe.

2. **C. boreale** Fern. Northern Wild Comfrey. Rather slender plants, once or twice forked above in the bractless inflorescence, with white spreading hairs; lower leaves with petioles, the upper with a broad clasping base; corolla pale blue.—Woods, south to Sawyer and Door counties, and along Lake Michigan to Racine County.

Fig. 467. *Lithospermum lati-*
folium, × .40.

Fig. 469. *Lithospermum canes-*
cens, × .40.

Fig. 468. *Lithospermum incisum,*
× .40.

3. Lithospermum   Puccoon; Gromwell

Annual or perennial herbs, mostly with reddish roots; leaves entire, sessile, pubescent; flowers in bracted cymes or sometimes also in the upper leaf axils, regular, funnel-shaped; anthers almost sessile; nutlets bony, often ivorylike, smooth or roughened, the scar nearly flat.

a. Corolla pale, about equaling the calyx *b*
   b. Midrib of leaf conspicuous, branch veins obscure . **1.** *L. arvense*
   b. Branch veins of leaf conspicuous *c*
      c. Leaves 0.6-1.2 cm broad . . . . **2.** *L. officinale*
      c. Leaves 1.2-4.5 cm broad . . . . **3.** *L. latifolium*
a. Corolla yellow or orange, several times exceeding calyx *d*
   d. Corolla lobes finely toothed . . . .**4.** *L. incisum*
   d. Corolla lobes not toothed *e*
      e. Plants with fine white silky hairs . . . .**5.** *L. canescens*
      e. Plants with bristly hairs which have swollen bases . . . .
      . . . . **6.** *L. caroliniense*

1. L. arvense L. Corn Gromwell. Corolla white or yellowish, the lobes glabrous on outer surface.—Naturalized from Europe; rare, Sheboygan County.

2. L. officinale L. Gromwell. Corolla greenish, 4.5 mm or less long, pubescent on lobes.—Naturalized from Europe; rare, Fond du Lac to Brown counties.

3. L. latifolium Michx. (Fig. 467). Corolla yellow, 4.5-4.8 mm long, pubescent on lobes.—Woods and open ground, north to Brown and Pierce counties.

4. L. incisum Lehm. Puccoon (Fig. 468). Leaves ribbonlike, whitened with fine hairs; corolla yellow.—Dry sand, north to Barron, Portage, and Sheboygan counties.

5. L. canescens (Michx.) Lehm. Hoary Puccoon (Fig. 469). Corolla orange yellow, the tube naked at base within. —Prairies and sandy woods, throughout most of the state.

6. **L. caroliniense** (Walt.) MacM. Corolla light yellow or orange, the tube hairy at base within.—Dry sandy and prairie soils, throughout the state.

4. Mertensia   Bluebells; Lungwort

Perennial herbs, with entire leaves; inflorescence a panicled or racemelike cyme, without bracts; flowers showy, corolla bell-shaped; stamens and elongate style within or slightly projecting beyond the corolla tube; nutlets ovoid, fleshy, with small scar, the surface dull and roughish.

a. Plants glabrous, glaucous; corolla to 25 mm long, lacking appendages . . . . 1. *M. virginica*
a. Plants pubescent, green; corolla to 15 mm long, with appendages . . . . .2. *M. paniculata*

1. **M. virginica** (L.) Pers. Virginia Cowslip (Fig. 470). Plants 2-6 dm high; lower leaves long-petioled, the upper without petioles, ovate; corolla light blue, pinkish in the bud, trumpet-shaped, without scales in the throat.—Local in wooded river bottoms, along the Mississippi, St. Croix, Wisconsin, Kickapoo, Rock, and Pecatonica rivers; escaped from cultivation, Dane and Milwaukee counties.

2. **M. paniculata** (Ait.) G. Don. Tall Lungwort. Plants sparsely pubescent; basal leaves long-petioled, upper leaves sessile, acuminate; corolla blue, pink in bud; the throat with conspicuous crests; stamens within the corolla tube, filaments broad and short.—Rich woods; northern Wisconsin, from Douglas to Florence counties.

Fig. 470. *Mertensia virginica*, × .60.

Fig. 471. *Lappula echinata,*
× .55.

Fig. 472. *Hackelia americana,*
× .30.

5. Lappula   Stickseed

Corolla with a slender tube and spreading lobes, the throat closed by 5 scales; stamens included; nutlets attached by their sides, erect, the margins or backs armed with barbed prickles; whole plant covered with minute stiff appressed hairs. Mature fruit is almost indispensable for the determination of species.

a. Fruit with a double row of bristles along the margin 1. *L. echinata*
a. Fruit with a single row of bristles along the margin 2. *L. redowskii*

1. **L. echinata** Gilib. (Fig. 471). Stems 1.5-6 dm high; leaves very rough to the touch; flowers blue.—Naturalized from Europe; pastures, along railroads, and open woodlands; mostly in southern and eastern Wisconsin.

2. **L. redowskii** (Hornem.) Greene. Similar to No. 1; flowers blue.—Along railroads, probably adventive from farther west; Iowa, Grant, Juneau, and Marquette counties.

6. Hackelia   Nodding Stickseed

Perennial or biennial herbs, with paired, naked or bracted,

one-sided inflorescences; flowers small, the throat of the corolla usually closed by the appendages; stamens and style shorter than the corolla; fruit 4 globose nutlets covered over the back with hooked bristles.

H. americana (Gray) Fern. (Fig. 472). Stems 0.3-1 m high, much branched above; basal leaves narrowly ovate-lanceolate; stem leaves narrowly to broadly lanceolate, mostly more than 1 cm wide; flowers white.—Occasional in rocky woods, in scattered localities in southern and eastern Wisconsin.

7. Lycopsis Bugloss

Annuals; inflorescence of bracteate cymes; corolla blue, lobes slightly unequal, tube bent; stamens included; nutlets wrinkled, erect, with a hollow scar where attached.

L. arvensis L. Plants 1-6 dm high, with copious spreading white flattened hairs; corolla funnel-shaped, somewhat curved and irregular.—Adventive from Europe; Ozaukee to Manitowoc counties.

8. Myosotis Forget-me-not

Small herbs; stem leaves without petioles, not toothed; upper flowers without bracts; corolla narrowly tubular below, abruptly spreading, with blunt appendages opposite each of the 5 rounded lobes; stamens included in the throat, nutlets compressed, smooth.

a. Calyx not 2-lipped; flowers usually blue *b*
   b. Calyx with straight close hairs *c*
      c. Calyx lobes shorter than the tube . . . . 1. *M. scorpioides*
      c. Calyx lobes about equaling the tube . . . . 2. *M. laxa*
   b. Calyx with curved spreading hairs *d*
      d. Annuals or biennials; corolla 4 mm or less broad . . . .
      . . . . 3. *M. arvensis*
      d. Perennials; corolla showy, 5-9 mm broad . . . . 4. *M. sylvatica*
a. Calyx 2-lipped; flowers white . . . . 5. *M. verna*

1. **M. scorpioides** L. Corolla 5-9 mm broad, blue with yellow center.—Naturalized from Europe; wet ground, throughout the state.

2. **M. laxa** Lehm. Corolla 5 mm or less broad, pale blue.—Wet ground along the lower Wisconsin River, Grant to Sauk counties, also Chippewa County.

3. **M. arvensis** (L.) Hill. Corolla 1.5-4 mm broad, light blue or occasionally white.—Introduced from Europe; well-drained soil, rare; Bayfield to Sheboygan counties.

4. **M. sylvatica** Hoffm. Garden Forget-Me-Not. Corolla

Fig. 473. *Myosotis verna*, × .25.

blue (pink or white), with yellow center, abruptly spreading, 5-9 mm broad.—Escaping from cultivation; Bayfield to Door counties, also Racine County.

5. **M. verna** Nutt. (Fig. 473). Corolla 1-2 mm broad; calyx hairs hooked or gland-tipped.—Dry open places; lower Wisconsin River and eastward.

LAMIACEAE (LABIATAE)   Mint Family

Herbs; stems usually square; leaves with volatile aromatic oils, opposite or sometimes whorled, simple; flowers perfect, regular to strongly 2-lipped; sepals 5, fused; petals 5, fused; stamens 4 (2 long, 2 short) or 2, borne on the corolla tube; style 1, coming from between the 4 lobes of the ovary and becoming 2-branched at the tip; fruit of 4 nutlets, each 1-seeded.

a. Calyx with a protuberance on the upper side of the tube . . . .
. . . . 1. *Scutellaria*
a. Calyx without a protuberance on the tube *b*
  b. Stamens 2 *c*
    c. Flowers axillary, in loose few-flowered whorls subtended by ordinary foliage leaves; corolla 6 mm or less long . . . .
. . . . 2. *Hedeoma*
    c. Inflorescence in a terminal raceme or head subtended by bracteal leaves much reduced in size; corolla 6-20 mm long *d*
      d. Inflorescence an elongate raceme, with more than 5 whorls in each inflorescence, whorls with less than 12 flowers . . . . 3. *Salvia*
      d. Inflorescence a dense head, with 1-5 whorls, whorls with many flowers . . . . 4. *Blephilia*

b. Stamens 4 *e*
  e. Plants creeping or only slightly erect . . . . 5. *Glechoma*
  e. Plants erect *f*
    f. Corolla 1-lipped (upper lobe very shallow, lower lobes extended into a lip) . . . . 6. *Ajuga*
    f. Corolla either 2-lipped or almost regular *g*
      g. Calyx lobes with stiff, bristlelike, spiny tips; leaves palmately veined . . . . 7. *Leonurus*
      g. Calyx lobes neither stiff nor spiny; leaves pinnately veined *h*
        h. Flowers in dense terminal heads; if subtended by leaves or bracts, the bracts much reduced in size . . . . .8. *Prunella*
        h. Flowers axillary, subtended by leafy bracts only slightly differentiated from the regular leaves *i*
          i. Leaves broadly ovate or roundish, not much longer than broad . . . . .9. *Lamium*
          i. Leaves lanceolate, more than twice as long as broad . . . . 10. *Dracocephalum*

1. Scutellaria   Skullcap

Mostly rhizomatous perennials; flowers solitary in the axils or in terminal or axillary racemes; calyx 2-lipped, entire, upper lip becoming enlarged with a transverse ridge at time of fruiting; corolla tube elongated, curved-ascending, dilated at the throat, the upper lip barely notched, the lateral lobes connected to the upper lip, the lower lip convex, notched; anthers bearded, those of the 2 upper stamens 2-celled and heart-shaped, the lower pair 1-celled and oval; nutlets pebbled.

a. Corolla 1.5-2 cm long; leaves 2-4 times as long as wide . . . .
  . . . . .1. *S. galericulata*
a. Corolla 6-12 mm long; leaves less than twice as long as wide *b*
  b. Plants minutely puberulent or glabrous; leaves entire, often revolute . . . . 2. *S. leonardii*
  b. Plants pilose; some leaves sparingly toothed, plane . 3. *S. parvula*

1. S. galericulata L. Common Skullcap. Plants very slender, 2-8 dm high with retrorsely appressed pubescence; leaves to 5 cm long, pinnately veined; flowers solitary, blue marked with white, corolla 1.5-2 cm long.—Common, throughout the state in wet places.

2. S. leonardii Epl. Smooth Small Skullcap (Fig. 474). Slender plants, 1-2 dm tall, mostly glabrous; the rhizome forming tubers like beads; leaves to 2 cm long; flowers axillary, blue, 7-9 mm long.—Dry sandy soil, especially on bluffs; mostly south of the Tension Zone, from St. Croix to Sheboygan counties southward; also Burnett and Price counties.

3. S. parvula Michx. Small Skullcap. Similar to No. 2, but

Fig. 474. *Scutellaria leonardii,*   Fig. 475. *Hedeoma hispida,*
× .40.   × .55.

finely woolly; leaves about 1 cm long; flowers 5-10 mm
long.—Rare on dry prairies; Pierce, St. Croix, and Rock
counties.

2. Hedeoma   Mock Pennyroyal

Plants mostly small and very strongly scented, with many
whorls of bluish flowers borne along nearly the entire length
of the plant; stems with downwardly pointing bristles; calyx
2-lipped; corolla weakly 2-lipped, about the same length as
the calyx tube; stamens 2; nutlets ovoid, smooth.

H. hispida Pursh (Fig. 475). Leaves linear, entire; calyx
bristly.—Dry sandy or gravelly soil, throughout most of the
state but less common northward.

3. Salvia   Sage

Plants with mostly showy, terminal spikelike inflorescences; calyx 2-lipped; corolla strongly 2-lipped, the upper lip straight or arched, the lower lip spreading and 3-lobed; stamens 2, filaments short.

a. Leaves mostly basal, long-petioled; corolla 15-20 mm long . . . .
. . . . .1. *S. pratensis*
a. Leaves cauline, short-petioled to sessile; corolla 9-12 mm long . . . .
. . . . .2. *S. nemorosa*

1. **S. pratensis** L. Stems to 1 m tall, somewhat hairy; lower leaves petioled, the blades rounded to slightly heart-shaped at base, irregularly and somewhat doubly toothed; stem leaves without petioles; corolla blue, 15-20 mm long, the upper lip strongly arching.—Escaping from cultivation, from Racine to Fond du Lac counties.

2. **S. nemorosa** L. Stems leafy to summit; leaves lanceo-late-ovate to lanceolate-triangular, margins crenate, cordate at base, underside pubescent; corolla blue, 9-12 mm long, the tube about the same length as the calyx.—Escape from cultivation; Dane, Green, Waukesha, and Pierce counties.

4. Blephilia   Wood Mint

Perennial herbs, with flowers densely crowded in the axils of the upper leaves; calyx distinctly 2-lipped, the upper lip longer; corolla 2-lipped, woolly, the upper lip straight, con-cave, and entire, the lower lip of 3 lobes and spreading, the 2 lateral lobes broader than the oblong middle lobe; stamens 2.

Fig. 476 *Blephilia ciliata,* × .30.

Fig. 477. *Glechoma hederacea,*   Fig. 478. *Ajuga genevensis,* × .25.
var. *parviflora,* × .40.

B. ciliata (L.) Benth. (Fig. 476). Stems pubescent, with short recurved hairs; leaves lanceolate; corolla purple.—Low places, in southeastern Wisconsin from Dane and Rock counties eastward.

5. Glechoma   Gill-over-the-Ground; Creeping Charlie

Plants mostly with creeping stems; leaves round or kidney-shaped; flowers in leaf axils, about 3 per axil; calyx lobes sharp-pointed; corolla 2-lipped, the upper lip with 2 shallow lobes, the lower lip larger with large middle lobe and 2 shorter lateral lobes; stamens 4.

G. hederacea L. Extensively creeping; corolla blue.

Var. **hederacea.** Leaves green; corolla 15-23 mm long.—Rare in Wisconsin.

Var. **parviflora** (Benth.) Druce (Fig. 477). Leaves reddish; corolla 9-15 mm long.—Native of Europe; very common in wet or shady spots, mostly near dwellings or in disturbed habitats.

6. Ajuga   Bugleweed

Perennials; foliage leaves dentate or lobed; inflorescence a leafy spike, the flowers in whorls in the axils of bracteal leaves; calyx lobes about as long as the tube; upper lip of corolla with 2 short lobes, the lateral lobes shorter than the lower 2-cleft elongate lip; stamens 4.

A. genevensis L. (Fig. 478). Flowers blue, 1.5 cm long, several in the axil of each broad 3-lobed bract.—Escaping

Fig. 479. *Leonurus cardiaca,* × .50.

Fig. 480. *Prunella vulgaris,* × .50.

from cultivation and locally abundant in Waukesha, Walworth, and Milwaukee counties.

7. **Leonurus**   Motherwort

Tall perennial herbs; leaves deeply 3-cleft and coarsely toothed; inflorescence a long terminal spike; calyx lobes long and needlelike; corolla 2-lipped, densely hairy, the upper lip erect, concave, the lower lip spreading; stamens 4.

L. cardiaca L. (Fig. 479). Leaves all petioled, the lower leaves palmately lobed and veined, the upper smaller and less cut; corolla pink or purple.—Naturalized from Europe; waste places and pastured woods; scattered throughout the state but less common northward.

8. **Prunella**   Self-heal

Perennial herbs; inflorescence a dense bracteate terminal spike, the bracts very different from the leaves; calyx lipped, the upper lip shallowly 3-toothed, the lower lip deeply cleft into 2 narrow segments; corolla with the upper lip concave, entire, and arched over the stamens, the 3-lobed lower lip spreading; stamens 4, the lower pair longer than the upper.

P. vulgaris L. (Fig. 480). Stems simple or a little branched; leaves ovate-oblong, narrowed at base, about one-third as broad as long; flowers 3 in a cluster, many clusters making up a dense cylindrical short spike; calyx green or purple; corolla commonly bluish, sometimes pink or white.—Naturalized from Eurasia; frequent weed in lawns, woods, and along roadsides; throughout Wisconsin.

9. **Lamium**   Dead Nettle

Low perennial herbs, with mostly heart-shaped leaves and whorled flower clusters; calyx 5-lobed, the lobes almost equal; corolla erect, 2-lipped, the upper lip concave, constricted at base, the lateral lobes reduced, the lower lip constricted at base, entire or 2 lobes deeply cleft; stamens 4, the 2 lower longer than the 2 upper; nutlets 3-sided, truncate at tip.

a. Upper leaves sessile and clasping, lower leaves petioled . . . .
. . . . 1. *L. amplexicaule*
a. All leaves petioled . . . . 2. *L. maculatum*

1. **L. amplexicaule** L. Henbit. Lower leaves petioled, the upper leaves sessile and clasping; corolla 12-18 mm long, the upper lip 3-5 mm long.—Roadsides and waste places, from Racine to Sheboygan counties along Lake Michigan, also Dane County.

2. **L. maculatum** L. Dead Nettle. Leaves petioled, frequently marked with a white spot on the upper side; corolla 20-25 mm long, the upper lip 7-11 mm long.—Introduced from Europe and escaping from cultivation; Grant, Wood, Rock, and Manitowoc counties.

10. **Dracocephalum**   Dragonhead

Herbs, with serrate leaves and flowers in whorls in leafy or bracted heads or spikes; calyx tubular, lipped; corolla tube gradually widened, weakly 2-lipped, the upper lip straight and 2-lobed, the lower lip deflexed and 3-lobed with the middle lobe notched; stamens 4, the upper pair slightly longer, the pollen sacs divergent; nutlets smooth.

Fig. 481. *Dracocephalum parviflorum,* × .40.

D. parviflorum Nutt. (Fig. 481). Plants 1.5-8 dm high; leaf blades ovate-lanceolate; flowers bluish, in the axils of crowded awn-toothed or fringed leafy bracts.—Not common, scattered localities in Wisconsin.

SOLANACEAE   Nightshade Family

Our species herbs or shrubs, with alternate leaves; flowers perfect, mostly regular; sepals 5, fused; petals 5, fused; stamens 5, borne on the corolla and alternate with the lobes; style 1; fruit a berry in our spring-flowering species.

a. Plants herbaceous *b*
    b. Flowers blue .... 1. *Solanum*
    b. Flowers cream, yellow, or white .... 2. *Physalis*
a. Plants shrubby .... 3. *Lycium*

1. Solanum   Nightshade

Our species herbs or slightly woody vines; corolla broadly spreading; stamens with short filaments and anthers united around the style, opening by terminal pores or slits; fruit a many-seeded berry.

S. dulcamara L. Bittersweet; Nightshade (Fig. 482). Somewhat climbing; leaf blades heart-shaped, sometimes with small lateral leaflets; flowers about 1 cm in diameter, blue, several in an open cluster; berries red.—Naturalized from Europe; wet thickets, shores, old fencerows, and roadsides; common in the southeastern half of Wisconsin, becoming rare northwestward.

2. Physalis   Ground Cherry

Annual or perennial herbs, bearing single or few flowers at the leaf nodes; calyx small in flower but enlarging and nearly enclosing the fruit at maturity, appearing inflated; corolla wide-spreading, entire or shallowly lobed, white to

Fig. 482. *Solanum dulcamara,* × .10.

Fig. 483. *Physalis heterophylla,*   Fig. 484. *Physalis virginiana,*
× .40.   × .30.

yellowish with a darker center; anthers opening longitudinally; fruit a many-seeded berry.

a. Annuals; corolla white with pale yellow center ... **1**. *P. grandiflora*
a. Deep-rooted perennials; corolla yellow *b*
    b. Plants sticky-hairy .... **2**. *P. heterophylla*
    b. Plants hairy but not sticky .... **3**. *P. virginiana*

1. **P. grandiflora** Hook. Large White Ground Cherry.
Annuals, with hairy and sticky stems; leaves about half as
broad as long; corolla 3-4 cm wide; fruiting calyx closely
fitting the berry and open at the throat.—Clearings, northern
Wisconsin.

2. **P. heterophylla** Nees. Clammy Ground Cherry (Fig.
483). Perennials; leaf blades rounded or heart-shaped at base,
nearly as broad as long, hairy on both sides; corolla 1.5-2 cm
wide; fruiting calyx much larger than the berry inside.—
Sandy soil, throughout the state.

3. **P. virginiana** Mill. Virginia Ground Cherry (Fig. 484).
Perennials, with forked stems; leaf blades several times as
long as broad; corolla 1-2 cm wide; fruiting calyx sunken at
base, much larger than the berry.—Sandy soil, throughout
the state.

3. Lycium   Matrimony Vine

Shrubs with long drooping branches; leaves lanceolate,
short-petioled; calyx 3-6-lobed; corolla 4-7-lobed, funnel-
shaped; anthers opening longitudinally; berry somewhat
compressed.

L. halimifolium Mill. Sparsely thorny gray green shrubs;

flowers greenish purple; berry red, 1-2 cm long.—Introduced from Europe; occasional escape, north to Dunn to Outagamie counties.

SCROPHULARIACEAE   Figwort Family

Herbs (our species); flowers perfect, mostly zygomorphic; sepals and petals 4-5, mostly fused; corolla usually 2-lipped; stamens usually 4 (rarely 5), in 2 pairs, or flower with only 2 stamens; fruit a many-seeded capsule, formed of 2 carpels.

a. Corolla nearly regular *b*
   b. Flowers 2 cm or more in diameter . . . . 1. *Verbascum*
   b. Flowers less than 1 cm in diameter *c*
      c. Corolla lobes much longer than the tube; leaves opposite . . .
      . . . . .2. *Veronica*
      c. Corolla lobes much shorter than the tube; leaves whorled . . .
      . . . . 3. *Veronicastrum*
a. Corolla 2-lipped *d*
   d. Corolla with a spur *e*
      e. Flowers in terminal racemes . . . . 4. *Linaria*
      e. Flowers single, in the leaf axils . . . . 5. *Chaenorrhinum*
   d. Corolla without a spur *f*
      f. Anther-bearing stamens 4 *g*
         g. Leaves not lobed *h*
            h. Stamens shorter than the corolla (Figs. 494-95) *i*
               i. Leaves with petioles . . . . 6. *Scrophularia*
               i. At least the upper leaves without petioles *j*
                  j. Flowers blue and white, 6-12 mm long . . . .
                  . . . . .7. *Collinsia*
                  j. Flowers pink, purple, or whitish, 15-20 mm
                  long . . . . 8. *Penstemon*
            h. Stamens longer than the corolla (Fig. 497) .9. *Besseya*
         g. Leaves lobed at least halfway to the midrib *k*
            k. Lobes ribbonlike (Fig. 498) . . . . 10. *Castilleja*
            k. Lobes rounded, not over twice as long as broad (Fig.
            500) . . . . .11. *Pedicularis*
      f. Stamens 2 *l*
         l. Leaves alternate . . . . .9. *Besseya*
         l. Leaves opposite . . . . .12. *Gratiola*

1. Verbascum   Mullein

Biennial herbs, the first-year rosette of basal leaves producing a stem the second year; leaves alternate; inflorescence an elongate raceme; flowers nearly regular; corolla saucer-shaped, yellow, pink, or white; stamens 5; stigma flattened.

V. blattaria L. Moth Mullein. Stems tall and stout, about 1 m high; lower stem leaves with lobed margins, sessile; corolla yellow or white, 2.5 cm wide.—Adventive from Europe; southern Wisconsin, also La Crosse and Portage counties.

2. **Veronica**   Speedwell

Herbs; foliage leaves opposite, those in the inflorescence mostly alternate and scattered; flowers small; calyx 4-lobed; corolla tube shorter than the lobes, the 4 lobes wide-spreading; stamens 2; style exserted; fruit short, often flattened and notched at the apex, tipped by the persistent style.

a. Flowers in long racemes in the axils of leaves *b*
  b. Stems pubescent *c*
    c. Inflorescence spikelike, the pedicels shorter than the bracts
    . . . . 1. *V. officinalis*
    c. Inflorescence racemose, the pedicels equaling or exceeding
    the bracts . . . . 2. *V. teucrium*
  b. Stems almost or quite without hairs *d*
    d. Leaves all petioled (Fig. 485) . . . . 3. *V. americana*
    d. Leaves, at least those of the flowering stems, without
    petioles *e*
      e. Leaves lanceolate, clasping the stem at the broad base
      (Fig. 486) . . . . 4. *V. anagallis-aquatica*
      e. Leaves nearly linear, narrowed to the base (Fig. 487) . . .
      . . . . .5. *V. scutellata*
a. Flowers along the stem in axils of foliage leaves *f*
  f. Stems creeping . . . . .6. *V. serpyllifolia*
  f. Stems erect *g*
    g. Corolla whitish . . . . .7. *V. peregrina*
    g. Corolla deep violet blue *h*
      h. Leaves toothed or entire . . . . 8. *V. arvensis*
      h. Leaves pinnatifid *i*
        i. Style surpassing the summit of the capsule . . . .
        . . . . .9. *V. dillenii*
        i. Style barely reaching the summit of the capsule . . . . .
        . . . . .10. *V. verna*

1. **V. officinalis** L. Common Speedwell. Stems creeping; leaves oval, toothed, hairy, with very short petioles or none; racemes 20-30-flowered, the pedicels shorter than the flowers; flowers pale lavender, with blue lines.—In eastern Wisconsin west to Dane, Lincoln, and Iron counties.

2. **V. teucrium** L. Stems erect, 3-8 dm high, pubescent; leaves opposite, serrate, sessile; racemes densely flowered, becoming loose in fruiting; corolla 1 cm wide, blue.—Native of Europe, escaping from cultivation; Outagamie County.

3. **V. americana** (Raf.) Schwein. American Brooklime (Fig. 485). Plants creeping only at base, with erect stems; leaf blades broadest near the base, toothed; racemes 10-25-flowered, several times as long as their subtending leaves; pedicels slender, several times as long as the pale blue purple-striped flowers.—Wet shores, over much of the state but not southeastern Wisconsin.

Fig. 485. *Veronica americana,* × .40.

Fig. 486. *Veronica anagallis-aquatica,* × .40.

Fig. 487. *Veronica scutellata,* × .40.

Fig. 488. *Veronica serpyllifolia,* × .35.

Fig. 489. *Veronica peregrina,* × .40.

Fig. 490. *Veronica arvensis,* × .40.

4. **V. anagallis-aquatica** L. Water Speedwell (Fig. 486). Stems mostly in the water; leaves obscurely toothed, mostly 1 cm or more wide; racemes 3-4 times as long as their subtending leaves, coarse, 15-30-flowered, the pedicels 3-6 mm long; corolla pale blue.—North to Door, Adams, and Pierce counties.

Var. **glaberrima** (Pennell) Fassett has pedicels without stalked glands.

5. **V. scutellata** L. Marsh Speedwell (Fig. 487). Stems slender and weak, occasionally with some weak hairs; leaves very narrow, long-pointed, often with minute distinct teeth; racemes 5-20-flowered, seldom twice as long as their subtending leaves; pedicels very slender, 6-10 mm long; flowers blue.—Occasional in wet places, throughout the state.

6. **V. serpyllifolia** L. Thyme-leaved Speedwell (Fig. 488). Leaves short-petioled, the blades oval, obtuse, obscurely toothed, mostly 1 cm or less long; corolla 3-4 mm wide, whitish or pale blue with deeper stripes.—Open woods and grassy places, throughout the state.

7. **V. peregrina** L. Purslane Speedwell (Fig. 489). Lowest leaves petioled, with ovate blades, the upper oval to linear, without petioles; flowers very small, shorter than their bracts, whitish.—A roadside and garden weed, southern half of the state, also Door and St. Croix counties.

Var. **xalapensis** (HBK.) St. John & Warren is covered with gland-tipped hairs.

8. **V. arvensis** L. Corn Speedwell (Fig. 490). Leaves mostly short-petioled, the blades oval, crenate; stems with spreading hairs; flowers violet blue.—Naturalized from Europe; occasional in fields, southern Wisconsin and north to Monroe to Door counties.

9. **V. dillenii** Crantz. Annuals, 1-2 dm tall; leaves 1-2 cm long, deeply pinnatifid; corolla blue, shorter than the calyx lobes; capsule obcordate, the persistent style extending beyond the lobes of the capsule.—La Crosse, Monroe, and Jackson counties.

10. **V. verna** L. Similar to *V. dillenii*, but persistent style of the capsule shorter than the capsule lobes.—Manitowoc and Columbia counties.

3. Veronicastrum   Culver's Root

Tall perennials, with leaves mostly in whorls of 3-7; inflorescence 1 or more terminal spikes of pale lavender or white

Fig. 491. *Veronicastrum virginicum,* × .25.

flowers; calyx 4-5-lobed; corolla tubular, the lobes much shorter than the tube, nearly regular; stamens 2, long-exserted; style about equaling the stamens; capsule narrowly ovoid.

V. virginicum (L.) Farw. (Fig. 491). Plants to 2 m tall; leaves lanceolate, acuminate, sharply serrate; spikes erect, 5-15 cm long; corolla 7-9 mm long.—Dry upland woods and prairies, throughout Wisconsin except for the very northern tier of counties.

4. Linaria   Toadflax

Annual to perennial herbs; leaves alternate, ribbonlike; flowers in racemes; corolla very irregular, spurred at base, strongly 2-lipped, the upper lip erect and 2-lobed, the lower lip 3-lobed, the throat open or closed by an elevated palate; stamens 4.

a. Flowers yellow to orange *b*
   b. Leaves linear, narrowed at the base . . . . 1. *L. vulgaris*
   b. Leaves lanceolate, broadest at the sessile base . . 2. *L. dalmatica*
a. Flowers blue . . . . 3. *L. canadensis*

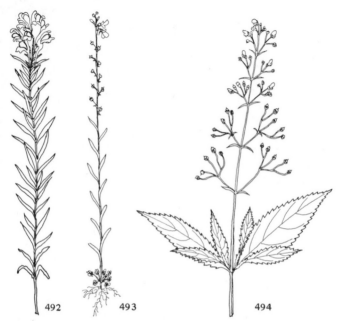

Fig. 492. *Linaria vulgaris,* × .25. Fig. 494. *Scrophularia lanceolata,*
Fig. 493. *Linaria canadensis,*   × .25.
× .20.

1. **L. vulgaris** Hill. Butter-and-Eggs (Fig. 492). Colonial
perennials; leaves numerous, pale green; corolla 2-3 cm long
(including the spur), the orange palate contrasting with the
rest of the flower.—Naturalized from Europe; common weed,
throughout the state.

2. **L. dalmatica** (L.) Mill. Stem leaves broad-lanceolate,
acuminate, sessile, leaves of the inflorescence narrower and
oblong; corolla 2.5-4 cm long.—Escape from cultivation; scattered localities in Green, Dane, Sauk, Bayfield, Sawyer, and
Vilas counties.

3. **L. canadensis** (L.) Dumont. Old-field Toadflax (Fig.
493). Stems usually simple or nearly so, very slender; corolla
1 cm or less long.—Sandy ground, often in disturbed soil;
north to Eau Claire and Waupaca counties.

5. Chaenorrhinum   Dwarf Snapdragon

Annuals or perennials; leaves alternate, narrow, entire; flowers solitary, in the leaf axils; calyx 5-lobed; corolla strongly 2-lipped, spurred at base; stamens 4; capsule ovoid.

C. minus (L.) Lange. Annuals 1-3 dm tall, glandular-pubescent; leaves 1-2 cm long; corolla 5-6 mm long, blue purple.—A rather uncommon weed along railroads; southern half of state north to Buffalo and Brown counties.

6. Scrophularia   Figwort

Coarse perennial herbs, 1-1.7 m high; leaf blades heart-shaped or rounded at base, sharply toothed; petioles long; branches of the inflorescence with copious gland-tipped hairs; corolla yellowish or purplish green, with a scale inside the upper lip representing a fifth stamen (Fig. 494), 2-lipped, the upper lip extended forward, the lower lip also forward with the middle lobe deflexed.

a. Inflorescence loosely branched; rudimentary stamen purplish brown . . . . 1. *S. marilandica*
a. Inflorescence cylindrical; rudimentary stamen yellowish green . . . . .2. *S. lanceolata*

1. S. marilandica L. Leaves serrate; inflorescence pyramidal; corolla 5-8 mm long; capsule shiny. Flowers later than the next, usually in July.—Oak woods, north to Dunn, Sauk, and Outagamie counties.

2. S. lanceolata Pursh (Fig. 494). Leaves coarsely serrate to incised; corolla 7-11 mm long; capsule dull brown.—Common on roadsides and rocky banks, throughout the state.

7. Collinsia   Blue-eyed Mary

Annuals or winter annuals; stem leaves opposite, the uppermost whorled; flowers axillary, in whorls of 2-8 at upper nodes; calyx deeply cleft; corolla bell-shaped, the lobes longer than the tube, 2-lipped, the lower lip deflexed; stamens 4; style elongate.

C. verna Nutt. Slender herbs; leaves clasping the stem, broadest at the base, toothed; flowers about 6 in a whorl; corolla deeply 2-lipped, blue and white.—Rock County.

8. Penstemon   Beard-tongue

Perennial herbs; stems usually several in a clump, not branched except in the inflorescence; stem leaves opposite,

without petioles, often somewhat clasping; corolla tubular,
dilated above, more or less 2-lipped, the upper lip 2-lobed,
the lower 3-cleft (Fig. 495), frequently streaked by a second
color (guidelines); fertile stamens 4, 1 sterile stamen usually
bearded at the tip (characterizing the genus); style elongate.

a. Leaves glaucous, entire, succulent; inflorescence unbranched . . . . .
. . . . 1. *P. bradburii*
a. Leaves green, margins entire to dentate, not succulent; inflores-
cence branched *b*
   b. Plants rather robust; leaf margins entire to slightly serrate;
     corolla tube abruptly inflated; sterile stamen bearded at tip
     with long flexuous hairs . . . . 2. *P. digitalis*
   b. Plants more slender; leaf margins usually serrate; corolla tube
     gradually inflated; sterile stamen bearded half its length by
     short stiff hairs *c*
      c. Lower lip of corolla upcurving to close the throat, corolla
        lacking guidelines, 2-3 cm long . . . . 3. *P. hirsutus*
      c. Lower lip of corolla not upcurving, the throat open; corolla
        with guidelines, 1.5-2.5 cm long *d*
        d. Corolla pale violet . . . . 4. *P. gracilis*
        d. Corolla white . . . . 5. *P. pallidus*

1. **P. bradburii** Pursh (*P. grandiflorus* Nutt.). Plants 5-10
dm high, stout, entirely without hairs or glands, and with a
whitish bloom; leaves rounded or blunt at tip, the upper
nearly as broad as long; inflorescence unbranched; corolla
4-5 cm long, pink or purplish.—Prairies, from Rock to
Waushara counties westward, and along the Mississippi River
north to Polk County.

2. **P. digitalis** Nutt. Stems rather stout, about 1 m or
more high; lower stem leaves sometimes reaching 16 cm in
length and 4 cm in width; upper leaves triangular; inflorescence open, the lower branches usually many times exceeding their subtending leaves; corolla 2-3 cm long, white,
streaked with purple, abruptly dilated into a wide throat.—
Probably adventive or escaping from cultivation; fields, north
to Lincoln County.

3. **P. hirsutus** (L.) Willd. (Fig. 495). Stems usually 2-6 dm
high, hirsute the full length; basal leaves long-petiolate, stem
leaves sessile, weakly serrate; corolla bluish purple externally,
the throat bearded with white hairs; staminodium bearded
much of its length with yellow hairs, usually exserted.—Open
ground, Sheboygan and Green Lake counties, southeastward;
also Vilas and Marinette counties.

4. **P. gracilis** Nutt. (Fig. 496). Stems 2-6 dm high;
basal leaves short-petiolate, stem leaves sessile, gradually
reduced in size toward the inflorescence; corolla 17-21 mm
long, purplish violet, throat open, guidelines present;

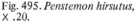

Fig. 495. *Penstemon hirsutus,*   Fig. 496. *Penstemon gracilis,*
× .20.   × .25.

staminodium bearded most of its length with yellow hairs, included in the tube.

Ssp. **wisconsinensis** (Pennell) Fassett. Leaves moderately to conspicuously pubescent.—Sandy places, from Polk and Lincoln counties, south to the lower Wisconsin River.

Ssp. **gracilis**. Completely glabrous.—Western Wisconsin, from Iowa to Portage counties westward, and along the Mississippi River north to Burnett County.

5. **P. pallidus** Small. Stems 2-6 dm high, conspicuously pubescent; leaves pubescent, basal leaves petiolate, cauline leaves sessile, margins toothed to entire; corolla 17-22 mm long, gradually inflated, white, corolla throat open, bearded lightly, guidelines usually conspicuous; staminodium bearded most of its length, included in the corolla tube.—Central Wisconsin, from Sauk to Green Lake counties, north to Oneida and Price counties, also Manitowoc County.

9. Besseya

Perennial herbs, with a basal rosette and stems with reduced leaves bearing a dense terminal spikelike raceme; calyx deeply 4-lobed, nearly regular; corolla strongly 2-lipped, yellow; stamens 2, exserted; style as long as the stamens; capsule nearly round, somewhat flattened.

Fig. 497. *Besseya bullii*, × .40.

B. bullii (Eat.) Rydb. (*Synthyris, Wulfenia*) (Fig. 497). Basal leaves petioled, the blades rounded, covered with short white hairs; stems unbranched, with small, sessile hairy leaves; flowers in a dense spike.—Open woods and prairies; southern tier of counties, also Pierce to Polk counties.

10. Castilleja   Painted Cup

Annual or perennial herbs, parasitic on the roots of other plants and characterized by colorful bracts in the inflorescence often more showy than the corolla; leaves both basal and alternate on the flowering stem, cut into ribbonlike lobes; inflorescence a dense terminal spike with bracteal

Fig. 498. *Castilleja coccinea*, × .40.

Fig. 499. *Castilleja sessiliflora*, × .25.

leaves; calyx 2-cleft, about as long as the corolla tube; corolla tubular, 2-lipped; flattened laterally; stamens 4; capsule ovoid.

a. Floral leaves scarlet-tipped or yellow-tipped . . . . 1. *C. coccinea*
a. Floral leaves green . . . . 2. *C. sessiliflora*

1. **C. coccinea** (L.) Spreng. Scarlet Painted Cup (Fig. 498). Flowers pale yellow, less conspicuous than the scarlet-tipped or yellow-tipped leaves in whose axils they are borne.—Locally abundant, throughout the state.

2. **C. sessiliflora** Pursh. Downy Painted Cup (Fig. 499). Corolla tubular, 3-4 cm long; floral leaves green.—Limestone bluffs, high prairies, and sand, north to Pierce and Columbia counties; mostly south of the Tension Zone.

11. **Pedicularis** Wood Betony; Lousewort

Annual to perennial herbs; leaves opposite or alternate, toothed to bipinnatifid; inflorescence a terminal spike or raceme; calyx entire or lobed; corolla strongly 2-lipped, the upper lip arching, often laterally compressed; stamens 4; capsule flattened.

P. canadensis L. (Fig. 500). Perennials; stems simple, clustered, hairy, 1.5-4 dm high; leaves pinnate or pinnatifid, the divisions with rounded teeth; flowers in a short spike, yellow, rarely scarlet.—Dry sandy woods and fields, throughout the state.

12. **Gratiola** Hedge Hyssop

Annual to perennial herbs; leaves opposite, sessile; flowers single in the leaf axils; sepals 5; corolla tubular to bell-shaped, 2-lipped; stamens 2.

G. neglecta Torr. Annuals; stems 1-3 dm high; flowers on

Fig. 500. *Pedicularis canadensis,* × .50.

peduncles 1-2.5 cm long; corolla 8-15 mm long, pale yellow.—Muddy places, throughout most of state except southeastern Wisconsin.

OROBANCHACEAE   Broomrape Family

Herbs, not green, parasitic on the roots of other plants; stems stout, fleshy; leaves replaced by scales; corolla 2-lipped, with 2 folds in the throat; fruit a capsule, with 1 cavity, 2-4 parietal placentae.

a. Flowers crowded in a thick scaly spike . . . . 1. *Conopholis*
a. Flowers on long pedicels . . . . .2. *Orobanche*

1. **Conopholis**   Squawroot

Stout unbranched herbs, the stems mostly concealed by fleshy overlapping leaf scales; flowers crowded in a dense spike; calyx 4-5-parted; corolla tubular, 2-lipped; stamens (4) and style about equaling the corolla; capsule ovoid, tipped by a persistent style.

C. americana (L.) Wallr. (Fig. 501). Parasitic on the roots of trees, mostly oaks; plants yellow or chestnut-colored throughout, not branched, 0.5-2 dm high, looking something like a pine cone; flowers covered by thick scales.—Woods; locally abundant in southern Wisconsin, occasional northward to Bayfield, Vilas, and Door counties.

2. **Orobanche**   Broomrape

Parasitic on herbs; stems short, the above-ground portions bearing several small bracts; flowers with a curved tube, each

Fig. 501. *Conopholis americana,* × .40.

Fig. 502. *Orobanche uniflora,* × .40.

on a long, erect, naked pedicel from the axil of a bract, upper lip 2-lobed, lower 3-lobed; stamens 4; capsule with 4 placentae.

a. Flowers 1-5; pedicels longer than the stem . . . . .1. *O. uniflora*
a. Flowers 6-12; pedicels shorter than the stem . . . . . 2. *O. fasciculata*

1. **O. uniflora** L. Cancerroot (Fig. 502). Flowers 1-5; pedicels longer than the short stem, which is only 1-7 cm tall; flowers lavender, fading to yellowish or white.—Sandy prairies, thickets, or moist woods, southwestern Wisconsin and Door County; rare.

2. **O. fasciculata** Nutt. Flowers 6-12, purplish; pedicels not longer than the elongate stem which is 5-10 cm high.—Sandy soil, Green, Iowa, Sheboygan, and Ozaukee counties; rare.

LENTIBULARIACEAE   Bladderwort Family

Aquatics or plants of moist ground; calyx 5-parted; corolla irregular, usually 2-lipped, having a short tube with a spur near its base, the throat partly closed by a palate; stamens 2; ovary of 2 carpels; fruit a many-seeded capsule.

Utricularia   Bladderwort

Plants submerged or on wet soil, bearing bladders which catch small animals; leaves divided into hairlike or linear segments; flowers on an emersed racemiform scape, yellow.

a. Leaves and bladders on the same branches; leaf segments rounded, threadlike . . . . .1. *U. vulgaris*
a. Leaves and bladders on separate branches; leaf segments flat . . . . .
. . . . 2. *U. intermedia*

1. **U. vulgaris** L. (Fig. 503). Leaves 2-3 times pinnately parted, bearing bladders.—Common in shallow water, throughout the state.

Fig. 503. *Utricularia vulgaris,* × .35.

2. **U. intermedia** Hayne. Leaves 4-5 times forked; bladders on separate leafless branches.—Peat bogs; less common, but throughout the state.

PLANTAGINACEAE   Plantain Family

Leaves all basal, simple, strongly veined; flowers in a slender spike; sepals 4, papery; corolla papery, 4-lobed; stamens 4; fruit a many-seeded capsule that opens in a line around the middle (circumscissile).

Plantago   Plantain

Characters as for family.

a. Leaf blades more than half as broad as long *b*
  b. Lateral veins of the leaf arising along the midrib . . 1. *P. cordata*
  b. Veins of the leaf parallel, free to base of the blade *c*
    c. Capsule circumscissile about the middle, 6-20-seeded; seeds smooth . . . . 2. *P. major*
    c. Capsule circumscissile about the base, 2-9-seeded; seeds netted . . . . 3. *P. rugelii*
a. Leaf blades many times longer than broad *d*
  d. Leaf blades lanceolate (Fig. 506) or oblanceolate *e*
    e. Leaves lanceolate, hairless; bracts and sepals hairless; spikes short; seeds brown . . . . 4. *P. lanceolata*
    e. Leaves oblanceolate, often toothed, hairy; bracts and sepals short-hairy; spikes long; seeds red . . . . 5. *P. virginica*
  d. Leaf blades ribbonlike (Fig. 508) *f*
    f. Bracts shorter than the flowers . . . . 6. *P. purshii*
    f. Bracts much longer than the flowers . . . . 7. *P. aristata*

1. **P. cordata** Lam. (Fig. 504). Leaf blades heart-shaped at base, on long petioles, base of petioles usually green; plants usually pubescent; roots fleshy; capsule 4-5 mm long.—Along streams; Kenosha, Racine, Milwaukee, Brown and Outagamie counties.

2. **P. major** L. Common Plantain (Fig. 505). Leaf blades broadly oval, thick and leathery, margins entire, tapered to the petiole or cordate; petioles usually green; capsule 2-4 mm long.—Common, throughout the state.

3. **P. rugelii** Dcne. Leaves broadly oval, glabrous, margins with 5-9 small teeth; base of petioles usually purple; capsule 6.5-8 mm long.—Common, throughout the state.

4. **P. lanceolata** L. Ribgrass (Fig. 506). Leaves long and narrow, strongly ribbed; spikes short, lengthening as the fruit matures, on scapes 2-7 dm high.—Naturalized from Europe; a weed in grasslands, throughout the state.

5. **P. virginica** L. Dwarf Plantain (Fig. 507). Hairy, short-lived annuals; leaves oblanceolate; corolla persistent after

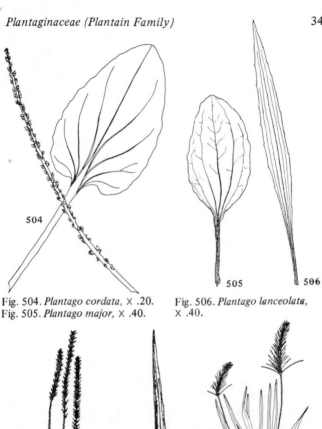

Fig. 504. *Plantago cordata*, × .20.
Fig. 505. *Plantago major*, × .40.

Fig. 506. *Plantago lanceolata*,
× .40.

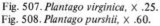

Fig. 507. *Plantago virginica*, × .25.
Fig. 508. *Plantago purshii*, × .60.

Fig. 509. *Plantago aristata*, × .25.

anthesis.—Grant, Richland, Iowa, Dane, Green, and Milwaukee counties.

6. **P. purshii** R & S. (Fig. 508). Whole plant silvery-woolly; leaves much shorter than the spike.—Sandy areas of central Wisconsin, from Richland and Monroe counties to Waushara County; also Burnett, Sawyer, Rock, and Waukesha counties.

7. **P. aristata** Michx. Bracted Plantain (Fig. 509). Similar to No. 6, but greener and less woolly, with loose hairs; bracts subtending the flowers awned.—Rare in dry places, north to Sheboygan, Lincoln, and Washburn counties.

RUBIACEAE   Madder Family

Herbs or shrubs; leaves opposite and with stipules, or appearing whorled; flowers perfect, mostly regular; sepals 4 or 5, free or fused; petals 4 or 5, fused, the lobes with their edges touching each other when in bud; stamens borne on the corolla, as many as the corolla lobes and alternate with them; ovary inferior, of 2-4 carpels, the calyx adherent to its wall.

a. Leaves whorled (Figs. 510-17) . . . . 1. *Galium*
a. Leaves opposite *b*
  b. Stems creeping . . . . 2. *Mitchella*
  b. Stems erect . . . . 3. *Houstonia*

1. Galium   Bedstraw

Slender herbs, with square stems; leaves whorled; inflorescences in cymes; flowers small, white to greenish white or yellowish; calyx teeth lacking; stamens (3) 4; styles 2; fruit dry, globular, of 2 nearly separate lobes.

a. Leaves mostly 4 in a whorl, gradually pointed or rounded, not with a short abrupt point (Figs. 510-14) *b*
  b. Flowers without pedicels (Fig. 510) *c*
    c. Corolla hairy . . . . .1. *G. circaezans*
    c. Corolla not hairy . . . . 2. *G. lanceolatum*
  b. Flowers with pedicels *d*
    d. Flowers in much-branched terminal clusters (Fig. 512) . . . . .
    . . . . 3. *G. boreale*
    d. Peduncles unbranched or with 2-4 branches, in the axils of foliage leaves *e*
      e. Corolla with 3 lobes *f*
        f. Flowers mostly solitary, on roughened curved pedicels
        . . . . 4. *G. trifidum*
        f. Flowers in 2s or 3s, on smooth straight pedicels . . . .
        . . . . 5. *G. tinctorium*
      e. Corolla with 4 lobes *g*
        g. Leaves ascending or spreading, lanceolate . . . .
        . . . . 6. *G. obtusum*

    g. Leaves pointing downward, linear . . 7. *G. labradoricum*
a. Leaves 5-8 in a whorl, with short abrupt rigid tips *h*
  h. Flowers in a terminal inflorescence that forks several times; leaf
    margins smooth or with a few bristles; fruit smooth . . . .
    . . . . 8. *G. concinnum*
  h. Flowers on axillary peduncles that are 3-flowered or slightly
    forking; leaf margins copiously hairy or bristly; fruit bristly *i*
    i. Leaves 5-8 mm broad, 3-4 times as long . . . . 9. *G. triflorum*
    i. Leaves 2-4 mm broad, about 10 times as long . . . .
    . . . . 10. *G. aparine*

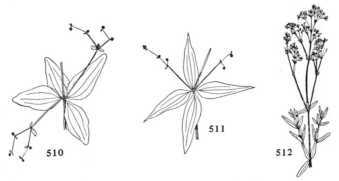

Fig. 510. *Galium circaezans,*
× .40.

Fig. 511. *Galium lanceolatum,*
× .40.
Fig. 512. *Galium boreale.* × .40.

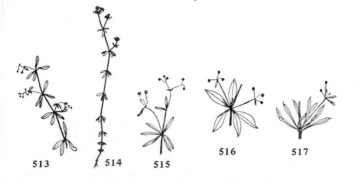

Fig. 513. *Galium tinctorium,*
× .40.
Fig. 514. *Galium labradoricum,*
× .30.

Fig. 515. *Galium concinnum,*
× .40.
Fig. 516. *Galium triflorum,* × .40.
Fig. 517. *Galium aparine,* × .40

1. **G.** circaezans Michx. Wild Licorice (Fig. 510). Stems erect, smooth, 3 dm high; leaves 1.5-4.5 cm long, one-third to one-half as wide, obtuse, their margins fringed with fine hairs; corolla greenish; fruit bristly.—Rich woods in southern Wisconsin; rare in localities north to Vernon, Waupaca, and Sheboygan counties.

2. **G.** lanceolatum Torr. Wild Licorice (Fig. 511). Similar to No. 1; leaves 3-7.5 cm long, about one-fourth as wide, long-pointed; corolla yellowish, turning dark purple; fruit bristly.—Not common, but widely scattered in most of the state.

3. **G.** boreale L. Northern Bedstraw (Fig. 512). Stems erect, smooth, 3-9 dm high; leaves long and narrow, roughened on the margins; flowers very many, white; a variable species.—Railroad embankments, bluffs, meadows, and along streams; common, throughout the state.

4. **G.** trifidum L. Stems slender, weak, reclining, roughened; leaves obtuse; flowers white; fruit smooth.—Throughout Wisconsin.

5. **G.** tinctorium L. (Fig. 513). Similar to No. 4, but stouter.—Marshlands; common, throughout the state.

6. **G.** obtusum Bigel. Stems erect, smooth; leaves 1-2 cm long, obtuse, slightly roughened on the margins and midrib; flowers white; fruit smooth.—Moist ground; frequent in southern Wisconsin, rare northward.

7. **G.** labradoricum Wieg. (Fig. 514). Similar to No. 6, but weak and much more slender, generally hidden among grasses.—Swales and swamps; throughout most of Wisconsin, but most frequent in the eastern half.

8. **G.** concinnum T. & G. (Fig. 515). Stems low and slender, 1.5-3 dm high, smooth or minutely roughened on the angles; leaves in 6s, narrow, the margins slightly roughened upwardly; flowers white; fruit smooth.—Woods, north to Polk, Clark, and Adams, and Brown counties, also Forest County.

9. **G.** triflorum Michx. Sweet-scented Bedstraw (Fig. 516). Stems often reclining, 3-10 dm long; peduncles 3-flowered; flowers greenish; fruit with hooked prickles.—Common in woods, throughout the state.

10. **G.** aparine L. Cleavers; Goosegrass (Fig. 517). Stems weak and reclining; leaves usually 8 in a whorl; flowers white; fruit bristly.—Moist woods; common in the southern half of Wisconsin.

2. Mitchella   Partridge Berry

Smooth trailing evergreen herbs, with opposite leaves; flowers in pairs, mostly terminal, white, bearded within, tinged with purple; style 1; stigmas 4; berry red, formed of 2 coherent ovaries and their calyces.

M. repens L. (Fig. 518). Stems rooting at the nodes, forming mats; leaf blades nearly round, 1-2 cm long.–Sandy soil in shady woods, throughout Wisconsin.

3. Houstonia

Small herbs, with basal rosettes of leaves; stem leaves with their bases connected by stipules; calyx and corolla 4-lobed; style 1; stigmas 2; fruit a 2-locular capsule.

a. Plants with single stems, from slender rhizomes; inflorescence single or few-flowered; stamens included in the corolla . . . . .1. *H. caerulea*
a. Plants with several stems, or caespitose; inflorescence cymose; stamens exserted . . . . .2. *H. longifolia*

1. **H. caerulea** L. Bluets; Innocence (Fig. 519). Plants slender, delicate, 0.5-2 dm high; leaves 6-9 mm long; flowers few on each stem, bluish or white.–Moist grassy places, north to Dane County.

2. **H. longifolia** Gaertn. (Fig. 520). Plants stouter and coarser, 1-2.5 dm high; leaves 1.5-2.5 cm long; flowers

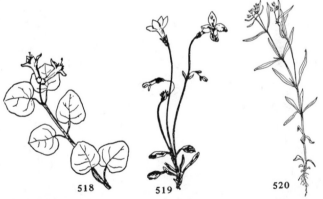

Fig. 518. *Mitchella repens*, × .75.
Fig. 519. *Houstonia caerulea*, × .75.

Fig. 520. *Houstonia longifolia*, × .40.

several in a branched inflorescence, pale purple or nearly white.—Gravelly and sandy ground, mostly northward.

CAPRIFOLIACEAE   Honeysuckle Family

Shrubs, vines, or slightly woody herbs, with opposite leaves, usually without stipules (sometimes with appendages at the base of the petiole, representing stipules in *Viburnum*); flowers perfect; calyx mostly 5-lobed, the lobes successively overlapping when in bud; stamens borne on the corolla tube and alternate with the lobes; style 1 or stigma sessile; ovary inferior; fruit a capsule, a berry, or a drupe.

a. Corolla tubular or bell-shaped; leaves simple *b*
  b. Shrubs *c*
    c. Leaves taper-pointed (Fig. 521) . . . . .1. *Diervilla*
    c. Leaves not taper-pointed *d*
      d. Corolla long-tubular, somewhat irregular . . . . .2. *Lonicera*
      d. Corolla short-bell-shaped, nearly regular . . . .
        . . . . 3. *Symphoricarpos*
  b. Herbs, or plants slightly woody at base *e*
    e. Plants delicate; stems creeping and slightly woody (Fig. 526)
      . . . . 4. *Linnaea*
    e. Plants coarse, erect, herbaceous (Fig. 527) . . . . 5. *Triosteum*
a. Corolla spreading, open *f*
  f. Leaves simple . . . . 6. *Viburnum*
  f. Leaves pinnately compound . . . . .7. *Sambucus*

1. Diervilla   Bush Honeysuckle

Low shrubs; leaves petioled, toothed; flowers 3 or more in the upper leaf axils or in a terminal cluster; fruit a slender beaked capsule.

D. lonicera Mill. (Fig. 521). Plants to 12 dm high; leaves acuminate, often fringed with hairs; corolla yellowish, some-

Fig. 521. *Diervilla lonicera,* × .55.

times marked with reddish, somewhat irregular.—Woods, throughout the state.

Var. **hypomalaca** Fern. Leaves velvety beneath.—Door, Bayfield, and Iron counties.

2. **Lonicera**   Honeysuckle

Erect or vinelike shrubs, with entire leaves; flowers in terminal clusters or in axillary pairs, each pair usually subtended by 2 bracts and 4 bractlets; calyx teeth very short; corolla tubular to funnel-shaped, often 2-lipped, the tube sometimes pouched or spurred at base; fruit a berry.

a. Erect shrubs *b*
  b. Bracts below the flowers small, scalelike or linear *c*
    c. Pith of twigs hollow *d*
      d. Leaves and twigs glabrous . . . . .1. *L. tatarica*
      d. Leaves and twigs more or less pubescent *e*
        e. Flowers pale yellow; filaments pubescent . . . .
        . . . . .2. *L. xylosteum*
        e. Flowers pink or white (fading to yellow); filaments glabrous *f*
          f. Leaves densely pubescent beneath . . 3. *L. morrowii*
          f. Leaves sparsely pubescent, with a few hairs on the veins beneath .4. *L. ✕bella*
    c. Pith of twigs solid *g*
      g. Peduncles rarely more than 1 cm long, usually shorter than the flowers; ovaries wholly united; fruit blue . . . . . .
      . . . . .5. *L. villosa*
      g. Peduncles mostly 2-4 cm long, longer than the flowers; ovaries separate; fruit red *h*
        h. Lobes of the corolla much shorter than the tube . . . .
        . . . . 6. *L. canadensis*
        h. Lobes of the corolla nearly as long as the tube . . . . . .
        . . . . .7. *L. oblongifolia*
  b. Bracts below the flowers broadly oval and leaflike, almost enclosing the flowers . . . . .8. *L. involucrata*
a. Climbing vines *i*
  i. Flowers in pairs, in the axils of upper leaves . . . . .9. *L. japonica*
  i. Flowers mostly in terminal clusters *j*
    j. Leaves without hairs on the margins and upper surfaces *k*
      k. Connate upper leaves longer than broad, green or slightly glaucous above . . . . .10. *L. dioica*
      k. Connate upper leaves forming a nearly round disk; strongly whitened above . . . . 11. *L. prolifera*
    j. Leaves with hairs along the margin and more or less pubescent on the upper surfaces . . . . .12. *L. hirsuta*

1. **L. tatarica** L. Tartarian Honeysuckle. Shrubs 1.5-3 m high; leaves 2-4 cm long, rounded or obtuse at tip, with short petioles, glabrous; corolla 1.5 cm long, white or rose.—An aggressive weedy shrub, often escaping from cultivation in southern Wisconsin.

2. **L. xylosteum** L. European Fly Honeysuckle. Similar to

L. tatarica; the twigs and upper surfaces of the leaves more or less pubescent, the under surfaces of the leaves pubescent; corolla yellowish or whitish.—Escaping from cultivation, in scattered localities in Wisconsin.

3. **L. morrowii** Gray. Similar to *L. tatarica* and *L. xylosteum* but corolla pubescent, pink to white and fading to yellow after flowering.—Japanese; cultivated shrub, escaping to scattered localities in southern Wisconsin.

4. **L. ×bella** Zabel. A hybrid of *L. tatarica* and *L. morrowii*; corolla glabrous, but the leaves slightly hairy beneath.—Scattered localities, north to Taylor and Lincoln counties.

5. **L. villosa** (Michx.) R. & S. Mountain Fly Honeysuckle (Fig. 522). Stems 3-9 dm high, with ascending branches; young twigs finely woolly and sometimes with sparse spreading hairs; leaf blades oval, somewhat hairy beneath; bracts fused, longer than the ovaries; corolla yellowish; berries blue.—Rare in bogs, northern and eastern Wisconsin.

6. **L. canadensis** Marsh. American Fly Honeysuckle (Fig. 523). Shrubs to 2 m tall; leaves triangular-ovate to oblong, the blades fringed with hairs; corolla pouched at base, yellowish, 1.5-2 cm long; fruit red.—Moist woods, north of the Tension Zone; rare in bogs and moist upland woods southeastward

7. **L. oblongifolia** (Goldie) Hook. Swamp Fly Honeysuckle. Shrubs to 2 m tall; leaves oblong to oblanceolate, the edges pubescent; corolla pouched at base, deeply 2-lipped, 1-1.5 cm long, yellowish white; fruit red.—Rare in swamps, east of a line from Rock to Columbia to Douglas counties.

8. **L. involucrata** (Richards.) Banks. Fly Honeysuckle. Shrubs to 3 m tall; leaves ovate to obovate; bracts oval, leaflike, almost enclosing the flowers; peduncles long, 2-4 cm; corolla pouched at base, yellow; fruit purple black.—Bayfield County.

9. **L. japonica** Thunb. Japanese Honeysuckle. Climbing pubescent vines; bracts foliaceous, bractlets round, ciliate; corolla strongly 2-lipped, whitish; fruit black.—Sporadic along riverbanks and edges of woods, southern Wisconsin; perhaps not hardy.

10. **L. dioica** L. (Fig. 524). Usually climbing; leaves without hairs, somewhat whitened beneath, the uppermost 1 or 2 pairs joined into a disk longer than broad, the tips obtuse;

Fig. 522. *Lonicera villosa,* × .35.

Fig. 523. *Lonicera canadensis,* × .40.

Fig. 524. *Lonicera dioica,* × .40.

Fig. 525. *Lonicera prolifera,* × .40.

corolla yellow to red or purple, without hairs, showy; filaments hairy; fruit red.—Oak woods; throughout the state, but less common northward.

Var. **glaucescens** (Rydb.) Butters. Leaves and corolla with spreading hairs.—Common, throughout the state.

11. **L. prolifera** (Kirch.) Rehd. Wild Honeysuckle (Fig. 525). Extensively climbing over shrubs; leaves strongly bluish whitened beneath, the uppermost pair joined into a nearly round disk, which is also strongly whitened above; corolla yellow; stamens conspicuous, the filaments without hairs; fruit red.—Common in fields and woods, north to Pierce and Waupaca counties.

12. **L. hirsuta** Eat. Hairy Honeysuckle. Climbing vines; leaves copiously fine-hairy on both surfaces, the uppermost forming a broad disk, the tips acuminate to mucronate; corolla yellow, filaments hairy; fruit red.—South to Barron, Clark, and Sheboygan counties.

3. Symphoricarpos Snowberry

Low shrubs; leaves entire to slightly lobed or toothed;

flowers in terminal short spikes or axillary clusters; corolla
bell-shaped; calyx short, 4-5-toothed; corolla 4-5-lobed; fruit
a drupe.

a. Corolla 5-6 mm long; style and stamens shorter than or equaling
    the corolla, not exserted; style 2-3 mm long, glabrous . . 1. *S. albus*
a. Corolla 6-9 mm long; style 4-8 mm long, exserted, pilose, rarely
    glabrous . . . . 2. *S. occidentalis*

1. **S. albus** (L.) Blake. Snowberry. Shrubs, to 1 m tall;
leaves ovate, usually hairy beneath, 2-3 cm long; flowers in
pairs on short pedicels or in few-flowered spikes; corolla
pink; berry white.—Throughout most of Wisconsin, south to
Vernon, Sauk, and Racine counties.

Var. **laevigatus** (Fern.) Blake. Coarser shrubs, to 2 m tall,
with the leaves glabrous to sparsely pilose on lower
surface.—A western variety which has been widely planted in
eastern North America and has become naturalized in scattered localities in northern and eastern Wisconsin.

2. **S. occidentalis** Hook. Wolfberry. Shrubs, to 1 m tall,
often producing stout unbranched canes; leaves ovate, often
coarsely crenate; flowers 4-10 per leaf axil, crowded in short
spikes; berry white.—Bluffs and dry prairies near the
Mississippi River, spreading eastward along railroads.

4. **Linnaea**   Twinflower

Slender creeping evergreen herbs, somewhat woody; leaves
short-petioled, with rounded blades; peduncles upright, with
2 (4 or 6) delicate pinkish, very fragrant flowers; corolla
bell-shaped; stamens 4; fruit dry, 1-seeded.

L. borealis, var. **americana** (Forbes) Rehd. (Fig. 526).
Leaves broadly oval, 1-2 cm; flowers in pairs on upright
stalks; corolla 8-15 mm long.—Shady woods northward;
rarely southward as far as Dane and Milwaukee counties

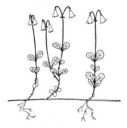

Fig. 526. *Linnaea borealis,* var. *americana,* × .25.

5. Triosteum   Horse Gentian; Wild Coffee

Coarse, usually dark green herbs, 0.5-1.2 m high; middle leaves with their bases uniting around the stem or only narrowing at the base, velvety beneath (or smooth in some forms); flowers clustered in axils of the leaves; corolla greenish yellow to purple; fruit orange or red, with 3 large seeds.

a. Principal leaves broadly clasping the stem; stem hairs of a short glandular type; flowers 3-4 in the leaf axils; style exserted 1.5-3 mm beyond the corolla lobes; fruit light orange, densely covered with short glandular hairs . . . . 1. *T. perfoliatum*
a. Principal leaves narrowed to the base (Fig. 527); stem hairs of a long nonglandular type, or of both long and short types; flowers 1-5 in each leaf axil; style included within the corolla lobes (exserted in No. 3); fruit bright orange or red *b*
  b. Style included; flowers 3-5 in each leaf axil; fruit bright orange; stem hairs of both long and short types . . . . . 2. *T. aurantiacum*
  b. Style exserted 1.5-2.5 mm; flowers 1-2 in each leaf axil; fruit red (sometimes purplish); stem hairs mostly of a long nonglandular type . . . . .3. *T. illinoense*

1. **T. perfoliatum** L. Tinker's Weed. Leaves thick, of a rugose texture, dark green above, light green below; corolla reddish purple, with cream-colored spots near the base.—On sandy, thin, or rocky soil in open woods and thickets; southern one-third of the state and in Dunn County.

2. **T. aurantiacum** Bickn. (Fig. 527). Leaves as in No. 1, except thinner and tapering to the base; corolla dull purplish red.—Throughout the state, in rich hilly or rocky woods and thickets.

3. **T. illinoense** (Wieg.) Rydb. Leaves pale green, shortened and more oval than the 2 species above; corolla dull reddish purple with cream colored spots.—In rich open woods, wooded ravines, and rocky wooded slopes near

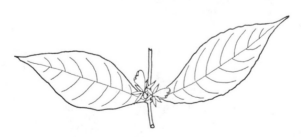

Fig. 527. *Triosteum aurantiacum,* × .25.

streams, usually in a limestone area; collected in the south-western corner of the state, from Vernon to Rock counties; rare.

6. Viburnum   Viburnum

Simple-leaved shrubs; leaf buds naked or with a pair of scales; flowers white, in cymes, either with all of the flowers alike and small or with the center flowers regular, small, and fertile, and the marginal flowers irregular, large, and sterile; calyx lobes, corolla lobes, and stamens 5; style lacking; stigmas 3, sessile, on a short stylopodium; fruit a 1-seeded drupe.

a. Leaves palmately veined and sometimes lobed *b*
  b. Leaves slightly hairy, with simple hairs, or glabrous beneath *c*
    c. Petioles with stipules at base and glands at the summit; leaves distinctly 3- or 5-lobed *d*
      d. Topmost pair of leaves with elongate pointed lobes, margins entire to obscurely toothed; glands at tips of petioles stalked and round-topped . . . . .1. *V. trilobum*
      d. Topmost pair of leaves with short broad lobes, margins obviously toothed; glands at tips of petioles short-stalked or sessile and saucer-shaped . . . . 2. *V. opulus*
    c. Petioles without stipules or glands; upper leaves sometimes not lobed or only slightly lobed . . . . 3. *V. edule*
  b. Leaves stellate-pubescent beneath . . . . .4. *V. acerifolium*
a. Leaves pinnately veined, not lobed *e*
  e. Leaves entire or finely toothed, their lateral veins branching near the margins *f*
    f. Inflorescence a stalked cyme, the branches arising from a central stalk *g*
      g. Leaves smooth beneath . . . . .5. *V. cassinoides*
      g. Leaves stellate-pubescent beneath . . . . 6. *V. lantana*
    f. Inflorescence a sessile cyme, no central stalk present *h*
      h. Leaves mostly with sharply acuminate tips; petioles with winged margins . . . . 7. *V. lentago*
      h. Leaves mostly with acute, obtuse, or rounded tips; petioles slightly or not wing-margined . .8. *V. prunifolium*
  e. Leaves coarsely toothed, their lateral veins straight or with only 1-2 branches, terminating in the teeth . . . .9. *V. rafinesquianum*

1. **V. trilobum** Marsh. Highbush Cranberry (Fig. 528). Shrubs, 1-4 m high; leaves 3-lobed, the lobes somewhat toothed; flowers in a flat-topped inflorescence, the marginal flowers with a large white corolla and no stamens or pistils, the inner fertile and with a small corolla; fruit bright red.—Moist thickets, throughout the state.

2. **V. opulus** L (Fig. 529). Similar to *V. trilobum*. Leaves often 5-lobed and with sessile, concave-topped glands at the top of the petiole.—European; escaping from cultivation, and naturalized in southeastern Wisconsin from Grant to Winnebago counties.

Fig. 528. *Viburnum trilobum,*
× .40.

Fig. 529. *Viburnum opulus,*
× .40.

530

Fig. 530. *Viburnum acerifolium,*
× .40.

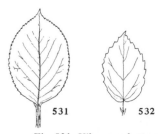

531   532

Fig. 531. *Viburnum lentago,*
× .40.
Fig. 532. *Viburnum rafines-*
quianum, × .40.

3. **V. edule** (Michx.) Raf. Squashberry. Blades shallowly lobed or not lobed; flowers uniform.—Barron County.

4. **V. acerifolium** L. Arrowwood (Fig. 530). Shrubs 1-1.5 m high; leaves maplelike; flowers all alike and fertile; fruit crimson, turning black.—Woods, throughout the state.

5. **V. cassinoides** L. Withe Rod. Leaf margins with obscure and irregular shallow teeth, or none; petioles scarcely winged.—Marinette and Oconto counties.

6. **V. lantana** L. Wayfaring Tree. Shrubs, to 3 m tall; leaves finely serrate; inflorescence with a short peduncle; fruit red, turning dark.—Widely planted as an ornamental, and sometimes escaping to edges of woods and along railroad embankments.

7. **V. lentago** L. Nannyberry (Fig. 531). Shrubs or trees,

reaching 9 m in height; leaf blades ovate, with 25-40 knobbed teeth on each side; petioles winged; midrib red-dotted beneath; fruit 1-1.5 cm long, blue black.—Woods and thickets, throughout the state.

8. **V. prunifolium** L. Black Haw. Shrubs or small trees; leaves glabrous to sparsely brown-scurfy when young, finely serrate, the petioles often slightly winged; cyme sessile; fruit blue black.—Milwaukee and Racine counties.

9. **V. rafinesquianum** Schultes. Downy Arrowwood (Fig. 532). Shrubs 6-14 dm high; leaves with 6-11 teeth on each side, velvety beneath; petioles 3-7 mm long.

Var. **affine** (Bush) House. Leaves not downy beneath; petioles 5-12 mm long. Between these 2 varieties occur occasional intermediates, with short petioles and smooth leaves, or with long petioles and leaves downy beneath.—Both varieties occur throughout the state, but var. *affine* is less common.

7. Sambucus   Elder

Shrubs (our species); leaves pinnately compound, leaflets serrate; inflorescence a large terminal cyme with numerous small flowers; corolla lobes and stamens 3-5, stamens exserted; style 3-5-lobed, very short; fruit a berry.

a. Pith white; inflorescence a flat cyme; berries purple black. . . . .
. . . . 1. *S. canadensis*
a. Pith brown; inflorescence a narrow ovoid cyme; berries red . . . . . .
. . . . 2. *S. pubens*

1. **S. canadensis** L. Common Elder (Fig. 533). Shrubs, 1-4 m high; pith white; lower leaflets often 2-3-parted; flowers white, in a broad flat-topped cyme; fruit purple black,

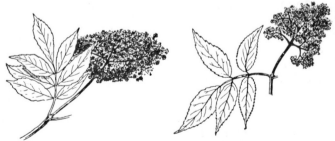

Fig. 533. *Sambucus canadensis,*
X .20.

Fig. 534. *Sambucus pubens,*
X .20.

edible.—Moist ground; common, throughout the state.

2. **S. pubens** Michx. Red-berried Elder (Fig. 534). Shrubs, 0.5-3.5 m high; twigs hairy, pith brown; flowers yellowish, white, or tinged with purple, in a narrow ovoid cyme; fruit red, poisonous.—Mostly northern, but occasionally found southward in the Driftless Area and in cool wooded ravines in southeastern Wisconsin.

ADOXACEAE　Moschatel Family

Delicate herbs, with basal leaves and 2 stem leaves, all ternately compound; inflorescence a few-flowered cyme; flowers perfect; petals 4-6, fused, the corolla wide-spreading; stamens borne on the corolla in pairs at the lobe sinuses; calyx tube not reaching to top of ovary; style 3-5-parted; fruit a small dry drupe.

Adoxa　Moschatel

Characters as for the family.

A. moschatellina L. (Fig. 535). Herbs, with a musky odor, 5-20 cm tall; leaflets obovate; flowers greenish or yellowish, 5-8 mm wide.—Rare in southwestern Wisconsin; on moist sandstone cliffs Vernon, Sauk, Grant, and Richland counties, and in rich woods in Pierce and Rock counties.

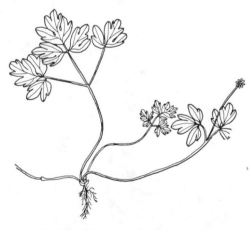

Fig. 535. *Adoxa moschatellina*, × .40.

VALERIANACEAE   Valerian Family

Perennials or annuals, often with an unpleasant odor; inflorescences and flowers diverse, the latter regular or zygomorphic, perfect or unisexual, mostly with 5 parts; stamens 3, borne on the corolla tube; style with a bilobed or trilobed stigma; fruit dry, indehiscent.

Valeriana   Valerian

Dioecious perennials; leaves in a rosette at base of stem, and opposite on the stem, without stipules, closely fringed with short hairs; inflorescence paniculate or corymbose; sepals and petals borne at the summit of the ovary; calyx with plumed bristles, these closely coiled in flower but expanding as the fruit matures.

a. Leaves thickish, parallel-veined . . . . 1. *V. edulis*, var. *ciliata*
a. Leaves thin, pinnately veined *b*
  b. Basal leaves simple, cauline leaves pinnatifid . . . .
   . . . . 2. *V. sitchensis*, ssp. *uliginosa*
  b. Basal and cauline leaves pinnately compound . . .3. *V. officinalis*

1. **V. edulis** Nutt., var. **ciliata** (T. & G.) Cronquist (Fig. 536). Roots large and deep; leaves thickish, the upper usually pinnate, with 3-7 ribbonlike divisions; flowers creamy white.—Moist meadows and prairies, north to Dane and Waupaca counties.

Fig. 536. *Valeriana edulis*, var. *ciliata*, ✕ .25.

2. **V. sitchensis** Borg, ssp. **uliginosa** (T. & G.) F. G. Meyer. Roots fibrous; leaves thin, the upper with 7-15 lanceolate divisions; flowers white.—Moist meadows; Outagamie, Portage, Waupaca, and Sheboygan counties.

3. **V. officinalis** L. Garden Heliotrope. Roots fibrous; stems hairy at least at the nodes; leaves divided into 11-21 leaflets.—An escape or remnant of cultivation; scattered localities in Wisconsin.

CAMPANULACEAE   Bluebell Family

Herbs, with milky juice; leaves alternate; corolla regular, 5-lobed; stamens 5, the anthers nearly free from the corolla tube; ovary inferior; style 1; stigmas 3; fruit a capsule, opening by pores or slits.

a. Leaves mostly longer than broad; corolla bell-shaped . 1. *Campanula*
a. Leaves as broad as long; corolla rotate . . . . 2. *Triodanis*

1. **Campanula**   Bluebell; Bellflower

Perennial herbs; sepals 5; corolla bell-shaped to rotate, 5-lobed, blue; stamens attached to base of the corolla; fruit short, usually strongly ribbed, opening by lateral pores.

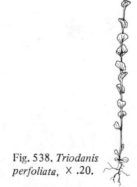

Fig. 537. *Campanula rotundifolia,* × .20.

Fig. 538. *Triodanis perfoliata,* × .20.

C. rotundifolia L. Bluebell; Harebell (Fig. 537). Stems 1-5 cm high; basal leaves (rarely present on flowering stems) long-petioled, with roundish blades; stem leaves narrowly ribbonlike; flowers several, long-peduncled; corolla bell-shaped.—Sandy and rocky places, throughout the state.

2. **Triodanis**   Venus' Looking-glass

Annuals; leaves toothed; flowers nearly sessile in the middle and upper leaf axils, lower flowers with reduced corolla or none, corolla rotate; stamens free; stigma lobed; capsules opening by pores above the middle.

T. perfoliata (L.) Nieuwl. (Fig. 538). Stems usually unbranched, 1-9 dm high; leaves sessile, roundish, less than 1 cm long, heart-shaped at clasping base; flowers not pediceled, in the upper leaf axils; corolla pale blue.—Sterile open ground and dry sandstone bluffs; southwestern half of Wisconsin, north to St. Croix and Waushara counties, also Racine County.

LOBELIACEAE   Lobelia Family

Herbs, often with a poisonous, milky juice; leaves simple, alternate; flowers irregular; corolla 5-lobed; stamens attached

Fig. 539. *Lobelia spicata*, × .25.

to base of the corolla tube and alternate with the lobes; anthers united into a tube around the style; fruit a capsule (our species).

Lobelia   Lobelia

Annuals or perennials, mostly with showy flowers in spikelike racemes; corolla split to the base on the upper side, 2-lipped, the upper lobes erect, the lower usually spreading; anthers colored, the lower 2 bearded at the tip; capsule opening at the top.

L. spicata Lam. (Fig. 539). Perennials, 3-10 dm tall; leaves lanceolate-oblong to obovate, 5-10 cm long, hairy, the upper leaves reduced to bracts; sepals often with conspicuous basal appendages; corolla blue to white, the lower lip bearded at the base.—Open sandy soils; southern half of Wisconsin, north to Polk to Marinette counties.

ASTERACEAE (COMPOSITAE)   Aster Family

Annual to perennial herbs, sometimes with milky juice; flowers many, in a dense head simulating a single flower, surrounded by an involucre of several close, often-overlapping bracts (phyllaries) the individual flowers often subtended by small bracts (chaff, pales); the flowers of 2 sorts, either or both present on a single head—(1) tubular flowers with a tubular corolla of fused petals, usually lobed, (2) ray or ligulate flowers with a strap-shaped corolla split down 1 side and appearing as a single petal; the calyx represented by a pappus of scales, bristles, teeth, or pappus lacking; stamens borne on the corolla, mostly with elongate anthers united into a tube around the style; style 1, usually 2-cleft; fruit an achene, often bearing the persistent pappus at the apex.

A single head of a member of this family may have its flowers arranged in 1 of 3 general types: (1) the head composed entirely of tubular flowers (discoid head), (2) the head entirely of ligulate flowers (radiate head), or (3) the head comprised of tubular flowers in the center surrounded by a series of ligulate flowers (head radiate with a disk). If all the flowers are tubular they are usually all perfect, or they may be unisexual, with pistillate or staminate heads on separate plants; if all the flowers on the same head are ligulate, they are generally all perfect; if the head is radiate, the ray flowers are either pistillate or without reproductive parts and the central tubular flowers are either perfect or functionally staminate.

a. Heads either with the inner flowers with a tubular corolla surrounded by ray flowers simulating petals, or heads with all tubular
flowers; juice not milky *b* . . . . Series I   Tubuliflorae
 b. Ray flowers present *c*
  c. Ray flowers yellow *d*
   d. Leaves opposite *e*
    e. Ray flowers pistillate; involucral bracts in 1 or 2
rows, all alike . . . . 1. *Heliopsis*
    e. Ray flowers neutral; involucral bracts in 2 rows, the
outer greener than the longer inner row. . 2. *Coreopsis*
   d. Leaves alternate or mainly basal *f*
    f. Leaves entire or little dissected *g*
     g. Foliage leaves basal, appearing after the flowers . . .
. . . . 3. *Tussilago*
     g. Leaves present on flowering stems *h*
      h. Plants rough-hairy . . . . 4. *Heterotheca*
      h. Plants smooth or finely pubescent . . . 5. *Senecio*
    f. Leaves finely dissected . . . . 6. *Anthemis*
  c. Ray flowers not yellow *i*
   i. Plants with bracted flowering scapes appearing before the
large basal leaves . . . . 7. *Petasites*
   i. Flowering stems leafy *j*
    j. Leaves finely divided into small segments; bruised
leaves aromatic *k*
     k. Heads many, in a flat-topped inflorescence . . . .
. . . . 8. *Achillea*
     k. Heads few, terminating short leafy branches *l*
      l. Receptacle naked, achenes 5-ribbed, smooth . . .
. . . . 9. *Matricaria*
      l. Receptacle with chaffy scales; achenes 10-
ribbed, glandular-tuberculate . . . . . . 6. *Anthemis*
    j. Leaves not finely divided; stem leaves simple, toothed
or lobed; bruised leaves not aromatic *m*
     m. Involucral bracts with spinescent tips; ray flowers
reflexed; plants coarsely hairy . . . . . . 10. *Echinacea*
     m. Involucral bracts not spinescent-tipped; ray flowers
spreading or ascending when flowers fully open *n*
      n. Plants softly hairy . . . . 11. *Erigeron*
      n. Plants glabrous . . . . 12. *Chrysanthemum*
 b. Ray flowers lacking (only tubular flowers present) *o*
  o. Involucral bracts imbricate *p*
   p. Plants lacking spines or bristles *q*
    q. Leaves finely dissected; heads on leafy branches;
plants not woolly . . . . 9. *Matricaria*
    q. Leaves entire, basal; heads on bracteate scapes; plants
quite woolly . . . . 13. *Antennaria*
   p. Plants very spiny and bristly . . . . 14. *Carduus*
  o. Involucral bracts in 1 row . . . . 15. *Cacalia*
a. Flowers all strap-shaped; juice milky *r* . . . . . Series II   Liguliflorae
 r. Plants smooth or woolly, without long hairs *s*
  s. Leaves rather broad, the sides not parallel *t*
   t. Heads 1-several on each scape; leaves entire or shallowly
toothed *u*
    u. Involucral bracts smooth . . . . 16. *Krigia*
    u. Involucral bracts woolly . . . . 17. *Crepis*
   t. Heads solitary on each scape; leaves pinnately lobed
nearly to the midrib . . . . 18. *Taraxacum*

1. Heliopsis   Oxeye

Perennial herbs, with opposite leaves; heads radiate, the rays yellow; involucral bracts in 1 or 2 series; receptacle conic, chaffy bracts concave and subtending both the ray and disk flowers; pappus lacking, or of a few teeth.

H. helianthoides (L.) Sweet (Fig. 540). Plants 5-15 dm tall; leaves ovate or lanceolate-ovate, serrate; disk 1-2.5 cm wide, 8-15 rays, 1.5-4 cm long.—Dry woods and prairies, throughout the state.

2. Coreopis   Tickseed

Herbs, with opposite leaves; heads radiate, the rays neutral, yellow; disk flowers perfect; involucral bracts in 2 rows of different shapes (dimorphic), joined at the base; receptacle bracts thin and flat; achenes usually winged, the pappus of 2 short awns or teeth, a minute crown, or lacking.

Fig. 540. *Heliopsis helianthoides,*
× .20.

Fig. 541. *Coreopsis palmata,*
× .45.

1. **C. palmata** Nutt. (Fig. 541). Perennials, 5-9 dm tall; leaves sessile, deeply trilobed, the lobes linear; heads mostly solitary, on short peduncles; rays 1.5-3 cm long.—Dry prairies; mostly southern and western Wisconsin, east to Waupaca and Winnebago counties, and north to Douglas and Bayfield counties.

2. **C. lanceolata** L., var. **lanceolata**. Glabrous perennials, 2-6 dm tall; leaves mostly basal or on lower half of stems, linear to lanceolate; heads on long naked peduncles; rays showy, 1.5-3 cm long and more than 1 cm wide.—Dunes of Lake Michigan, Sheboygan to Door counties.

Var. **villosa** Michx. Stems and leaves pubescent.—Rarely escaping from cultivation; scattered localities, throughout Wisconsin.

3. **C. grandiflora** Hogg. Coreopsis. Similar to *C. lanceolata*, but stems leafy to the summit; leaves pinnatifid into linear segments; peduncles not so elongate.—Escaping from cultivation, naturalized in Green County.

3. Tussilago   Coltsfoot

Perennial herbs; leaves large, basal, appearing after flowering; heads solitary, on scaly bracted stems; involucral bracts in 1 row; heads radiate, receptacle naked; pappus of numerous capillary bristles.

T. farfara L. Low perennials, to 5 dm tall; leaves long-petioled, round in outline, shallowly lobed, cordate at base; stems hairy; bracts about 1 cm long; rays yellow.—European; Winnebago County.

4. Heterotheca   Golden Aster

Herbs, with alternate, mostly entire leaves; heads radiate, both ray and disk flowers fertile, yellow; involucral bracts imbricate; receptacle naked; pappus double, the inner of bristles, the outer of short coarse bristles or scales.

H. villosa (Pursh) Shinners (Fig. 542). Taprooted perennials, 2-10 dm tall, rough-pubescent; stems leafy, the leaves mostly alike, 2-7 cm long, entire; bracts pubescent, sometimes glandular.—Sandy prairies and bluffs; St. Croix, Pierce, and Eau Claire counties in western Wisconsin; Adams, Marquette, and Waushara counties in central Wisconsin.

5. Senecio   Ragwort

Leaves alternate, toothed, only the basal ones with petioles; heads usually several; both ray and tubular flowers

Fig. 542. *Heterotheca villosa,* × .60.

fertile, yellow to orange or reddish, ray flowers sometimes lacking; receptacle naked; pappus of white bristles.

a. Stems leafy to the inflorescence *b*
  b. Heads without ray flowers .... 1. *S. vulgaris*
  b. Heads with ray flowers ....2. *S. congestus*, vars.
a. Stems leafy below, above nearly naked or with much-reduced leaves *c*
  c. Blades of basal leaves heart-shaped at base ....3. *S. aureus*
  c. Blades of basal leaves narrowed to the petiole ....
  ....4. *S. pauperculus* complex

1. **S. vulgaris** L. Common Groundsel. Plants usually much branched near the base; leaves pinnately deeply cut, the divisions oblong, sharply toothed; short lowest bracts of the involucre black-tipped.—Naturalized from Europe; eastern and extreme northern Wisconsin, and Dane and Sauk counties.

2. **S. congestus** (R. Br.) DC., var. **intonsus** Fern. Marsh Fleabane. Stems stout, not branched below; leaves lanceolate, without petioles, variously toothed, clasping the stem at base. The inflorescence in var. **palustris** (L.) Fern. is long-hairy; var. **tonsus** Fern. is smoother, with few or no long hairs.—Northwestern Wisconsin, Polk to Bayfield counties.

3. **S. aureus** L. Golden Ragwort. Basal leaves long-petioled, the blades rounded, toothed; stem leaves pinnate or pinnately deeply cleft, the uppermost bractlike.—Wet meadows, moist cliffs, and low wet woods, throughout the state.

4. **S. pauperculus** complex (Fig. 543). Preliminary evidence indicates that the *Senecio pauperculus* complex in Wisconsin contains 3 cytological races that are morphologically similar: a slender-leaved tetraploid (n=44) of damp ground, commonest in the north (*S. pauperculus* Michx.); a

Fig. 543. *Senecio pauperculus,*   Fig. 544. *Anthemis cotula,* × .50.
× .25.

wider-leaved diploid (n=22) of dry bluffs, pastures, savannas, commonest in the south (possibly *S. balsamitae* Muhl.); and a wider-leaved, often hairy tetraploid (n=46) of dry prairielike habitats, rare in the south (*S. plattensis* Nutt.).

6. Anthemis   Dog Fennel; Chamomile

Herbs, usually aromatic; leaves alternate, incised to much dissected; heads radiate; involucral bracts usually imbricate, dry, with scarious margins; pappus a short crown or lacking.

a Rays white; pappus lacking . . . . 1. *A. cotula*
a. Rays yellow; pappus a low crown . . . . 2. *A. tinctoria*

1. **A. cotula** L. Chamomile (Fig. 544). Aromatic annuals, ill smelling; leaves 2-3 times pinnatifid, segments narrow; disk ovoid; receptacle bracts narrow, firm; rays 10-20, sterile.—Naturalized from Europe; now in scattered localities, throughout the state, especially common southward.

2. **A. tinctoria** L. Yellow Chamomile. Perennials, 3-7 cm tall; leaves pinnatifid, rachis winged, underside villous; heads solitary or on long peduncles; receptacle bracts with firm, yellow awn tips equaling the disk flowers.—Native of Europe; scattered localities in Wisconsin.

7. **Petasites** Sweet Coltsfoot

Perennials, with large basal leaves; stems stout, bearing 10 or more purplish or greenish bracts, the lower somewhat inflated and clasping; heads in a raceme; involucral bracts in a single row; receptacle naked; flowers white, fragrant; pappus of numerous white slender hairs.

a. Leaves palmately veined, lobed, rounded or heart-shaped in outline
....1. *P. palmatus*
a. Leaves pinnately veined, dentate, arrow-shaped.... 2. *P. sagittatus*

1. **P. palmatus** (Ait.) Gray (Fig. 545). Stems with short, crinkly, somewhat glandular hairs; leaves with rounded and deeply 5-7-lobed blades, coming from the rootstock only as the fruit matures.—Swamps, from Douglas to Manitowoc counties.

2. **P. sagittatus** (Pursh) Gray. Stems densely white-woolly; leaf blades evenly toothed, broadly triangular to arrow-shaped.—Rare, Douglas County.

8. **Achillea** Yarrow; Milfoil

Perennial plants, with alternate, simple to compound leaves; heads small, many together in a flat-topped inflorescence; involucral bracts with scarious margins, in several rows; heads with 3-13 marginal ray flowers; pappus lacking.

Fig. 545. *Petasites palmatus,*
× .25.

Fig. 546. *Achillea millefolium,*
× .30.

Fig. 547. *Matricaria matri-carioides,* × .60.

Fig. 548. *Echinacea pallida,* × .25.

A. millefolium L. (Fig. 546). Plants 3-10 dm high; stems simple, forked in the inflorescence; leaves aromatic when bruised, pinnate, the leaflets deeply divided; flowers white.—Mostly native, but in part naturalized from Europe; common in open ground, throughout Wisconsin.

Forma **rubicunda** (Farw.) Farw. Rays in shades of pink.

9. Matricaria Wild Chamomile

Aromatic plants, with finely and pinnately dissected leaves; heads with radiate or with all flowers tubular; receptacle domed; pappus a short crown (*M. matricarioides*) or lacking.

a. Heads with ray flowers .... 1. *M. chamomilla*
a. Heads lacking ray flowers .... 2. *M. matricarioides*

1. M. chamomilla L. Aromatic annuals, with finely dissected leaves; ray flowers 10-20, white; disk 6-10 mm wide.—Roadsides and waste places, scattered localities in Wisconsin.

2. M. matricarioides (Less.) Porter. Pineapple Weed (Fig. 547). Leaves with a pineapple odor when bruised; heads rayless; the disk 5-9 mm wide; flowers, yellow.—Roadsides, city streets, and gardens, throughout Wisconsin; naturalized, probably from the Pacific slope.

10. Echinacea   Purple Coneflower

Perennial herbs; leaves alternate, mostly 3-veined; heads radiate, single, on long peduncles; involucral bracts in 2-4 rows; receptacle conic; bracts conspicuous, exceeding the disk flowers; ray flowers persistent; pappus a short-toothed crown.

E. pallida Nutt. (Fig. 548). Taprooted hairy perennials; leaves linear to lanceolate; rays purple.—Dry prairies; Dane, Green, Rock, and Grant counties.

11. Erigeron   Fleabane

Annuals or perennials; leaves alternate; heads 1-many; marginal ray flowers fertile, white, purple, blue, or pink; tubular flowers yellow; the bracts of the involucre subequal in 1 row, or with a long inner series and a very short outer one; pappus a single row of bristles, with short ones inter- mixed, or with a ring of short bristles outside the long ones; achenes flattened, 2-nerved, usually pubescent.

Fig. 549. *Erigeron pulchellus,* × .20.
Fig. 550. *Erigeron philadel-phicus,* × .20.

Fig. 551. *Erigeron strigosus,* × .25.

a. Heads mainly 2-3 cm broad (including the rays), solitary to about 10 (except *E. philadelphicus* occasionally more) *b*
  b. Perennials (remnants of previous year visible), with woody stems; lower leaves entire-margined, rather stiff, nearly glabrous; pappus with a ring of short bristles outside the elongate ones . . . . .1. *E. glabellus*
  b. Short-lived perennials, from rosettes with short or long basal offshoots; lower leaves rounded, toothed, soft-hairy; pappus with only elongate bristles *c*
    c. Stolons elongate; ray flowers about 50-100, 1 mm wide . . . . .2. *E. pulchellus*
    c. Stolons short or none; ray flowers more than 100, less than 1 mm wide, hairlike . . . . .3. *E. philadelphicus*
a. Heads mainly 1-1.5 cm broad, many in a corymb *d*
  d. Plants with the hairs appressed . . . . 4. *E. strigosus*
  d. Plants with the hairs spreading . . . . 5. *E. annuus*

1. E. **glabellus** Nutt. Biennials or perennials, nearly glabrous; lower leaves oblanceolate, upper leaves linear to lanceolate; ray flowers mostly blue or pink, 8-15 mm long.—Sandy barrens; extreme northwestern Wisconsin.

2. E. **pulchellus** Michx. Robin's Plantain (Fig. 549). Plants with soft white hairs; stems producing offshoots from the base, simple or slightly branched above; stem leaves few and rather small; rays light bluish purple.—Dry soil in open woods, north to Jackson, Sauk, and Milwaukee counties.

3. E. **philadelphicus** L. (Fig. 550). Plants hairy; stems somewhat branched; stem leaves larger than in the preceding; rays rose purple or flesh color.—Fields and woods, throughout the state.

4. E. **strigosus** Muhl. Daisy Fleabane (Fig. 551). Hairy annuals, the minute hairs mostly appressed; stem leaves entire; heads many in a corymb; rays white.—Common weed, throughout the state.

5. E. **annuus** (L.) Pers. Daisy Fleabane. Very similar to the preceding, but the sparse stiff hairs spreading and the stem leaves coarsely toothed.— Common weed, throughout the state.

12. Chrysanthemum   Chrysanthemum

Annuals or perennials, with alternate leaves; heads 1-many, with ray flowers; involucral bracts with scarious margins; receptacle bracts lacking; achenes 5-10-ribbed; pappus a short crown or lacking.

C. **leucanthemum** L., var. **pinnatifidum** Lecoq & Lamotte. Oxeye Daisy (Fig. 552). Stems simple or forked above; basal leaves long-petioled, the blades coarsely toothed or deeply cleft; stem leaves deeply toothed; rays white, tubular flowers

Fig. 552. *Chrysanthemum leucanthemum,*
var. *pinnatifidum,* × .40.

yellow.—Naturalized from Europe; locally abundant, throughout the state.

13. Antennaria   Ladies' Tobacco; Pussy-toes

Dioecious perennials, with stolons from the basal rosette,
the stolons producing side tufts of leaves; stems and often
the leaves covered with cobwebby hairs; heads 1-many, the
inflorescence usually compact at flowering and becoming
corymbose to racemose when fruiting; involucral bracts thin
and papery; pappus of abundant white hairy bristles. A
single patch often consists of only one plant of one sex;
staminate plants of some species either rare or unknown.

a. Rosette leaves with 1-3 prominent veins; the lateral veins, if
   present, rarely prominent beyond broadest part of leaf *b*
  b. Stolons prostrate, elongate *c*
    c. Cauline leaves with scarious appendages......1. *A. neglecta*
    c. Cauline leaves without scarious appendages . 2. *A. petaloidea*
  b. Stolons short, ascending, stiff *d*
    d. Upper cauline leaves with scarious appendages; leaves
       glabrous above ....3. *A. canadensis*
    d. Cauline leaves without scarious appendages; leaves glabrous
       to pubescent above ....4. *A. neodioica*
a. Rosette leaves with 3-7 prominent veins, the 2 main lateral veins
   converging toward and nearly reaching the tip *e*
  e. Stolons elongate, prostrate ....5. *A. munda*
  e. Stolons short, ascending, stiff *f*
    f. Involucre 4-7 mm high; pistillate corolla 4-6 mm long; sta-
       minate corolla 3-4.5 mm long; nodes on flowering stem 3-5
       ....6. *A. plantaginifolia*
    f. Involucre 6-9 mm high; pistillate corolla 5-8 mm long; sta-
       minate corolla 4-5.5 mm long; nodes on flowering stem
       5-12 *g*
      g. Rosette leaves tomentose above ....7. *A. fallax*
      g. Rosette leaves glabrous above; stem often purple-glandu-
         lar .... 8. *A. parlinii*

1. **A. neglecta** Greene (Fig. 553). Rosette leaves narrow,
usually 1-veined; stolons spreading, long and flexuous;
flowering stems 4-30 cm high; heads 1-7· involucral bracts

Fig. 553. *Antennaria neglecta,*
× .70.

Fig. 554. *Antennaria plantaginifolia,* × .20.

white to purple, often with scarious tips; both pistillate and
staminate plants abundant.—Common in dry habitats in
southern Wisconsin, less common northward to Douglas
County.

2. **A. petaloidea** Fern. Similar to No. 1; involucral bracts
green, brown, or purple, tips not scarious.—Pine woodlands,
prairies, and pastures; widespread in Wisconsin. Staminate
plants very rare.

3. **A. canadensis** Greene. Stolons erect, short, and stiff,
with large terminal rosettes developing early; rosette leaves
usually 1-veined, glabrous above; flowering stems 5-30 cm
high, the upper stem leaves with scarious appendages.—Pine
barrens, dry woodlands, and prairies; scattered localities in
northern Wisconsin, and south to Sauk and Adams counties.

4. **A. neodioica** Greene. Stolons erect, short, and stiff,
with early-developing rosettes; leaves 1-3-veined, tomentose
to glabrous above; flowering stems 15-40 cm high, stem
leaves lacking scarious appendages; bracts green, purple, or
brown, with scarious or white tips.—All plants in Wisconsin
female; sandy soil, throughout the state.

5. **A. munda** Fern. Stolons elongated; rosette leaves
3-5-veined, the lateral veins approaching the blade tip;
flowering stems 10-45 cm high; heads in a dense
corymb.—Wisconsin specimens all female; mostly on prairies
in south-central Wisconsin, but in very scattered locations
elsewhere throughout the state.

6. **A. plantaginifolia** (L.) Richards. (Fig. 554). Stolons
short, ascending; rosette leaves 3-7-veined; flowering stems

10-40 cm high; heads 4-20; bracts often purple at base, with white tips.—Male and female plants in about equal numbers; oak woods and pastures, most common in southern Wisconsin, and northward, especially along the Mississippi River, to Douglas County.

7. **A. fallax** Greene. Stolons short, ascending; rosette leaves 3-7-veined; flowering stems 15-45 cm high, their leaves lacking scarious appendages.—Dry habitats; female plants common throughout Wisconsin, male plants rarer.

8. **A. parlinii** Fern. Stolons short, ascending; rosette leaves 3-7-veined; glabrous and bright green above; flowering stems sometimes purple-glandular, pistillate stems 15-45 cm tall, staminate stems 3-30 cm tall; stem leaves lacking scarious appendages.—Oak woods and grazed prairies; both male and female plants common in southern Wisconsin, and in scattered localities northward to Douglas to Door counties.

14. Carduus   Thistle

Spiny herbs; leaves alternate, serrate to pinnatifid, leaf bases usually decurrent; heads with all flowers tubular, clustered or solitary; involucral bracts imbricate; flowers perfect, corollas purple, reddish to white, or rarely yellow; achenes glabrous; pappus of rough capillary bristles, deciduous in a ring.

Fig. 555. *Carduus nutans,* × .20.

Fig. 556. *Cacalia atriplicifolia,* × .20.

C. nutans L. Nodding Thistle; Musk Thistle (Fig. 555). Stout, very spinescent biennials; leaves deeply lobed; heads mostly solitary, nodding, usually large (4-8 cm wide); flowers purple, showy.—Native of Europe; roadsides, pastures, and waste places, southeastern Wisconsin from Kenosha and Racine counties to Dane and Rock counties.

15. Cacalia   Indian Plantain

Herbs, with alternate leaves; heads discoid, cylindrical; involucral bracts in 1 row, partly herbaceous, sometimes with scarious margins; flowers perfect, white or pale; receptacle naked; pappus of capillary bristles.

C. atriplicifolia L. (Fig. 556). Tall perennials, with glaucous stems; leaves palmately veined, shallowly lobed, glaucous on the underside.—Moist places, from Rock and Dane counties eastward, and north to Manitowoc County; also Trempealeau County.

16. Krigia   Dwarf Dandelion

Herbs, with milky juice; leaves all in a basal rosette, entire or pinnatifid; heads on long naked or nearly naked scapes; involucral bracts campanulate, 5-18 in a single row; corollas yellow; pappus of several scales and 5-40 longer bristles.

Fig. 557. *Krigia virginica*, × .30.   Fig. 558. *Krigia biflora*, × .50.

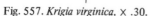

a. Flowering scapes naked; heads single . . . . .1. *K. virginica*
a. Flowering scapes with 1-3 bracts; heads 2 or more . . . 2. *K. biflora*

1. **K.** virginica (L.) Willd. (Fig. 557). Scapes naked, terminated by a single head.—Sandy soil along the Wisconsin River; Richland, Sauk, Iowa, and Dane counties.

2. **K.** biflora (Walt.) Blake (Fig. 558). Plants very smooth and somewhat bluish-powdery; scapes with 1-3 bracts near the summit, enclosing the bases of several peduncles.—Common in open sandy ground, in prairies, and oak woods, mostly south of the Tension Zone, restricted to pine woods in northern Wisconsin.

17. Crepis   Hawk's-beard

Annuals (our species), with basal rosettes or leafy stems, glabrous to hispid; involucral bracts in 2 series, the outer becoming thickened at the base; flowers all ligulate; corollas yellow; achenes fusiform; pappus of white hairs.

C. tectorum L. (Fig. 559). Basal rosette of coarsely and pinnately toothed leaves; stems furrowed, somewhat whitened with a fine wool, branched.—Sandy ground, locally abundant, especially northwestward; adventive from Europe.

18. Taraxacum   Dandelion

Plants with milky juice; leaves all basal, entire to deeply cut; scapes hollow; flowers all ligulate, perfect; corollas yellow; involucral bracts in 2 rows, the outer row reflexed or erect, the inner row erect; the achenes with an elongate tip, topped by a pappus of numerous capillary bristles.

Fig. 559. *Crepis tectorum,*
× .20.

Fig. 560. *Taraxacum officinale,*
× .80, flower detail × 2.

a. Mature achenes tan to olivaceous, not red; leaves various, deeply
lobed to entire .... 1. *T. officinale*
a. Mature achenes reddish to deep brown or purplish; leaves generally
deeply lobed or cut to midrib ....2. *T. erythrospermum*

1. T. officinale Weber. Common Dandelion (Fig. 560).
Leaves variously lobed to entire, usually somewhat hairy on
underside and on the midrib; involucral bracts erect and
becoming reflexed in age; achenes pale gray brown to olive
brown.—Naturalized from Europe; abundant, throughout the
state.

2. T. erythrospermum Andrz. Red-seeded Dandelion.
Leaves deeply cut the whole length, the lobes narrow; inner
bracts usually with a hooded appendage at the tip; achenes
bright red or reddish purple.—Native of Eurasia; throughout
much of Wisconsin, but much less common than the other
species.

19. Tragopogon   Goat's-beard

Tall annuals, biennials, or perennials, with a taproot;
stems smooth, juice milky; leaves grasslike, smooth, somewhat clasping at base; heads large, solitary; involucral bracts
in a single row; flowers all ligulate; pappus of plumose
bristles united at base, the plume branches interwebbed.

Fig. 561. *Tragopogon pratensis,*
× .30.

Fig. 562. *Tragopogon dubius,*
× .25.

a. Flowers yellow b
    b. Involucral bracts shorter than the flowers . . . . . . 1. *T. pratensis*
    b. Involucral bracts equaling or longer than the flowers . . . .
    . . . . 2. *T. dubius*
a. Flowers purple . . . . 3. *T. porrifolius*

1. **T. pratensis** L. (Fig. 561). Stems scarcely enlarged below the head; bracts of involucre shorter than the flowers; flowers yellow; pappus white.—Naturalized from Europe; fields and roadsides in scattered localities, throughout the state, probably less common than No. 2.

2. **T. dubius** Scop. (Fig. 562). Stems conspicuously enlarged upward below the head; bracts of involucre longer than the flowers; flowers yellow; pappus white.—Naturalized from Europe; fields and roadsides, throughout the state.

3. **T. porrifolius** L. Salsify; Oyster Plant. Biennials; peduncles enlarged under the head; involucral bracts exceeding the flowers; flowers purple; pappus brownish.— Escape from cultivation and naturalized in scattered localities.

20. Microseris

Taprooted herbs, with milky juice; leaves all basal, grass-like, with woolly margins; heads solitary on naked scapes; involucre campanulate; flowers all ligulate and perfect;

achenes columnar; pappus of numerous capillary bristles and scales.

M. cuspidata (Pursh) Schultz-Bip. (Fig. 563). Scapes to 5 dm tall, woolly below the head; flowers yellow.—Rare on prairies in southwestern Wisconsin, Grant to Columbia and Rock counties, Pierce County.

21. Hieracium   Hawkweed

Fibrous-rooted perennials from rhizomes, with milky juice, often pubescent and glandular; leaves in a basal rosette or alternate; flowers all ligulate and perfect, corollas yellow to red orange; achenes cylindrical; pappus of whitish or brownish capillary bristles.

a. Flowers bright red orange....................1. *H. aurantiacum*
a. Flowers yellow *b*
    b. Stolons present; leaves pubescent ....2. *H. caespitosum*
    b. Stolons lacking; leaves essentially glabrous ... 3. *H. florentinum*

1. H. aurantiacum L. Orange Hawkweed (Fig. 564). Leaves mostly basal, with rough brown hairs; scapes 2-6 dm high, usually with 1 or 2 small leaves; heads several; flowers orange red; involucres densely covered with long black and short glandular hairs.—Naturalized from Europe, a noxious weed in the north; fields and roadsides, throughout most of Wisconsin except the southern tier of counties and the extreme western section.

2. H. caespitosum Dumort. (*H. pratense* Tausch). King Devil. Glandular hairy perennials, 4-6 dm tall, with short erect rootstocks, often with arched or erect leafy stolons;

Fig. 563. *Microseris cuspidata,*
× .25.

Fig. 564. *Hieracium aurantiacum,*
× .50.

leaves all basal, entire; involucres cylindrical, the bracts dark green with a black midrib.—Native of Eurasia; scattered localities in northern Wisconsin, south to La Crosse to Juneau counties.

3. H. florentinum All. Yellow Devil's Paintbrush. Similar to No. 2, but lacking stolons, essentially glabrous, and with smaller heads.—Usually found with *H. aurantiacum* in north-central Wisconsin, from Clarke to Oconto counties northward, where the 2 species make a spectacle of color; native of Europe.

Glossary

Acaulescent. Lacking an upright stem.

Achene. A small, dry, one-seeded, indehiscent fruit.

Acuminate. With a long, tapering point (Fig. 571*a*).

Acute. With a sharp point (Fig. 571*c*).

Adventive. Imperfectly naturalized.

Alternate (of leaves). Borne along a stem, with only one at each level.

Anther. The upper, usually enlarged part of a stamen, bearing the pollen (Figs. 565, 566).

Anthesis. The period during which a flower is fully expanded and functional.

Apex. The end farthest from the central stem.

Appressed. Lying close to a part, such as a hair on a stem or leaf.

Aril. A fleshy appendage to a seed of Bittersweet or Wahoo, derived from the growth of the seed stalk.

Auricle. An earlike lobe (Fig. 571*c*).

Auriculate. With ear-shaped lobes or appendages.

Awn. An elongate hard projection or bristle.

Axil. The angle formed by a leaf with the stem.

Axillary. Arising from a node in the axil of a leaf (see flowers in Figs. 569*b* and 569*c*, or the inflorescence in Figs. 330 and 486).

Beak. A terminal appendage on certain seeds or fruits.

Berry. A juicy fruit, with several compartments (carpels).

Biconvex. Convex on two surfaces.

Bipinnate. Pinnate, with the leaflets themselves pinnate.

Biternate. With each of three divisions again divided into three.

Blade. The expanded, flattened part of a leaf.

Bloom. A white or bluish, powdery or waxy covering.

Bract. A more or less modified, usually small and scalelike leaf, subtending a flower or on a stem.

Branchlet. A twig or small branch.

Bristle. A stiff hair or small spine.

Bulb. An underground bud with scaly or fleshy layers.

Caespitose. Growing in dense tufts.

Calyx. The outer, usually green, whorl of a flower, made up of sepals (Figs. 565-67). Plural: calyces.

Calyx lobe. The free part of a sepal which is united below with other sepals (Fig. 567).

Calyx teeth. Bristlelike or very short calyx lobes.

Calyx tube. The lower tubular portion of a calyx formed of united sepals (Fig. 567).

Campanulate. Bell-shaped.

Canescent. With finely hoary pubescence.

Capitate. Headlike (usually referring to a stigma).

Capsule. A dry fruit, with 2 or more united carpels, opening at maturity.

Carpel. A macrosporophyll; a modified, seed-bearing leaf, of which from one to several make up a pistil, the number often determined by observing the number of stigmas, or of compartments in the ovary (Fig. 568).

Catkin. A spike of flowers, closely aggregated without pedicels but with conspicuous scales, the whole flexible and drooping at maturity.

Caulescent. With a well-developed stem.

Cauline. Pertaining to the stem.

Cell. A compartment, or carpel (when used of fruit).

Chaff. Fine bristles or scales on the receptacle subtending the flowers in the heads of certain Asteraceae.

Cilia. Marginal hairs.

Ciliate. With marginal hairs.

Circumscissile. Opening by a line extending around the middle.

Clavate. Club-shaped and increasing in diameter toward the summit.

Claw. The narrow basal portion of some sepals and some petals.

Cleft. Cut halfway or more to the base.

Cleistogamous. Describes a flower which does not open, yet self-fertilizes and produces seeds.

Compound. Divided into several leaflets, each of which is distinct and not joined to the other leaflets by any leaflike tissue (Figs. 570c, 570d).

Connate. Grown together or attached.

Convolute. Said of sepals or petals in bud stage when one overlaps the next along one edge in a slanting twisting arrangement.

Cordate. With 2 broadly rounded lobes at the base of a leaf and referring either to the base or the entire heart-shaped leaf.

Corm. Enlarged, fleshy, solid base of a stem.

Corolla. The whorl of a flower just within the calyx, usually of white, or colored, separate or united petals (Figs. 565-67).

Corolla lobe. The free part of a petal which is united below with other petals (Fig. 567).

Corolla tube. The lower tubular portion of a corolla, formed of united petals (Fig. 567).

Corona. An appendage toward the base of the blades of petals forming an extra crownlike or petallike structure.

Corymb. A shortened raceme with the lower pedicels long, and the upper short to form a broad, flat-topped inflorescence (Fig. 569*e*).

Corymbose. In a corymb.

Cotyledon. The leaf of an embryo plant as found in the seed.

Crenate. Scalloped; with rounded teeth (Fig. 571*d*).

Culm. The flowering stem of a grass or sedge.

Cuneate. Narrowly triangular with the acute angle downward; wedge-shaped.

Cyathium. A cuplike involucre of fused bracts surrounding the individual inflorescences in *Euphorbia* (Fig. 344).

Cyme. A type of inflorescence in which a central flower terminates the stem and opens first, a pair of branches below this continues the growth, and these in turn are terminated by flowers. At the base of each flower stalk the branching is again repeated by dichotomies, resulting in a more or less flat-topped cluster (Fig. 569*f*).

Decompound. Describing a leaf repeatedly and irregularly divided into many leaflets; more than once compound.

Decurrent. Forming wing-like extension of a leaf continuing down the stem.

Deflexed. Bent downward.

Dehiscent. Opening to discharge the contents, usually referring to a fruit or anther.

Dimorphic. Of two kinds or shapes.

Dioecious. With the staminate (male) and pistillate (female) flowers on different plants.

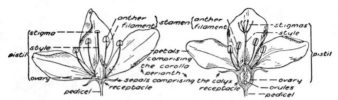

Fig. 565. Hypogynous flower, ovary superior.

Fig. 566. Epigynous flower, ovary inferior.

Fig. 567. Flower with fused perianth.

Fig. 568. Pistil of 3 carpels.

Diploid. Containing twice the number of chromosomes possessed by the male or female gametes.

Discoid. Disklike, used to describe a head of tubular flowers in the Asteraceae.

Dissected. Cut into numerous segments.

Divided. Lobed or segmented to the base.

Downy. Covered with short fine hairs.

Drupe. A fruit with epidermal, fleshy and stony layers (pit) enclosing a seed, e.g., cherry or peach.

Elliptical. Narrowed equally at each end (Fig. 571*d*).

Embryo. The little dormant plant in a seed.

*Entire.*Without teeth or lobes (Fig. 571*e*).

Exserted. Extending beyond the corolla lobes.

Falcate. Sickle-shaped.

Fibril. Tiny fiber.

Fibrillose. With many fibers.

Filament. The usually threadlike stalk of a stamen (Figs. 565, 566).

Filiform. Threadlike.

Fimbriate. Fringed.

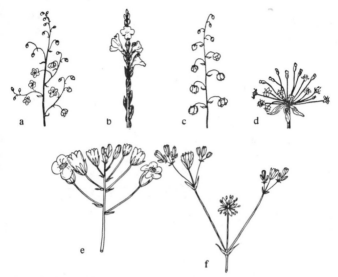

Fig. 569. Types of inflorescences: *a.* panicle; *b.* spike; *c.* raceme; *d.* umbel; *e.* corymb; *f.* cyme.

Flexuous. Bent alternately in opposite directions; zigzag.
Floccose. Woolly.
Floricane. The second-year stem of a biennial woody plant bearing the flowers, characteristic of *Rubus*.
Foliate. With leaflets.
Follicle. A dry dehiscent fruit opening along one suture.
Fusiform. Thick near the middle and tapering at both ends.

Glabrate. Almost glabrous, with but few scattered hairs.
Glabrous. Lacking hairs; usually smooth.
Globose. Round; globe-shaped.
Glomerule. Compact, headlike cluster.
Glume. One of the 2 lower scales of a grass spikelet (Fig. 3*c*).
Guideline. Colored lines on petals supposedly guiding insects toward the nectary.
Gynostegium. The compound structure formed by the uniting of the stamens and pistil in the Orchidaceae and Asclepiadaceae.

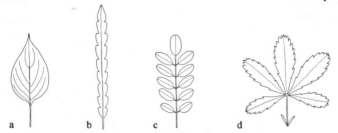

Fig. 570. Types of leaves: *a.* simple; *b.* pinnatifid; *c.* pinnately
compound; *d.* palmately compound with stipules.

Head. A dense cluster of flowers without pedicels, sometimes
with and sometimes without an involucre.
Herb. A plant with nonwoody stems, which die back to the
ground level each winter.
Herbaceous. Not woody, of soft texture and usually green.
Heterostylous. With styles of 2 or 3 lengths and the stamens
also of corresponding lengths but in different flowers.
Hirsute. Coarsely hairy.
Hispid. Bristly.
Hyaline. Rather transparent.
Hygroscopic. Sensitive to moisture changes.
Hypanthium. A flat to cup-shaped structure originating
below the ovary and formed from the fusion of the bases
of stamens, petals, and sepals, or from a cuplike exten-
sion of the receptacle and bearing the sepals, petals, and
stamens.

Imbricate. Overlapping like shingles.
Impressed. Furrowed.
Included. Within; referring to stamens in a corolla tube.
Indehiscent. Usually referring to fruits which do not open at
maturity.
Inferior. Referring to an ovary which is below the position
of attachment of the other flower parts.
Inflorescence. The flowering portion of a plant
Infrastipular. Beneath the stipules, as in some of the spines
in *Rosa.*
Internode. The portion of a stem between 2 nodes.
Introduced. Brought intentionally from another region.
Involucre. A collection of bracts around a flower or cluster
of flowers (Fig. 569*d*), especially in the Asteraceae (Figs.
563, 564).

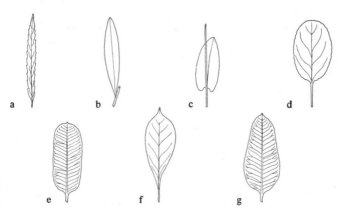

Fig. 571. Leaf shapes and characters: *a.* linear with serrate margin and acuminate tip; *b.* lanceolate, sessile (petiole lacking); *c.* sessile with auricles; *d.* elliptical with crenate margin; *e.* oblong with entire margin; *f.* obovate with acute tip; *g.* ovate.

Keel. A ridge like the keel of a boat.

Lanceolate. Long and tapering, and broadest near the base (Fig. 570*b*).

Lax. Of loose arrangement; weak and reclining.

Leaf axil. See *Axil.*

Leaflet. A single division of a compound leaf (Figs. 570*c*, 570*d*).

Legume. A dry, simple fruit, usually dehiscing along 2 sutures.

Lemma. The outer of 2 bracts enclosing the stamens and pistil of a grass flower (Fig. 3*c*).

Ligule. A collarlike, variously modified structure between the leaf blade and the sheath in a grass.

Linear. Long and narrow, and with parallel margins (Fig. 571*a*).

Lip. One of the petals in the Orchidaceae, enlarged and often cleft or inflated; 1 of the 2 divisions into which the corolla is sometimes cut when it is composed of united petals (Fig. 496)

Lobed. With divisions (Figs. 140, 244, 369, 482).

Loculus (Locule). One compartment of an ovary or a stamen.

Loculicidal. Describing a capsule which dehisces along the outer median line of each locule.

Midrib. The main vein of a leaf.
Mm. A millimeter, 1/10 of a cm (about 1/25 of an inch).
Monadelphous. Referring to stamens in which the filaments are united into a single tube.
Monoecious. Bearing staminate and pistillate flowers on the same plant.
Mucronate. Tipped with a short, sharp point.

Naturalized. Thoroughly established.
Nectary. A nectar-producing organ.
Nerve. A small vein.
Net-veined. With the veins branching, and sometimes re-uniting toward the edge of the leaf; or netted veins in a seed.
Node. The level on a stem where one or more leaves are borne.

Obconic. Inversely conic; conical but attached at the broader end.
Obcordate. Inversely cordate with the point basal.
Oblanceolate. Lanceolate with the broadest part toward the apex.
Oblong. Longer than broad, with nearly parallel sides (Fig. 571*e*).
Obovate or *obovoid.* Egg-shaped, with the broadest part toward the apex (Fig. 571*f*).
Obtuse. Blunt or rounded at the end (Fig. 571*e*).
Ocrea. A sheathing structure composed of fused stipules at the attachment of leaf to stem in the Polygonaceae.
Opposite (of leaves). Arranged in pairs on the stem, 2 at each node.
Orbicular. Circular.
Oval. Broadly elliptical.
Ovary. The lower, usually swollen, part of a pistil, where the seeds develop (Figs. 565, 566).
Ovate or *ovoid.* Egg-shaped, with the broadest part toward the base (Fig. 571*g*).

Pale. A chaffy scale around the achene in certain Asteraceae.
Palea. The inner of 2 bracts enclosing the stamens and pistil in a grass flower.
Palmate. Radially lobed or divided (Fig. 570*d*).

Panicle. A loose, several-times-branched inflorescence, with stalked flowers (Fig. 569*a*).

Papilionaceous. Having a standard, wings, and keel; the petal arrangement of many Fabaceae.

Pappus. Hairs, scales, or bristles growing from the summit of an achene in the Asteraceae and representing the modi fied sepals.

Parallel (of veins). All running from the base to the tip of the leaf without branching (Figs. 62, 74).

Parasite. A plant growing on and deriving nourishment from another living plant or animal.

Parietal placentation. Ovules borne on the inner side of the exterior wall of an ovary.

Pedicel. The stalk of a single flower (Figs. 565, 566).

Peduncle. A primary flower stalk, bearing a cluster or a solitary flower.

Peltate. Attached toward the center instead of the margin.

Perennial. Living for several years.

Perfect. Referring to flowers with both stamens and pistils present.

Perfoliate. Referring to the fused bases of opposite leaves which form a cup where they are attached to the stem, appearing as if the stem passes through the leaves (Fig. 524).

Perianth. Calyx, or calyx and corolla (Figs. 565-67).

Perianth tube. A tube formed of united sepals and petals.

Perigynium. The saclike structure enclosing the achene of a sedge.

Petal. A division of the corolla (Figs. 565, 566).

Petaloid. Petallike.

Petiole. The stalk of a leaf.

Petiolule. The stalk of a leaflet in a compound leaf.

Phyllary. A bract of the involucre in the Asteraceae.

Pilose. With very long hairs.

Pinnate. Compound, with the leaflets on the opposite sides of a common axis (Fig. 570*c*).

Pinnatifid. Pinnately deeply cleft, but not into distinct leaflets (Fig. 570*b*).

Pistil. The seed-bearing (female) part of a flower, having typically style, stigma, and ovary (Figs. 565, 566), formed of one to several carpels (Fig. 568).

Plumose. With fine spreading hairs, the hairs feathery-branched.

Pod. A dry fruit which splits open.

Pollinium. A structure formed of many fused pollen grains

all discharged as a group in the Orchidaceae and Asclepiadaceae. Plural: pollinia.

Polygamous. Bearing partly perfect and partly unisexual flowers.

Pome. A fleshy fruit of which the apple is typical, formed from an inferior ovary and the hypanthium and with an inner cartilaginous layer around the seeds.

Prickle. A small sharp outgrowth.

Primocane. A first-year nonflowering stem of a biennial woody plant, characteristic of *Rubus.*

Puberulent. Slightly hairy.

Pubescent. With soft short hairs.

Raceme. An inflorescence with stalked flowers borne on a common axis (Fig. 569c).

Rachis. The axis of a spike or compound leaf.

Receptacle. The often somewhat enlarged tip of a stem, bearing the sepals, petals, stamens, and pistils (Figs. 565, 566, where the receptacle is indicated by vertical shading).

Reflexed. Bent back or downward.

Regular. Symmetrical on several radii.

Reniform. Kidney-shaped.

Replum. The false partition in some fruits of the Brassicaceae.

Retrorse. Directed backward.

Revolute. Inrolled, such as the margins of certain leaves.

Rhizome. A horizontal underground stem.

Rootstock. A rhizome.

Rosette. A close cluster of leaves in a circular form, as in the Dandelion and in *Antennaria* (Fig. 553).

Rotate. Wheel-shaped; referring to a widely spreading corolla.

Sagittate. Arrowhead-shaped.

Salverform. Having corolla with a long narrow tube which abruptly flares out at the top.

Samara. A fruit with an indehiscent winged appendage.

Saprophyte. A plant growing on and deriving nourishment from dead plant or animal matter.

Scabrous. Rough to the touch.

Scale. A thin brownish body, usually a degenerate leaf.

Scape. A leafless flowering stem arising from the ground, as in a Dandelion.

Scarious. Thin, papery, and translucent; not green.

Schizocarp. A capsule which splits into sections at the septa and the portions containing some seeds dispersed in separate units.

Sepal. A division of a calyx (Figs. 565, 566).

Septate. Divided by partitions or septa.

Serrate. With sharp teeth pointing forward.

Serrulate. Finely serrate.

Sessile. Arising directly from the stem, without any stalk (Figs. 571*b*, 571*c*).

Sheath. A tubular envelope, usually about a stem and formed by the base of a leaf.

Silicle. A short silique, usually less than 3 times as long as broad.

Silique. An elongate capsule of the Brassicaceae in which the 2 valves are separated by a longitudinal partition and which is more than 3 times as long as broad.

Simple. Not compound; of one piece (Fig. 570*a*).

Sinus. The cleft between 2 lobes.

Spadix. A spike with a fleshy axis.

Spathe. A large bract or pair of bracts enclosing an inflorescence.

Spatulate. Like a spatula, somewhat broadened at the tip.

Spicule. A fine point.

Spike. An inflorescence of flowers without pedicels on an elongated common axis (Fig. 569*b*).

Spikelet. A small spike, especially as in the Poaceae and Cyperaceae.

Spine. A sharp, rigid, woody outgrowth from a stem.

Spur. A hollow tubular or saclike extension from some part of a flower.

Stamen. The pollen-bearing (male) organ of a flower (Figs. 565, 566).

Staminodium. A sterile structure, usually petaloid, in the position of stamens. Plural: staminodia.

Standard. The upper petal of a papilionaceous flower.

Stellate. Starlike.

Stem. Upright part of a plant bearing secondary stems, leaves, or the inflorescence.

Sterile. Unproductive; without stamens or pistils.

Stigma. The usually sticky tip of the style, on which the pollen lodges; it may be single (Fig. 565), or there may be as many stigmas as there are carpels in the pistil (Figs. 566, 568).

Stipe. A stalk, below the ovary and above the other flower parts, as in some Capparidaceae.

Stipitate. With a short stalk.

Stipule. An appendage, usually paired, at the base of a petiole
    (Fig. 570*d*).
Stolon. A basal branch which loops over the soil and roots at
    intervals.
Stoloniferous. Bearing stolons.
Stone. The hard inner layer (pit) of the fruit of a drupe and
    containing the seed.
Style. The slender part of a pistil (Figs. 565, 566).
Stylopodium. A disklike expansion at the base of the style, as
    in the Apiaceae.
Sub. A prefix meaning somewhat, or slightly less than.
Superior. Referring to an ovary which is attached on the
    receptacle above the other flower parts.
Syncarp. A multiple fruit formed by the fusion of a number of
    flowers, as in *Morus* and *Maclura.*

Taproot. The main, vertical root of a plant.
Tendril. A stem or leaf modified to twist around a support.
Tepal. A term used to describe the sepals and petals when
    they resemble each other and are difficult to distinguish.
Terete. Round in cross section; cylindrical.
Terminal. Borne at the tip of the stem (as the flower in Fig.
    373, and the inflorescence in Figs. 343, 345).
Tetraploid. Containing 4 times the chromosome count of the
    male or female gamete.
Thorn. A short, stiff, sharp, modified branch.
Tomentose. Woolly; the hairs branching and matted.
Toothed. Notched or jagged on the edge. Singly toothed:
    with all teeth alike (Figs. 121, 531); doubly toothed:
    with small teeth on the larger ones (Figs. 129, 133).
Trichome. A plant hair.
Triternate (leaf). A compound leaf in which the leaf is
    divided 3 times, then branched 3 times, and then each
    division is further divided into 3 leaflets.
Tuber. A short, underground swelling of a stem.
Tubercle. A warty outgrowth.

Umbel. An inflorescence in which the flower stalks all arise
    from one point (Fig. 569*d*).
Umbellet. One of the small umbels in a compound umbel.
Unisexual. Of one sex, bearing either stamens or pistils.
Utricle. A small, thin-walled and inflated capsule or achene.

Vascular bundle. A strand of conductive tissue.

Vein. A thread of vascular tissue in a leaf.
Versatile. Referring to the attachment of anthers of grass to the filament, at the center instead of the base (Fig. 3*c*).
Villous. Softly long-hairy.

Whorl. The arrangement of organs in a circle about a common axis.
Whorled (of leaves). With several borne at one node.

Zygomorphic. Referring to irregular or bisymmetrical flowers which may be divided in only one plane of symmetry.

Selected References

Through the years, the Wisconsin Academy of Sciences, Arts and Letters has published in its *Transactions* a series of significant papers entitled "Preliminary Reports on the Flora of Wisconsin." As a convenience to the reader, these reports are gathered together and listed below under "Wisc. Acad. Sci. Arts Letters," according to the number of the report and the family or group treated.

Curtis, John T. 1959. The vegetation of Wisconsin. Univ. Wis. Press, Madison.

Davis, H. A., A. M. Fuller, and Tyreeca Davis. 1967-70. Contributions toward the revision of the Eubati of eastern North America. *Castanea* 32: 20-37; II. Setosi. *Castanea* 33: 50-76; III. Flagellares. *Castanea* 33: 206-41; IV. Verotriviales. *Castanea* 34: 157-79; V. Arguti. *Castanea* 34: 235-66; VI. Cuneifolii. *Castanea* 35: 176-94.

Fassett, Norman C. 1951. Grasses of Wisconsin. Univ. Wis. Press, Madison.

Fassett, Norman C. 1961. Leguminous plants of Wisconsin. Univ. Wis. Press, Madison.

Fernald, M. L. 1950. Gray's manual of botany. American Book Co., New York.

Fuller, Albert M. 1933. Orchidaceae of Wisconsin. Bull. Milwaukee Public Mus. 14(1): 1-284.

Gleason, H. A., and Arthur Cronquist. 1963. Manual of vascular plants of northeastern United States and Canada. Van Nostrand Reinhold Co., New York.

Greene, H. C., and J. T. Curtis. 1955. A bibliography of Wisconsin vegetation. Milwaukee Public Mus. Publ. Botany No. 1.

Morley, Thomas. 1969. Spring flora of Minnesota. Univ. Minn. Press, Minneapolis.

Musselman, Lytton, Theodore Cochrane, William Rice, and Marion Rice. 1971. The flora of Rock County, Wisconsin. *Michigan Botanist* 10(4): 147-93.

Seymour, F. C. 1960. Flora of Lincoln County, Wisconsin. P. F. Nolan, Taunton, Mass.

Shenefelt, R. D. 1970. A key to Wisconsin trees. Wis. Dept. Nat. Res., Madison.

Voss, Edward G. 1972. Michigan flora. Part I. Cranbrook Institute of Science, Bloomfield Hills, Mich.

Wisc. Acad. Sci. Arts Letters. *Transactions.* Preliminary reports on the flora of Wisconsin.

    1. Juncaginaceae, Alismaceae. 1929. Norman C. Fassett. 24: 249-56.

    2. Ericaceae. 1929. Norman C. Fassett. 24: 257-68.

    3. Lobeliaceae, Campanulaceae, Cucurbitaceae. 1929. Kenneth L. Mahony. 24: 357-61.

    4. Lycopodiaceae, Selaginellaceae. 1930. Leonard R. Wilson. 25: 169-75.

    5. Coniferales: Taxaceae, Pinaceae. 1930. Norman C. Fassett. 25: 177-82.

    6. Pandanales: Typhaceae, Sparganiaceae. 1930. Norman C. Fassett. 25: 183-87.

    7. Betulaceae. 1930. Norman C. Fassett. 25: 189-94.

    8. Aceraceae. 1930. Norman C. Fassett. 25: 195-97.

    9. Elatinaceae. 1930. Norman C. Fassett. 25: 199-200.

    10. Haloragidaceae. 1930. Norman C. Fassett. 25: 201-3.

    11. Ranunculaceae. 1930. Lois Almon. 25: 205-14.

    12. Polypodiaceae. 1931. Edith W. Breakey and Ruth I. Walker. 26: 263-73.

    13. Fagaceae. 1931. David F. Costello. 26: 275-79.

    14. Hypericaceae. 1931. Willard T. McLaughlin. 26: 281-88.

    15. Polygonaceae. 1932. Kenneth L. Mahony. 27: 207-25.

    16. Xyridales: Eriocaulaceae, Xyridaceae, Commelinaceae, Pontederiaceae. 1932. Norman C. Fassett. 27: 227-30.

    17. Myricaceae: Juglandaceae. 1932. Norman C. Fassett. 27: 231-34.

    18. Sarraceniales: Sarraceniaceae, Droseraceae. 1932. Florence B. Livergood. 27: 235-36.

    19. Saxifragaceae. 1932. Norman C. Fassett. 27: 237-46.

    20. Malvales: Malvaceae, Tiliaceae. 1932. Alice M. Hagen. 27: 247-49.

    21. Geraniales: Linaceae, Oxalidaceae, Geraniaceae, Rutaceae, Polygalaceae, Euphorbiaceae, Callitrichaceae. 1933. Norman C. Fassett. 28: 171-86.

22. Cornaceae. 1933. A. A. Drescher. 28: 187-90.
23. Urticaceae. 1933. David F. Costello. 28: 191-96.
24. Salicaceae. 1935. David F. Costello. 29: 299-318.
25. Arales: Araceae, Lemnaceae (Lemnaceae in collaboration with Lawrence E. Hicks). 1937. Norman C. Fassett. 30: 17-20.
26. Convolvulaceae. 1937. Sidney O. Fogelberg. 30: 21-25.
27. Lentibulariaceae. 1940. John W. Thomson. 32: 85-89.
28. Caprifoliaceae. 1940. Dorothy R. Wade and Douglas E. Wade. 32: 91-101.
29. Anacardiaceae. 1940. Norman C. Fassett. 32: 103-6.
30. Rhamnales: Rhamnaceae, Vitaceae. 1940. Richard W. Pohl. 32: 107-11.
31a. Solanaceae. 1943. Norman C. Fassett. 35: 105-12.
31b. Boraginaceae. 1944. Emil P. Kruschke. 36: 273-90.
33a. Ranunculaceae, Nymphaeaceae, Ceratophyllaceae, Berberidaceae, Menispermaceae, Lauraceae. 1947. Norman C. Fassett. 38: 189-209.
33b. Najadaceae. 1951. James G. Ross and Barbara M. Calhoun. 40: 93-110.
34. Liliales: Juncaceae, Dioscoreaceae, Liliaceae, Amaryllidaceae, Iridaceae. 1950. Joan A. McIntosh. 40: 215-42.
35. Araliaceae. 1950. Norman C. Fassett and H. J. Elser. 40(1): 83-85.
36. Scrophulariaceae. 1951. Peter J. Salamun. 40(2): 111-38.
37. Cyperaceae. Part 1 (exclusive of *Carex*). 1953. H. C. Greene. 42: 47-67.
38. Rubiaceae. 1958. Emil K. Urban and Hugh H. Iltis. 46: 91-104.
39. Phrymaceae. 1958. Hugh H. Iltis. 46: 105.
40. Asclepiadaceae. 1958. Gottlieb K. Noamesi and Hugh H. Iltis. 46: 107-14.
41. Labiatae. 1958. Robert C. Koeppen. 46: 115-40.
42. Rosaceae I. 1959. Harriet Gale Mason and Hugh H. Iltis. 47: 65-97.
43. Primulaceae. 1960. Hugh H. Iltis and Winslow M. Shaughnessy. 49: 113-35.
44. Cruciferae. 1961. Jacqueline P. Patman and Hugh H. Iltis. 50: 17-73.
45. Amaranthaceae. 1961. Jonathan Sauer and Robert Davidson. 50: 75-87.
46. Caryophyllaceae. 1961. Robert A. Schlising and Hugh H. Iltis. 50: 89-139.

47. Thymelaeales, Myrtales, and Cactales. 1962. Donald Ugent. 51: 83-134.
48. Compositae I. Tribes Eupatorieae, Vernonieae, Cynarieae, and Cichorieae. 1963. Miles F. Johnson and Hugh H. Iltis. 52: 255-342.
49. Compositae II. The genus *Senecio*. 1963. T. M. Barkley. 52: 343-52.
50. Compositae III. The genus *Solidago*. 1963. Peter J. Salamun. 52: 353-82.
51. Salicaceae. The genus *Salix*. 1964. George W. Argus. 53: 217-72.
52. Gentiana hybrids in Wisconsin. 1964. James S. Pringle. 53: 273-81.
53. Gentianaceae and Menyanthaceae. 1965. Charles T. Mason, Jr., and Hugh H. Iltis. 54: 295-329.
54. Equisetaceae. 1965. Richard L. Hauke. 54: 331-46.
55. Compositae IV. Tribes Helenieae and Anthemideae. 1966. Carol J. Mickelson and Hugh H. Iltis. 55: 187-222.
56. Compositae V. Tribe Inuleae. 1966. Edward W. Beals and Ralph F. Peters. 55: 223-42.
57. Polemoniaceae. 1966. Dale M. Smith and Donald A. Levin. 55: 243-53.
58. Hydrophyllaceae. 1966. Jack W. Shields. 55: 255-59.
59. Plantaginaceae. 1967-68. Melvern F. Tessene. 56: 281-313.
60. Tiliaceae and Malvaceae. 1970. Fred H. Utech. 58: 301-23.
61. Hypericaceae. 1970. Fred H. Utech and Hugh H. Iltis. 58: 325-51.
62. Compositae VI. The genus *Ambrosia*. 1970. Willard W. Payne. 58: 353-71.
63. The genus *Trifolium*. 1973. John M. Gillett and Theodore S. Cochrane. 61: 59-74.
64. Adoxaceae. 1974. Theodore S. Cochrane and Peter J. Salamun. 62: 247-52.
65. Dipsacaceae. 1974. Peter J. Salamun and Theodore S. Cochrane. 62: 253-60.
66. Cyperaceae II. The genus *Cyperus*. 1974. Brian G. Marcks. 62: 261-84.

Index